Manjit S. Kang
Editor

Crop Improvement
Challenges in the
Twenty-First Century

More pre-publication
REVIEWS, COMMENTARIES, EVALUATIONS . . .

"This is a timely book on the challenges plant breeders will face in feeding the ten to twelve billion people on earth by the middle of the twenty-first century. Professor Kang has assembled authoritative contributions on this critical plant breeding issue from a team of diverse scientists of food and fiber crops. Dr Duvick, a well-known research administrator, envisions a global tripartite system of public, private, and participatory breeding to meet the challenge. World food laureate Dr. Khush and colleagues discuss the past and future of rice breeding, including biotechnology. Other important crops such as maize, sorghum, common bean, sugar beet, bananas and plaintains, and cotton are also covered, as well as issues applicable to all crops, such as abiotic stress and genotype-by-environment interactions. This will be a useful book for research workers and teachers of advanced plant breeding."

Bikram S. Gill
University Distinguished Professor,
The Wheat Genetics Resource Center,
Department of Plant Pathology,
Kansas State University,
Manhattan, Kansas

"This book provides a broad synoptic view of breeding achievements and future trends for some major crops. It includes recent developments on biotechnology, conservation of genetic resources, and statistics for handling genotypic responses in multi-environmental conditions. The contributors from around the world have interests ranging from genetics to breeding to physiology, and their experiences reflect a wide range of farming systems. It is amazing to find the contributions of so many distinguished and respected figures in genetics and breeding combined in this book by Dr. Kang, a prominent scientist in this field. The chapters are informative, comprehensive, and easy to read, and also provide many sources of literature for readers interested in a particular area. This book will serve as excellent educational material for geneticists, breeders, and graduate students."

Robert T. Magari, PhD
Statistician,
Beckman Coulter Corporation,
Miami, Florida

Food Products Press®
An Imprint of The Haworth Press, Inc.
New York • London • Oxford

Crop Improvement
Challenges in the Twenty-First Century

FOOD PRODUCTS PRESS
Seed Biology, Production, and Technology
Amarjit S. Basra, PhD
Senior Editor

Heterosis and Hybrid Seed Production in Agronomic Crops edited by Amarjit S. Basra

Seed Storage of Horticultural Crops by S. D. Doijode

New, Recent, and Forthcoming Titles of Related Interest:

Dictionary of Plant Genetics and Molecular Biology by Gurbachan S. Miglani

Advances in Hemp Research by Paolo Ranalli

Wheat: Ecology and Physiology of Yield Determination by Emilio H. Satorre and Gustavo A. Slafer

Mineral Nutrition of Crops: Fundamental Mechanisms and Implications by Zdenko Rengel

Conservation Tillage in U.S. Agriculture: Environmental, Economic, and Policy Issues by Noel D. Uri

Cotton Fibers: Developmental Biology, Quality Improvement, and Textile Processing edited by Amarjit S. Basra

Intensive Cropping: Efficient Use of Water, Nutrients, and Tillage by S. S. Prihar, P. R. Gajri, D. K. Benbi, and V. K. Arora

Plant Growth Regulators in Agriculture and Horticulture: Their Role and Commercial Uses edited by Amarjit S. Basra

Crop Responses and Adaptations to Temperature Stress edited by Amarjit S. Basra

Physiological Bases for Maize Improvement edited by María E. Otegui and Gustavo A. Slafer

Plant Viruses As Molecular Pathogens by Jawaid A. Khan and Jeanne Dijkstra

In Vitro Plant Breeding by Acram Taji, Prakash P. Kumar, and Prakash Lakshmanan

Crop Improvement: Challenges in the Twenty-First Century edited by Manjit S. Kang

Bacterial Disease Resistance in Plants: Molecular Biology and Biotechnological Applications by P. Vidhyasekaran

Tillage for Sustainable Cropping by P. R. Gajri, V. K. Arora, and S. S. Prihar

Crop Improvement
Challenges in the Twenty-First Century

Manjit S. Kang
Editor

Food Products Press®
An Imprint of The Haworth Press, Inc.
New York • London • Oxford

Published by

Food Products Press®, an imprint of The Haworth Press, Inc., 10 Alice Street, Binghamton, NY 13904-1580.

Cover design by Jennifer M. Gaska.

Library of Congress Cataloging-in-Publication Data

Crop improvement : challenges in the twenty-first century / Manjit S. Kang, editor.
    p. cm.
  Includes bibliographical references (p.).
  ISBN 1-56022-904-7 (alk. paper)—ISBN 1-56022-905-5 (pbk. : alk. paper)
   1. Crop improvement. I. Kang, Manjit S.

SB106.I47 C75 2001
631.5'23—dc21

                                                                2001040468

CONTENTS

ABOUT THE EDITOR

Manjit S. Kang, PhD, is Professor of Quantitative Genetics in the Department of Agronomy at Louisiana State University. He earned his BSc in agriculture and animal husbandry with honors from the Punjab Agricultural University in India, an MS in biological sciences, majoring in plant genetics from Southern Illinois University at Edwardsville, an MA in botany from Southern Illinois University at Carbondale, and a PhD in crop science (genetics and plant breeding) from the University of Missouri at Columbia.

Dr. Kang is the editor, author, or co-author of hundreds of articles, books, and book chapters. He enjoys an international reputation in genetics and plant breeding. He serves on the editorial boards of *Crop Science, Agronomy Journal,* and *Journal of New Seeds,* as well as the Haworth Food Products Press.

Dr. Kang is a member of Gamma Sigma Delta and Sigma Xi. He was elected a Fellow of the American Society of Agronomy and of the Crop Science Society of America. In 1999 he served as a Fulbright Senior Scholar in Malaysia.

Dr. Kang edited *Genotype-By-Environment Interaction and Plant Breeding* (1990), which resulted from an international symposium that he organized at Louisiana State University in February 1990. He is the author/publisher of *Applied Quantitative Genetics* (1994), which resulted from teaching a graduate level course on Quantitative Genetics in Plant Improvement. Another book, *Genotype-by-Environment Interaction,* edited by him and Hugh Gauch, Jr., was published by CRC Press in 1996. He edited *Crop Improvement for the 21st Century* in 1997 (Research Signpost, India). He is also Editor of *Quantitative Genetics, Genomics and Plant Breeding* (2002; CABI Publishing), which resulted from an international symposium that he organized in Baton Rouge in March 2001.

Dr. Kang's research interests are: genetics of resistance to aflatoxin, weevils, and herbicides in maize; genetics of grain dry-down rate and stalk quality in maize; genotype-by-environment interaction and crop adaptation; interorganismal genetics; and conservation and utilization of plant genetic resources.

Dr. Kang taught Plant Breeding and Plant Genetics courses at Southern Illinois University-Carbondale (1972-1974). He has been teaching a graduate level Applied Quantitative Genetics course at Louisiana State University since 1986. He developed and taught an intermediary plant genetics course in 1996 and team-taught an Advanced Plant Genetics course (1993-1995). He also taught an Advanced Plant Breeding course at LSU in 2000. He has directed six MS theses and six PhD dissertations. He has been a Full Professor in the Department of Agronomy at LSU since 1990. He has received many invitations to speak at international symposium relative to genetics and plant breeding.

Dr. Kang was recognized for his significant contributions to plant breeding and genetics by Punjab Agricultural University at Ludhiana at its 36th Foundation Day in 1997. He served as President (2000-2001) of the LSU Chapter of Sigma Xi—The Scientific Research Society. He was elected President of the Association of Agricultural Scientists of Indian Origin in 2001 for a two-year term. In addition, he serves as the Chairman of the American Society of Agronomy's Member Services and Retention Committee (2001-2004). Dr. Kang's biographical sketches have been included in *Marquis Who's Who in the South and Southwest, Who's Who in America, Who's Who in the World, Who's Who in Science and Engineering,* and *Who's Who in Medicine and Healthcare.*

CONTRIBUTORS

Monica Balzarini, PhD, is Professor, Department of Biometry, University of Cordoba, Argentina.

Larry G. Campbell, PhD, is Research Geneticist, USDA Agricultural Research Service, Northern Crop Science Laboratory, Fargo, North Dakota.

Donald N. Duvick, PhD, is Affiliate Professor of Plant Breeding, Iowa State University, and Senior Vice President of Research (retired), Pioneer Hi-Bred International, Inc., Johnston, Iowa.

Michael L. Gilbert, PhD, is Head of USA Field Seeds, Aventis Crop Science, Lubbock, Texas.

John Hartman, PhD, was Postdoctoral Fellow, Crop Improvement Division, International Institute of Tropical Agriculture, Ibadan, Nigeria.

Bingru Huang, PhD, is Associate Professor, Department of Plant Science, Rutgers University, New Brunswick, New Jersey.

Yiwei Jiang, PhD, is Postdoctoral Research Associate, Department of Crop and Soil Sciences, The University of Georgia, Griffing, Georgia.

M. Altaf Khan, PhD, is Postdoctoral Research Assistant, Plant and Soil Science Department, Mississippi State University, Starkville, Mississippi.

Gurdev S. Khush, PhD, is Principal Plant Breeder, Division of Plant Breeding, Genetics and Biochemistry, International Rice Research Institute, Makati City, Philippines.

Scott B. Milligan, PhD, is Chief Geneticist, U.S. Sugar Corporation, Clewiston, Florida.

Orlando J. Moreno, PhD, is National Coordinator of the Rice Breeding Program, National Institute of Agricultural Research of Venezuela (INIA), Venezuela.

Gerald O. Myers, PhD, is Associate Professor, Department of Agronomy, Louisiana State University, Baton Rouge, Louisiana.

Henry T. Nguyen, PhD, is Paul Whitfield Horn Distinguished Professor of Genetics, Plant Molecular Genetics Laboratory, Department of Plant and Soil Science, Texas Tech University, Lubbock, Texas.

Ripusudan L. Paliwal, PhD, is Director of Maize Program (retired), CIMMYT, Mexico, and currently a maize consultant, Owego, New York.

Hans-Peter Piepho, PhD, is Senior Lecturer in Biometrics, Institut für Nutzpflanzenkunde, Universität-Gesamthochschule Kassel, Witzenhausen, Germany.

Michael Pillay, PhD, is Associate Scientist (Cytogeneticist), Crop Improvement Division, International Institute of Tropical Agriculture, Ibadan, Nigeria.

Darrell T. Rosenow, PhD, is Professor of Sorghum Breeding, Texas A&M University Agricultural Reseach and Extension Center, Lubbock, Texas.

Shree P. Singh, PhD, is Bean Breeder, Kimberly Research and Extension, University of Idaho, Kimberly, Idaho.

Margaret E. Smith, PhD, is Associate Professor, Department of Plant Breeding, Cornell University, Ithaca, New York.

J. McD. Stewart, PhD, is Professor of Cotton Biotechnology and Chair, Althimer Laboratory, Department of Crops, Soils, and Environmental Sciences, University of Arkansas, Fayetteville, Arkansas.

Prasanta K. Subudhi, PhD, is Assistant Professor of Research, Department of Agronomy, Louisiana State University, Baton Rouge, Louisiana.

A. Tenkouano, PhD, is Scientist (Breeder/Geneticist), Crop Improvement Division, International Institute of Tropical Agriculture, Ibadan, Nigeria.

Fred A. van Eeuwijk, PhD, is Associate Professor of Plant Breeding/ Biometrician, Department of Plant Sciences, Wageningen University, Netherlands.

Parminder S. Virk, PhD, is Affiliate Scientist, Division of Plant Breeding, Genetics and Biochemistry, International Rice Research Institute, Makati City, Philippines.

Foreword

Crop Improvement: Challenges in the Twenty-First Century is an important work, with chapters authored by outstanding, well-placed scientists. With the population projected to double in the next 50 years, food and fiber production must be increased dramatically in order to feed and clothe the world's people. These food production challenges must be met, in large part, by making significant improvements in the major food and fiber crops—such as rice, maize, sorghum, and cotton—and then rapidly transferring this technology to all parts of the world. This timely volume will make a significant contribution toward educating people about recent technological advances with the major crops and then assisting them in meeting the great challenge of doubling crop production in the next half century.

The breadth and depth of the work presented here is impressive, with contributions reflecting an international perspective and both university and agribusiness/agency/research center viewpoints. This volume is edited by Manjit S. Kang, PhD, whose excellent background and experience, as well as his global perspective, have given him the proper credentials to oversee and edit this important work.

Donald M. Elkins, PhD
Dean, College of Agriculture and Human Ecology
Tennessee Technological University
<http://eagle.tntech.edu/www/acad/aghec>

Preface

In the year 2000, the human population of the world crossed the six-billion mark. Before 1825, world population was about one billion. It took about 100 years (1825 to 1925) to add the second billion, but the third billion was added in only about 35 years (1925 to 1960) (Evans, 1998). The time span needed to add each additional billion has been narrowing; for example, it took only about 15 (1960 to 1975), 10 (1975 to 1986), and 15 (1986 to 2000) years, respectively, to add the fourth, fifth, and sixth billion to the world population. Projections are that the world population will reach the 10-billion mark by the middle of the twenty-first century, which translates into a population increase of one billion per 10 or 12 years during the next 40 to 50 years. The challenge before us, therefore, is: Can we feed four billion more people by 2040?

Recently, a prominent journalist, David Brinkley, eloquently stated in television commercials on behalf of an agricultural company (paraphrased here): (1) from the standpoint of food, we must view the world as one civilization; and (2) until everyone has food to eat, there will be no peace or prosperity in the world. These statements are quite accurate; we know from genomics research that all human beings are 99.9 percent alike in their genetic makeup; all of the large and small differences reside in the remaining 0.1 percent of the human genome. It would be difficult to ignore the common humanity of the peoples of the world.

In March 2000, Nobel Laureate Norman E. Borlaug remarked,

> Agricultural researchers and farmers in Asia face the challenge during the next 25 years of developing and applying technology that can increase the cereal yields by 50-75 percent, and to do so in ways that are economically and environmentally sustainable. Much of the near-term yield gains will come from applying technology already on the shelf. But there will also be new research breakthroughs—especially in plant breeding to improve yield stability and, hopefully, maximize genetic yield potential—if science is permitted to work as it should be.

He further stated, "Genetic improvement—continued genetic improvement of food crops—using both conventional and biotechnology research tools is needed to shift the yield frontier higher and to increase stability of yield." Borlaug optimistically states, "I now say that the world has the technology—either available or well advanced in the research pipeline—to feed a population of 10 billion people." He cautions us, however, that world food production would have to double over the next 30 years and triple over the next 50 years (Borlaug, 1999).

The current book *Crop Improvement: Challenges in the Twenty-First Century* brings together contributions from expert agronomists—plant breeders/geneticists, agricultural statisticians, and crop physiologists. The purpose of the book is to provide state-of-the-art information to crop breeders throughout the world and transfer technology to all parts of the world. This book addresses several important crops that are grown globally. In 1997, a number of crops and issues relative to crop improvement were included in a previous volume that I edited (Kang, 1997). Many important crops and issues could not be included in that volume, however. Because of the continued need for gathering the latest, authoritative scientific information concerning crop breeding as well as related subjects and for transferring technology to wherever it is needed the most, this volume was conceived.

In the foreword, Donald M. Elkins, PhD, an eminent educator and administrator, forcefully puts forth concerns about food production in the twenty-first century. He exhorts, "These food production challenges must be met, in large part, by making significant improvements in the major food and fiber crops—such as rice, maize, sorghum, and cotton—and then rapidly transferring this technology to all parts of the world."

The book begins with a contribution by Donald Duvick, PhD, a distinguished geneticist and agricultural research administrator, titled "Crop Breeding in the Twenty-First Century." Duvick provides a thought-provoking analysis of the past and current state of plant breeding and predicts the future of plant breeding. In this excellent chapter, he envisions the evolution of a tripartite system involving the private sector, the public sector, and participatory plant breeding. He discusses alternative ways in which the system could go and how the different components could interact, pointing out the essential unity of the global plant breeding system. He concludes, "The global plant breeding system will operate most efficiently and serve its primary customers (the farmers) most adequately if all concerned understand this unity in diversity, this system in which all parts work together for the good of the whole."

The second part of the book includes chapters dealing with improvement of specific crops or problems: (a) *food crops:* rice, tropical maize, aflatoxin (a potent carcinogen) in maize, sorghum, common bean, sugarbeet, and bananas and plantains, and (b) *fiber crop:* cotton.

The chapter on rice (Chapter 2) is authored by World Food Laureate Gurdev S. Khush and his colleague, Parminder S. Virk, at the International Rice Research Institute (IRRI) in the Philippines. This authoritative chapter provides a thorough review of what has been accomplished and what needs to be accomplished in rice breeding to meet the needs of Asia (90 percent of rice production and consumption occurs in Asia) and the rest of the world. They discuss the traditional methods of rice improvement as well as those based on molecular-marker technology.

Ripusudan L. Paliwal and Margaret E. Smith, both with vast international experience in maize at CIMMYT and elsewhere, discuss new and innovative approaches—both conventional and molecular breeding—used to improve tropical maize in Chapter 3. They also provide their perspective on the public and private sector seed systems and indicate their concern about the weakening of the public maize research system. They clearly point out what must happen to successfully improve tropical maize, which is of increasing importance in developing countries.

Chapter 4, written by Manjit S. Kang and Orlando J. Moreno, provides a comprehensive review of the problem of aflatoxin contamination of maize grain. Aflatoxin contamination occurs not only in tropical countries but also in temperate regions of the world. We discuss the information that has become available, in the past 25 years, on the complex inheritance of resistance to aflatoxin accumulation and possible methods of combating this serious problem.

Chapter 5, on sorghum breeding, has been prepared by a team dedicated to sorghum research at Texas Tech University at Lubbock. The work reflects the vast experiences of Prasanta K. Subudhi, Henry T. Nguyen, Michael L. Gilbert, and Darrell T. Rosenow. They emphasize that the challenges for sorghum breeders will be the production of new hybrids with increased productivity and profitability. Genome mapping and molecular breeding as well as tissue culture and genetic engineering have been highlighted in this chapter.

Shree P. Singh, who has been providing leadership for a number of years to the common bean-breeding program at CIAT in Colombia, addresses in Chapter 6 the important issue of the narrow genetic base of commercial cultivars, presents strategies for overcoming biotic and abiotic stresses, and advocates the need for developing high-yielding, high-quality cultivars that are less dependent on water, fertilizer, pesticides, and manual labor. He points out that the importance of dry beans should increase worldwide because of the advantages attributed to consumption of dry beans with a high nutritive protein content and their role in preventing cancer and lowering cholesterol.

Larry G. Campbell presents a comprehensive treatment of issues relative to sugarbeet improvement Chapter 7. He clearly elucidates inheritance of important traits, hybridization and hybrid production, selection techniques and strategies employed, biotic and abiotic constraints, and the narrow genetic base of sugarbeet germplasm. This chapter should be an excellent resource for breeders of sugar crops throughout the world.

Michael Pillay, A. Tenkouano, and John Hartman of the International Institute of Tropical Agriculture (IITA) in Nigeria present an excellent overview of the importance of banana and plantain *(Musa)* in the developing countries in the tropics (Chapter 8). They enumerate the challenges that *Musa* breeders face. They also suggest that the greatest challenge in the near term is to breed for resistance to banana streak virus (BSV) because the viral sequences (DNA) are integrated into the *Musa* genome. They advocate that a better knowledge of *Musa* genomics and a better understanding of transmission genetics should enhance *Musa* breeding efforts.

While the previous chapters treated improvement of food crops, the next chapter on cotton improvement (Chapter 9) relates to clothing people. M. Altaf Khan, Gerald O. Myers, and J. McD. Stewart present their experiences with cotton improvement. They especially emphasize the roles of molecular markers and genomics in cotton improvement, pointing out that cotton has a narrow genetic base and that the application of biotechnology to cotton should assure a virtually unlimited gene pool.

The third part of the book includes three chapters dealing with issues that relate to all crops. First, Bingru Huang and Yiwei Jiang discuss the physiological and biochemical responses of plants to drought and heat stress in Chapter 10. A thorough understanding of physiological/biochemical traits should help plant breeders develop appropriate strategies for developing resistance to abiotic stresses. Biotic and abiotic stresses have been associated with genotype-by-environment interactions in various organisms.

Chapter 11, on stability analysis in crop performance trials by Hans-Peter Piepho and Fred A. van Eeuwijk, discusses how to handle genotype-by-environment interactions encountered in yield trials. They advocate the use of mixed-model methodology. They expound on the concept of risk assessment relative to variety selection. Mixed models involving stability parameters are indicated to be more flexible and more realistic than the commonly used mixed-model ANOVA.

Chapter 12 also relates to genotype-by-environment interactions in crop performance trials. It specifically addresses the use of best linear unbiased prediction (BLUP) in multienvironment trials. Monica Balzarini, Scott B. Milligan, and Manjit S. Kang describe a new application of BLUP that is intended to improve genotype performance prediction in multienvironment trials.

In the final analysis, it is hoped that the works presented in this volume will help plant breeders understand some of the most important contemporary issues and challenges encountered in different crops and enhance their ability to feed and clothe the burgeoning world population. The chapters are expected to be useful to practicing crop breeders as well as to students of plant breeding. The optimistic outlook presented by various authors offers hope that a measure of peace and prosperity will be shared by all citizens of the world.

REFERENCES

Borlaug, N.E. 1999. How to feed the 21st century? The answer is science and technology. In J.G. Coors and S. Pandey (Eds.), *Genetics and exploitation of heterosis in crops.* (pp. 509-519), ASA, CSSA, SSSA, Madison, WI.

Borlaug, N.E. 2000. Text of speech. <http://usinfo.state.gov/topical/global/biotech/00030701.htm>.

Evans, L.T. 1998. *Feeding the ten billion: Plants and population growth.* Cambridge University Press, Cambridge, UK.

Kang, M.S. (Ed.) 1997. *Crop improvement for the 21st century.* Research Signpost, Trivandrum, India.

PART I:
INTRODUCTION—VISION

Chapter 1

Crop Breeding in the Twenty-First Century

Donald N. Duvick

INTRODUCTION

Crop breeding is as old as domesticated plants. When our forebears nurtured and selected desirable plants from a few favored wild species and eventually formed the first landraces, they practiced crop breeding identical in its fundamentals to today's "scientific" crop breeding. During the past 10,000 years, farmer-breeders have developed untold numbers of landraces (farmer varieties), and thousands of them are still on hand, although the numbers are shrinking rapidly as professionally bred varieties answering the demands of the marketplace replace the landraces in many parts of the world.

Our only record of changes in plant breeding procedures during the past 10,000 years has occurred during the past couple of centuries. We imagine, without evidence, that chance outcrossing provided heterozygous new materials from which new varieties could be selected. But we do know that discovery of the nature of sexuality in plants in the eighteenth century fostered deliberate outcrossing as a basis for new variety formation. And we know that rediscovery of Mendel's principles of genetics at the end of the nineteenth century gave great impetus to development of plant breeding as a science, eventually providing full-time work to professional plant breeders.

Other discoveries, inventions, and technological advances also increased the power and speed of plant breeding. Statistical theory helped breeders create efficient field plot designs and methods for precise data analysis, thereby greatly improving the accuracy of selection procedures. Statistical theory, including quantitative genetics, also provided great assistance in the design of breeding programs, giving them better precision, speed, and direction. Mechanization of planting and harvesting machinery greatly increased the quantity of yield trial data that a breeder could generate. Even though the odds for success might not be increased, breeders operated on a numerical base that was many times greater than when all trials were conducted by hand.

3

PRESENT STATUS

Division of Responsibilities

The players in the game of plant breeding have differentiated and changed over the years. Full-time professional breeders using science to aid their empirical endeavors came on the scene about 100 years ago. They soon divided into two groups: publicly employed and privately employed. The public-sector breeders are employed by government and university institutions. The private-sector breeders are employed by seed companies or are self-employed. The general public pays the bills—via tax revenues—for public-sector breeding. Farmers pay the bills—via seed purchases—for private-sector breeding.

Both groups work together for a common cause—variety improvement to suit the needs of farmers. Public-sector breeders have tended to specialize in development of theory and basic research including germplasm development. They also are responsible for the education of future generations of plant breeders and researchers. Private-sector breeders specialize in the development and deployment of finished cultivars (varieties). But public-sector breeders also develop varieties for minor crops and for large-scale field crops in regions that are not conveniently served by private enterprise (for example, crops with low profit margins per unit of sales or a high proportion of "seed saving" and consequently a prohibitively small market size). Private-sector breeders increasingly engage in basic research, especially in aspects of molecular biology applied to plants. Together, the two sectors, public and private, comprise a complete plant breeding system. They depend on each other.

Private-sector plant breeding has less impact globally than might be supposed from its prevalence in industrialized countries. Commercial wheat varieties comprise about 4 percent of all wheat planted in developing countries as compared to about 30 percent in industrialized countries. Commercial soybean varieties comprise an estimated 30 to 60 percent of the planted soybean area in developing countries compared with 70 to 90 percent in industrialized countries. Commercial maize varieties (mostly hybrid) are planted on an estimated 15 percent of the total maize area in developing countries as compared with essentially 100 percent in industrialized countries (Heisey, 2000).

Nevertheless, it is a fact that in industrialized countries private-sector plant breeding predominates in most of the major field crops. In the United States, for example, approximately 70 percent of the field breeders for agronomic crops were employed by private industry in the mid-1990s (Frey, 1996). Private-sector breeders outnumbered those from the public sector in each of the major crops except wheat.

In developing countries, farmer selection practices are still an important, and sometimes the only, source of variety development and maintenance. Farmers and farm communities maintain and shape their own varieties, predominantly in regions that are least favorable for commercial crop production, and in crops that are unlikely to have dependable commercial markets. Farmer-breeders by definition are not full-time breeders, and they use little or no scientific theory. They depend instead on empirical methods developed by themselves and their ancestors. But there is a growing movement among some nongovernmental organizations (NGOs) to teach farmers how to use of some of the techniques of professional breeding, if they are appropriate to the farmers' capacities and needs. The new approach usually is called *participatory plant breeding,* although other terms are also used. Farmers and professional breeders jointly may try, for example, a refinement to simple mass selection (e.g., stratified mass selection) for improvement of maize varieties, or they may perform controlled hybridization to generate useful new variation for selection of new varieties in rice. Farmers can evaluate the usefulness of these procedures in their own circumstances and then adopt them, adapt them, or abandon them.

Technology and Science

Increasingly sophisticated genetic and laboratory techniques have helped plant breeding during recent decades. They include embryo culture to facilitate wide crosses, deep knowledge of the interactive (interorganismal) genetics of host-plant resistance and pathogen virulence, delicate but speedy laboratory analyses to quantify desirable chemical components of seeds, and laboratory culture of insects and diseases, coupled with artificial infestation techniques, to facilitate selection for pest resistance. These techniques help breeders continually to refine and speed up their work. Computers and computer science have made further additions to plant breeders' power and efficiency. Masses of data can be analyzed in detail and instantaneously transported worldwide. Computer-aided modeling can help breeders choose among numerous breeding approaches or gain better understanding of physiological interactions and their potential to affect yield or other important traits.

The past two decades have seen a landmark change in plant breeding potentials. Molecular biology and its offspring, biotechnology, have given breeders the opportunity to add useful new traits governed by genes that up to now simply were not available for plant breeding. And breeders are learning how to modify and amplify existing plant genetic systems in ways never before imagined. Genetic transformation, molecular markers, genomics, proteomics, and bioinformatics provide new tools for moving, understanding, regulating, and redesigning genes and their products.

Biotechnology has not and will not replace "classical" plant breeding, but it will take it to heights never before thought possible (although, perhaps, not as soon as some have expected). Most important, biotechnology will enable breeders to identify in increasing detail the interacting metabolic pathways that enable a plant to express its genetic potential. Simultaneously, breeders will identify and learn how to regulate the genes that control those pathways. With this knowledge, they will be able to analyze existing genotypes—existing cultivars—for strengths and weaknesses, metabolic and genetic. Then they will be able to redesign the genotypes to provide greater stability of performance, more drought or insect tolerance, higher yield, or whatever trait farmers (or other users further down the line) deem to be in greatest need of improvement.

Biotechnology also gives plant breeders the opportunity to develop plants that make entirely different kinds of products (for example, nonfood products, such as plastics or pharmaceuticals). Less spectacular but also new will be plants that produce oils, proteins, or carbohydrates of altered composition to provide more nutritious foods or to satisfy unique industrial needs. This latter class of opportunities from applications of biotechnology has greatly enhanced expectations for a growing new market for field crops. Crops, such as maize and soybeans, increasingly are regarded not only as commodities but also as potential specialty crops when bred to suit the specialized needs of animal feeders, grain millers, or food companies.

With or without biotechnology, breeders can develop cultivars with grain (or other plant parts) of altered chemical composition. Farmers no longer produce only commodities for sale on the open market. They also produce specialty crops, often on contract with (for example) swine feeders, starch millers, or food companies. These companies, in turn, are directly influenced by the ultimate end user, the public that consumes the food that is produced from the specialty crops. Breeders, therefore, now must breed not just for the farmers but also for the end users of the farmers' crops.

Organization

Not only science and technology have changed—the current organization of professional plant breeding is very different than it was only 20 years ago.

The Private Sector

The major seed companies no longer exist as separate entities. They have been purchased by large firms, often agrochemical, but sometimes with pharmaceutical interests. The primary impetus for this change was the lure of profits from biotechnology applied to plant breeding. The purchasers believed that they could use biotechnology to make vastly improved and desirable cultivars in a short time and that the improvements would be protected by patents and other kinds of intellectual property rights. Also important was the expectation, by some firms, that their product line of environmentally undesirable chemical pesticides could be replaced by a product line of cultivars with extraordinarily high levels of biological resistance to insects and diseases. The companies that moved the most swiftly and strongly would be the winners. A corollary to the urge to buy into the "new" field of (hopefully) highly profitable plant breeding was the inescapable fact that biotechnology applied to plant breeding requires high fixed costs in people and infrastructure. Most of the seed companies could not raise the needed funds on their own. They required the funding base of a larger corporation. So, at the turn of the twenty-first century, old-line companies have disappeared, or they exist in name only as subsidiaries of larger firms that have only secondary interest—and little experience—in plant breeding. But despite the consolidations and acquisitions of large seed companies, scores of small seed companies are still in operation, collectively filling an important niche in the global plant breeding and seed production complex.

A new category of small companies serving the needs of plant breeding has arisen during the past decade. These private companies, typically working on contract, specialize in various applications that loosely can be classified as part of genomics. They serve primarily the medical sciences, but because of the unity of biology at the molecular level, they also serve plant-breeding organizations. They specialize in, for example, gene discovery, directed genetic recombination, and generation of new kinds of genetic diversity. These companies add a new dimension to the greater plant-breeding community, increasing its diversity as well as its plant breeding capabilities.

The Public Sector

Public plant breeding research has changed, in part, because of increased emphasis on biotechnology, but also as part of a trend to devolve cultivar development to the private sector. A further impetus to change in public plant breeding has been the gradual loss of stable funding for long-term projects, such as cultivar development. Funding for public plant breeding primarily now comes from competitive short-term grants, typically three years. Public research in plant breeding has moved toward investigations that produce intellectual products (e.g., publications) rather than biological products (e.g., cultivars). An unfortunate consequence is that the public sector has fewer field-experienced breeders and, therefore, has less capacity to train field breeders in agricultural universities.

A second change in organization and funding of public plant breeding research has given public researchers further inducement to work for results in the short term. Administrators in the public sector advocate plant variety protection certificates, patents, and ensuing royalties as a way to supplement increasingly scarce research funds. Scientists, therefore, are motivated to do their work with the promise of producing patentable products or knowledge in the near term and have a diminished incentive to do chancy long-term research. For example, they have little incentive to do prebreeding ("germplasm enhancement") of exotic germplasm. Prebreeding is essential if one is to use exotic (and by definition unadapted) germplasm as a source of needed genetic diversity for breeders' elite but narrow-based breeding pools. But such breeding typically requires many years of sustained effort and, in the end, may have less than spectacular results even though the occasional successes may be groundbreaking and of great value. (Some breeders argue that such successes, even though rare, are the only sure way to ensure continuing gains in yield and other important traits over the long term.) Emphasis on patentable research also can diminish incentives for public-sector breeders to work on cultivar development in the minor crops (sometimes called "orphan crops"), such as oats or red clover. By definition, the minor crops have small seed markets and, therefore, small possibility for income generation via plant variety protection or patents.

During the past 30 years, a new kind of public sector for plant breeding has arisen, organized and funded on a new model. The Consultative Group on International Agricultural Research (CGIAR) is composed of several widely scattered semiautonomous research centers, each with responsibilities for breeding specific crops. Most of the centers are located in developing countries; they were organized and still operate primarily to serve the rural poor in developing countries. Plant breeding is at the heart of many of the centers. Self-built, the CGIAR and its centers depend on grants from public agencies and private foundations worldwide. They have no power of taxation, nor do they support themselves with product sales.

Socioeconomic Considerations

One more change, a new force, affects plant breeding. A global coalition of environmental and social action NGOs has attacked plant breeding that is assisted by biotechnology. They say its products (in particular, genetically engineered cultivars) are inherently dangerous to human health, to the environment, and to the well-being of society. A well-organized and well-financed campaign has succeeded in banning transgenic (genetically engineered) cultivars from use or cultivation in many countries of the world. The campaign is aimed primarily at the private sector and its use of biotechnology ("the biotechnology industry"). But the effects of the antibiotechnology campaign are beginning to have effects on plant breeding of any kind. Night raiders impartially destroy public and private plant breeders' field plantings and laboratory experiments, traditional as well as transgenic. Professional plant breeding of any kind is said to be reductionist and, therefore, bound to run afoul of nature that operates holistically. Professional plant breeding is said to serve the needs of an industrialized, globally organized food sector and, therefore, harms the cause of subsistence farmers and small-scale farm-to-market producers.

SUMMARY AND PREDICTIONS

Given today's setting of professional plant breeding with potentially greater scientific capability than ever before, with greater private sector involvement than ever before, with weakening public support of plant breeding, and with strong antagonism to biotechnology in plant breeding, what will happen in the future?

A Tripartite System

I see the potential for three parallel movements in crop breeding.

Private Sector

Commercial farmers who produce major commodity crops, worldwide, could be served primarily by the private sector. Breeders would use biotechnology as an auxiliary tool primarily to provide cultivars with more stress tolerance and more durable pest resistance. The emphasis would be on improving existing genetic systems rather than on introducing foreign genes. However, novel genes would be introduced to, for example, enable development of specialty cultivars that produce unique high-value products, such as pharmaceuticals, that provide foods with a sorely needed nutritional improvement, or that provide needed pest resistance that cannot be obtained from within the crop species or its cross-fertile relatives.

Public Sector

Growers of minor crops could be served primarily by the public sector in both industrialized and developing countries. In developing countries, the public sector also would be the primary provider of improved versions of self-pollinated, clonally propagated, and open-pollinated crop varieties; the private sector in those countries would concentrate on hybrid crops. Exceptions would be sectors of developing countries in which the technology and economics of commercial crop production closely resemble that in industrialized countries. In those sectors, private breeders would be the primary providers of improved varieties of the major commodity crops.

CGIAR centers would be especially important to the developing countries. They would furnish cultivars to the poorest countries and germplasm and breeding theory to the richer of the developing countries.

Crop breeding in the public sector may move toward a semicommercial status for some crops. Farmers may choose to support public-sector breeding with checkoff funds (often at levels well beyond present practice) in those cases where they can depend on neither public institutions nor private industry to fund breeding of the new varieties that they need. This action would be added to the current practice in which the public sector obtains intellectual property rights on its varieties and then licenses them for a fee to individual seed companies that wish to sell them. In either case, public breeders do "breeding for hire." Their customers are farmers or seed companies.

In addition to variety development, the public sector in all countries of the world, often in collaboration with the private sector, would concentrate on development of new breeding materials and new breeding theory supported by experiments to test the theories. The research findings of biologists in nonagricultural universities would be a significant supplement to this research. Their research in plant molecular biology, for example, would develop knowledge and products that could be used in practical plant breeding.

The public sector would continue to be the primary agent for training new plant breeders at its bases in agriculture-oriented universities. It seems likely that, in some instances, industry would need to help, because of the trend toward fewer field breeders at the universities. Industry could provide field experience for students and perhaps experience in specialized laboratory investigations. Internships or similar on-site training programs might allow students to earn credit while working as temporary employees in qualified seed companies.

Participatory Plant Breeding

Farmers who maintain their own germplasm would be aided in interactive fashion by a new kind of NGO. The new breed of NGO would provide advice and germplasm as needed and requested by farmers, in ways that would increase the farmers' effectiveness in maintaining and improving their varieties. The farmer-breeders (acting either as individuals or in associations such as communities) and their NGO partners would produce varieties with utility in farming systems that are not well served (or not served at all) by formal plant breeding, either public or private. The NGOs, like the present CGIAR centers, would need to be self-supporting via grants from public and private agencies. Some of the CGIAR centers themselves might support some of these activities.

Interactions

This tripartite division of plant breeding operations and services would call for more than improved science and technology. It would call for a broadly shared understanding of a wide range of product needs and how they fit into a diverse assemblage of farming systems and socioeconomic conditions. Most important, it would require an understanding of how the three systems and intergrades among them comprise a single plant breeding network.

What techniques, science, and organizational methods would be common to all three systems, or to any two of them? For example, all will share the need to generate segregating populations for selection of new improved genotypes. But will they all share the need—and appetite—for products of biotechnology, such as the new knowledge and power to be gained from genomics or transgenic organisms? If not, what problems would this present?

At what points can the systems help each other and when will they compete? For example, private industry has amassed considerable bioinformatics data, most of them treated as trade secrets or patented. This proprietary information can be used or acquired only by satisfying terms specified by the owners (e.g., through licensing, reciprocal exchanges). Subsistence farmers in developing countries hold and continually modify landraces—farmer varieties—which collectively are an invaluable global storehouse and generator of crop genetic diversity. But landraces now are considered as proprietary germplasm, more or less as though they were patented with no expiration date. The 1992 Convention on Biological Diversity states that they are not a public good but instead are subject to sovereign rights of states over their natural resources. They can be used or acquired only by satisfying terms specified by the owners—the farmers, villagers, or ultimately the country of origin. Can patented DNA sequences and "patented" landraces be shared or exchanged to the benefit of both classes of owner? Or is it a mistake to assume that these two categories of germplasm can be judged by the same standards, monetary or otherwise?

A Prediction

Finally, and of greatest importance for the future, how much support will there be for plant breeding, and what kinds of plant breeding will be supported or allowed?

On the one hand, we see a steady decline in funding for public-sector plant breeding, a direct result of declining public interest in plant breeding and its products. The decrease in monetary support is shared by the CGIAR centers as well as by government and university plant-breeding programs. Industrialized countries, because of their overproduction, see little reason to support research that will add to the surpluses. They see more reason to support biological research to improve environmental health or (especially in developing countries) to increase social justice.

On the other hand, forceful and eloquent alliances deprecate and hope to curtail private-sector plant breeding, particularly when it is aided by biotechnology and/or serves farmers in developing countries. Private-sector plant breeding has a further problem in that the recent acquisitions and mergers of the large seed companies have placed them in an uncomfortable state of reorganization and realignment. Straightforward cultivar development is hindered when people, systems, and research goals are changed too much or too often.

One can predict a future in which plant breeding of all kinds and by all sectors will be significantly scaled back, both scientifically and quantitatively. Biotechnology—or at least genetic engineering—will be forbidden for use as a tool in plant breeding, and private-sector breeding will be greatly reduced in scope and effectiveness. Alternatively, one can predict a future in which commercialism will overtake all sectors, such that all plant breeding research, public and private, will be aimed at producing salable (and proprietary) products in the short term—products such as cultivars, germplasm fragments, or patentable technical processes. Fundamental plant breeding research and long-term, risky germplasm development projects will be reduced to insignificant amounts, even lower than present inadequate levels.

But a more optimistic outlook would predict that an approximation of the tripartite division of responsibilities that was described earlier in this section eventually will prevail. An inevitable and massive increase in world population during at least the next several decades will force the production of more food. Since very little land remains that profitably and safely can be converted to farming, production per unit area must increase, and this is what plant breeders have demonstrated they can do by producing cultivars that make more yield per unit area. Water supplies for agriculture also will soon be limited, and, again, plant breeders have demonstrated their ability to select for tolerance to drought at any level of severity. Thus, it seems likely that the public eventually will realize the essential role of plant breeding in our global efforts to feed a burgeoning population. They will be forced to learn how to evaluate rationally the utility and safety of new technologies (including biotechnology) with potential to benefit the world's people and the environment in which we live. And they will support actions that increase plant breeding's capacity to help feed the world in ways that are environmentally and socially sound.

Gradual increases in the economic well-being of people in developing countries and accompanying growth in their urban populations will foster the development of a reasonably stable commercial agriculture in those countries. This, in turn, will mean that increasing numbers of their farmers will be able to afford and profit from the products of professional plant breeding, both public and private, in appropriate crops and growing regions.

The organization of commercial plant breeding will continue to evolve, as corporations divest plant breeding operations that are insufficiently profitable and new combinations arise. Probably a few plant breeding companies will dominate globally, as in the past, but they will be balanced by a large number of small companies, operating vigorously in developing as well as industrialized countries. The small companies collectively will provide cultivars for a significant share of the market and, thereby, hold back incipient tendencies toward monopoly or oligopoly. A viable public-sector plant breeding system will be an important means of support for the small-company segment. It will provide germplasm, cultivars, and other research results that the small companies individually cannot furnish for themselves.

A new class of professional breeders gradually will learn how to work as partners with a select category of farmer-breeders. Their partnership will bring the benefits of appropriate plant breeding technology and germplasm to those farmers and those regions of the world that need them but cannot access them through standard plant breeding institutions. In some cases, the farmer-breeders (as individuals or as community associations) will evolve into full-time for-profit breeders, serving local needs on a scale and at prices that cannot be matched by traditional commercial seed companies. They will know how to access advanced breeding institutions for techniques and germplasm appropriate to the special needs of their local farming communities. Such evolution toward commercialism on a small local scale may be the best way to ensure long-term continuity and independence of farmer/professional partnerships that serve the needs of farmers who fall outside the purview of traditional breeding institutions.

In Conclusion

This prediction of what might develop—of three sectors interacting, intermingling, and depending on each other—points out the essential unity of the global plant breeding system. Like the plants that it manages, it is composed of separate organs, each essential for the survival of the whole but each with a different function. The global plant breeding system will operate most efficiently and serve its primary customers (the farmers) most adequately if all concerned understand this unity in diversity, this system in which all parts work together for the good of the whole.

REFERENCES

Frey, K. J. 1996. National plant breeding study I: Human and financial resources devoted to plant breeding research and development in the United States in 1994. Iowa Agriculture and Home Economics Experiment Station, and Cooperative State Research, Education and Extension Service/USDA cooperating, No. 98.

Heisey, Paul. USDA/ERS. (Personal communication, March 6, 2000).

PART II:
CROP IMPROVEMENT

Chapter 2

Rice Improvement: Past, Present, and Future

Gurdev S. Khush
Parminder S. Virk

INTRODUCTION

Of the three major cereals (rice, wheat, and maize) that feed the world, rice (*Oryza sativa* L.) is the most important. In 1999, it was planted on 153 million hectares (ha) of land with a production of 587 million tons. In contrast, wheat and maize were planted on 214 and 139 million ha with a production of 578 and 600 million tons, respectively. Unlike maize and wheat that are used partly as animal feed, rice is consumed exclusively by human beings. It is a staple food of 40 percent of the world population; for another 20 percent, it is a major item of diet. In the developed nations where it is not a staple food, rice consumption is increasing due to immigration and diversification in the diets of inhabitants. For Asians, rice is the lifeline. More than 90 percent of the world rice is produced and consumed in Asia—the most densely populated region of the world.

Rice was first domesticated in southern China and northeastern India—probably independently—about 8,000 years ago (Khush, 1997). Until about 1900, farmers themselves were responsible for most rice improvement. The best known examples are the rono varieties, such as Shinriki in Japan (Khush, 1987b). These were shorter in stature and, therefore, responded to nutrient inputs with higher yields. Systematic rice breeding started in the early twentieth century with the establishment of breeding stations in China, India, and Japan. Rice breeders' initial activities were the purification of existing varieties through pure-line selection. The development of varieties through crossbreeding or hybridization had little impact until the 1950s. Varieties, such as Mahsuri and ADT 27, developed by hybridization between indica and japonica varieties under the auspices of the International Rice Commission in the early 1950s, are still popular among farmers in South and Southeast Asia. Taichung Native 1 (TN1), the first semidwarf rice variety, was released in 1956 by agricultural scientists in Taiwan. TN1 was derived from a cross between 'Dee-geo-woo-gen,' a dwarf rice from China, and a local variety 'Tsai-Yuan-Chung.' Because of its dwarf phenotype, it was resistant to lodging when nitrogenous fertilizers were applied.

Despite the impact of varieties such as Mahsuri, ADT 27, and TN1, rice farmers in tropical and subtropical Asia grew tens of thousands of unimproved or semi-improved varieties even in the early 1960s. These varieties were low yielding, tall and weak-stemmed, and late maturing. Their harvest index (grain:straw ratio) was very low (0.3) and when nitrogenous fertilizers were applied at rates >40 kg·ha^{-1}, many tillered profusely, grew excessively tall, lodged early, and yielded less than they would with lower fertilizer inputs. In addition, many were photoperiod or daylength sensitive and could be grown only in limited areas due to their specific adaptation (Khush, 1987a).

The International Rice Research Institute (IRRI) was established in 1960 with the objective of increasing production of rice through interdisciplinary research. It was recognized that to enhance the yield potential of tropical rice, it would be necessary to increase harvest index and lodging resistance, hence nitrogen responsiveness. This was accomplished by reducing the plant stature through the incorporation of a recessive gene, *sd-1*, for short stature from the Chinese variety Dee-geo-woo-gen. In 1962, IRRI scientists crossed Dee-geo-woo-gen with Peta (a vigorous variety from Indonesia), which led to the release of IR8 variety in 1966. It was widely accepted because of its superior yield potential. IR8 also had a number of other desirable features, such as profuse tillering, dark green and erect leaves, and sturdy stems. It responded to nitrogenous fertilizers much better than traditional varieties such as Peta (see Figure 2.1), had a harvest index of 0.5, and growth duration of 130 days. Being photoperiod insensitive, it could be planted at any time of the year in the tropics.

The success of IR8 in raising the yield potential of rice was so impressive and revolutionary that rice breeders the world over immediately initiated breeding programs to develop short-statured varieties. Most of the improved rice varieties developed after IR8 at IRRI and by National Agricultural Research Systems (NARS) have the short stature conditioned by *sd-1*.

POST–GREEN REVOLUTION VARIETIES

Apart from high yield potential, improved rice varieties have many desirable features that were lacking in pre–green revolution varieties. Some of these features are:

- Short growth duration
- Multiple disease and insect resistance

- Grain quality
- Tolerance to abiotic stresses, especially those related to problem soils

Short Growth Duration

Pre–green revolution varieties in tropical and subtropical Asia matured in 160 to 170 days and many were photoperiod sensitive. Because of their long duration, these were not suitable for multiple cropping systems. IR8 and subsequent varieties, such as IR20 and IR26, matured in about 130 days. Even these earlier maturing varieties were suitable for growing only one crop of rice in one rainy season, which meant that farmers could not grow another crop after rice or a second crop of rice in one rainy season. Therefore, a major emphasis was placed on developing improved varieties with even shorter growth durations. IR28 and IR30, released in 1974, and IR36, developed in 1976, mature in 110 days. Growth duration was further reduced to 105 days in IR50 and IR58. However, yield is a primary consideration in a variety development program. Therefore, during the selection process, only short-duration lines with yield potential matching that of medium-duration varieties were saved. This was achieved by selecting geno-

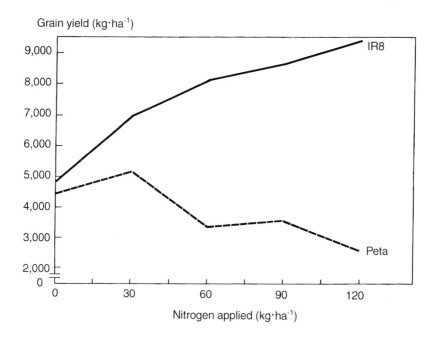

FIGURE 2.1. Nitrogen response of Peta and IR8 during 1966 dry season at IRRI.

types with rapid vegetative vigor at early growth stages (Khush, 1987a). Because of higher earlier growth rates at earlier stages, the short-duration varieties, such as IR36, IR64, and IR72, are able to produce the same amount of biomass in 100 to 115 days as the medium-duration varieties do in 130 to 135 days. Under most situations, yields of early- and medium-duration varieties are similar and their per day productivity is higher. For example, in replicated yield trials at the IRRI during 1997, short- and medium-duration varieties produced, on average, 53 kg/day vis-à-vis 35 kg/day grains from a late-maturing variety, IR74.

The availability of short-duration varieties has led to major changes in cropping patterns in Asia (Khush, 1995b). In the Philippines, many farmers grow an upland crop either before or after rice under rainfed conditions. In some areas, two crops of rice are regularly grown during the rainy season and, in irrigated areas, farmers grow three crops in one year. In Indonesia and Vietnam, where short-duration varieties have been widely adopted, the area under rice double-cropping has increased rapidly. Short-duration varieties are excellent for input economy, because they grow rapidly during the vegetative phase and are, thus, more competitive with weeds. Consequently, weed control costs are reduced. They also utilize less irrigation water, thus further reducing the production costs. The availability of short-duration varieties led to major increases in cropping intensity, greater on-farm employment, increased food supplies, and higher food security in Asian countries. Many national rice improvement programs have also developed short-duration varieties, such as Ratna, Cauvery, and TKM9 in India; BG34-7 and BG367-7 in Sri Lanka; and Chandina in Bangladesh.

Multiple Disease and Insect Resistance

The variety composition and cultural practices for rice have changed significantly during the post-IR8 era. A relatively small number of improved varieties have replaced thousands of traditional cultivars, thereby reducing the genetic variability of the crop (Khush, 1977). Farmers have adopted improved cultural practices, such as application of more fertilizers and establishment of higher plant populations per unit area. Availability of short-duration, photoperiod-insensitive varieties, coupled with the development of irrigation facilities, has enabled farmers in tropical Asia to grow successive crops of rice throughout the year. Reduced genetic variability because of large-scale adoption of short- or medium-duration varieties and continuous cropping with rice have increased the genetic vulnerability of the crop. Chemical control of diseases and insects for prolonged periods in tropical climate is very expensive and damaging to the environment. The use of host-plant resistance for disease and insect control is the logical approach to overcome these production constraints. Therefore, IRRI's rice improvement program has placed a major emphasis on developing germplasm with multiple resistance to major diseases and insects. Similarly, many national programs have given priority to developing varieties with multiple disease and insect resistance.

In most tropical and subtropical Asian countries, five diseases (blast, bacterial blight, sheath blight, tungro, and grassy stunt) and four insects (brown planthopper, green planthopper, stem borers, and gall midge) are of major importance. At IRRI, we continue to concentrate our efforts on developing germplasm with multiple resistance to these major diseases and insects. A large number of germplasm collections were screened and donors for resistance identified at IRRI (Khush, 1977), which led to the development of improved varieties with resistance to as many as four diseases and four insects (Khush, 1992). The first variety with multiple resistance, IR26, was released in 1973. Earlier IRRI varieties released in the Philippines (IR5, IR8, IR20, IR22, and IR24) were susceptible to most of the diseases and insects. Since 1973, many varieties with multiple resistance have been developed at IRRI (Table 2.1) and by many national programs. These varieties have as many as 20 different parents in their ancestry, which helped restore genetic diversity to some extent (Khush, 1995b).

Large-scale adoption of varieties with multiple resistance has helped stabilize world rice production. This can be illustrated with the following example. The yield of susceptible IR8 fluctuates from year to year much more than those of varieties with multiple resistance, e.g., IR36 and IR42 (see Figure 2.2), that have greater yield stability. However, resistant varieties do not remain resistant forever. They become susceptible due to the development of new races of pathogens and/or biotypes of insects. At IRRI, we have endeavored to identify diverse genes for resistance to each of the diseases and insects. When a variety with a particular gene breaks down, a new one with different genes is released (Khush, 1992). In this regard, we have transferred resistant genes from wild species of rice to the cultivated ones. For example, we have recently bred elite lines resistant to tungro by backcrossing IR64, a very popular variety, and a tungro-resistant accession of *O. rufipogon*.

Breeding for resistance to some diseases, such as sheath blight, remains elusive owing to nonavailability of resistance source(s) within the gene pools of rice. This is where genetic engineering holds much promise.

Grain Quality

Grain quality of rice is evaluated relative to several consumer-oriented criteria (Khush, 1995b). Most consumers in the tropics and subtropics prefer long or medium long, slender and translucent grains. However, in the temperate areas short, bold, and roundish grains are preferred by the consumers. Higher milling is a universal requirement and is, to some extent, dependent on the size, shape, and amount of chalkiness in the grains.

TABLE 2.1. Disease and Insect Reactions of IR Varieties of Rice

IR Variety	Blast	Bacterial	Grassy stunt	Tungro	GLH	BPH biotype			Stem borer	Gall Midge
						1	2	3		
R5	MR	S	S	S	R	S	S	S	MS	S
IR8	S	S	S	S	R	S	S	S	S	S
IR20	MR	R	S	MR	R	S	S	S	MR	S
IR22	S	R	S	S	S	S	S	S	S	S
IR24	S	S	S	S	W	S	S	S	S	S
IR26	MR	R	MR	MR	R	R	S	R	MR	S
IR28	R	R	R	R	R	R	S	R	MR	S
IR29	R	R	R	R	R	R	S	R	MR	S
IR30	MS	R	R	MR	R	R	S	R	MR	S
IR32	MR	R	R	MR	R	R	R	S	MR	R
IR34	R	R	R	R	R	R	S	R	MR	S
IR36	R	R	R	R	R	R	R	S	MR	R
IR38	R	R	R	R	R	R	R	S	MR	R
IR40	R	R	R	R	R	R	R	S	MR	R
IR42	R	R	R	R	R	R	R	S	MR	R
IR44	R	R	S	R	R	R	R	S	MR	S
IR46	R	R	S	MR	MR	R	S	R	MR	S
IR48	R	R	R	R	R	R	R	S	MR	-
IR50	MS	R	R	R	R	R	R	S	MR	-
IR52	MR	R	R	R	R	R	R	S	MR	-
IR54	MR	R	R	R	R	R	R	S	MR	-
IR56	R	R	R	R	R	R	R	R	MR	-
IR58	R	R	R	R	R	R	R	S	MR	-
IR60	R	R	R	R	R	R	R	R	MR	-
IR62	MR	R	R	R	R	R	R	R	MS	-
IR64	MR	R	R	R	R	R	MR	R	MR	-
IR65	R	R	R	R	R	R	R	S	MS	-
IR66	MR	R	R	R	R	R	R	R	MR	-
IR68	MR	R	R	R	R	R	R	R	MR	-
IR70	R	S	R	R	R	R	R	R	MS	-
IR72	MR	R	R	R	R	R	R	R	MR	-
IR74	R	S	R	R	R	R	R	R	MR	-

Notes: S = susceptible, MS = moderately susceptible, MR = moderately resistant. Reactions were based on tests conducted in the Philippines for all disease and insects except gall midge conducted in India.

Yield (t·ha⁻¹)

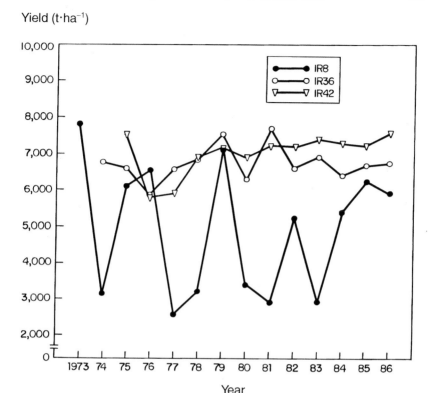

Year

FIGURE 2.2. Yields of IR8, IR36, and IR42 for dry-season, replicated yield trials at IRRI. Yields of multiple resistant IR36 and IR42 show little year-to-year variation; yield of susceptible IR8 fluctuates widely.

Cooking quality is another factor that is important to consumers and is determined largely by the amylose content and gelatinization temperature of rice starch. In temperate areas of China, Korea, and Japan, consumers prefer rice with a low amylose content and low gelatinization temperature, which gives a moist and sticky cooked product. On the other hand, in the tropics and subtropics, varieties with intermediate amylose content and intermediate gelatinization temperature are preferred.

Improvement of milling recovery and grain appearance received immediate attention in the early years of the breeding program at IRRI. The earliest IRRI varieties, such as IR5 and IR8, have poor grain quality. They have bold and chalky grains of poor appearance that frequently break during milling. In addition, they cook dry because of high amylose content and have poor consumer acceptance. All the varieties released after IR5 and IR8 have slender and translucent grains and have very good milling recovery. However, improvements in cooking quality were only slowly achieved primarily due to the fact that all the donors for disease and insect resistance used in the hybridization program had high amylose content and low gelatinization temperature. IR64 is the first IRRI variety, released in 1985, that has a desirable combination of intermediate amylose content and intermediate gelatinization temperature. Before the release of IR64, several varieties with improved rice quality were released, but they all lacked a desired combination of amylose content and gelatinization temperature. For example, IR48 has an intermediate amylose content but low gelatinization temperature. Similarly, IR20, IR32, IR36, and IR46 have intermediate gelitinization temperatures but high amylose content. Not surprisingly, therefore, IR64 has been widely accepted as a high-quality rice in the Philippines, India, Indonesia, Vietnam, Bhutan, and several other countries. In fact, IR64 is planted on about 9 million ha of rice land and is the most widely planted variety of rice in the world today.

Another important quality characteristic of rice is the presence of aroma. Aromatic rices fetch a premium price in the world market. None of the improved varieties is aromatic. However, several elite breeding lines with aromatic properties have been developed and are in the advanced stages of testing.

Tolerance to Abiotic Stresses

Large areas of land suitable for growing rice remain unplanted because of severe nutritional deficiencies and toxicities. A vast majority of rice soils has varying levels of alkalinity or salinity. Even well-managed rice lands suffer from mild nutritional deficiencies or toxicities. For example, zinc deficiency in rice soils is becoming a common concern in many countries. Several improved varieties possess moderate to high levels of tolerance to several nutritional deficiencies and toxicities. IR36, for example, has a tolerance to salinity, alkalinity, peatness, and iron and boron toxicities. It also tolerates zinc deficiency (IRRI, 1982). Similarly, IR42 has a broad spectrum of tolerance to many soil problems (see Table 2.2). These varieties also have a more stable yield performance across years (see Figure 2.2) and have been helpful for reclamation of degraded and marginal lands.

Combination of Desirable Traits

For widespread adoption, a variety must have a favorable combination of traits, such as high yield potential, desirable growth duration, multiple resistance to diseases and insects, superior grain quality, and tolerance to abiotic stresses. Many rice varieties with such desirable attributes have been developed. IR36 released in 1976 was the first such variety. It has a high harvest

TABLE 2.2. Reactions of Some IR Varieties to Adverse Soil Conditions

Variety	Toxicity				Deficiency		
	Salinity	Alkalinity	Peat	Iron	Boron	Phos.	Zinc
IR5	4	7	0	6	4	5	5
IR8	3	6	5	7	4	4	4
IR20	5	7	4	2	4	1	3
IR28	7	5	6	4	4	3	5
IR36	3	3	3	3	3	7	2
IR42	3	4	5	3	2	3	4
IR48	4	7	5	6	0	5	5
IR64	3	3	4	5	4	4	4

Notes: Scale of 0–9; 0–no information; 1–almost normal plant; 9–almost dead or dead plant.

index (0.55), long slender grains, and high yield potential and it matures in 110 days. It also has multiple resistance to major diseases and insects and is tolerant to abiotic stresses. Because of these desirable attributes, it was widely accepted and soon became the most widely planted variety of rice or any other crop the world has ever known. During the early 1980s, it was planted on 11 million ha of rice land all over the world. Subsequent releases, such as IR64, IR72, PSBRc18, and PSBRc52, have all the desirable attributes of IR36 but yield better or possess superior quality grains (Khush, 1995b). The stepwise improvements that have been incorporated into rice varieties are shown in Figure 2.3.

IMPACT OF MODERN VARIETIES

Modern varieties have almost completely replaced the traditional varieties grown under irrigated conditions in several countries, *viz.,* Japan, China, South Korea, the United States, and Egypt. The adoption rate is relatively low in some countries, such as Thailand, Myanmar, Cambodia, and Brazil, where a substantial portion of rice is grown under rainfed conditions, both in the uplands and lowlands. On average, more than 70 percent of the rice area of the world is planted with modern varieties.

In the Philippines, 46 IRRI inbred lines have been released, for irrigated and favorable rainfed lowland rice systems, as varieties (Table 2.3). The number of IRRI breeding lines named as varieties, in national programs of several rice-growing countries, has already surpassed 300. More than 1,000 improved varieties also have been developed by various national programs, 75 percent of them being the progenies of crosses with IRRI-bred varieties

FIGURE 2.3. Stepwise improvement in rice varieties. Right to left: Leb Mue Nahang, a tall traditional variety is photoperiod sensitive, susceptible to diseases and insects and has a harvest index of 0.3; Peta, a semi-improved variety is tall, photoperiod insensitive, matures in 155 days, is susceptible to diseases and insects and has a harvest index of 0.35. IR8, an improved short-statured variety, matures in 135 days, is susceptible to diseases and insects, and has a harvest index of 0.5. IR36, an improved short-statured variety, matures in 110 days, is resistant to diseases and insects and has a harvest index of 0.5.

or elite breeding lines (Khush, 1995b). A Yale economist, Bob Evenson, has computed the annual net worth of each released variety to be US $2.5 million, which means the annual net worth of IRRI-bred materials exceeds US $2.7 billion.

Large-scale adoption of modern varieties, coupled with improved management practices, has dramatically increased rice production in major rice-growing countries. Farmers get 5 to 7 t·ha[-1] of unmilled rice from high-yielding varieties (HYVs), compared with 1 to 3 t·ha[-1] from traditional varieties. Farmers have raised average yields by substituting HYVs for traditional ones, thereby increasing the proportion of the total area planted with HYVs. Since 1966, when the first HYV was released, rice-harvested area increased only marginally from 126 to 153 million ha (17 percent), whereas average rice yield has increased from 2.1 to 3.6 t·ha[-1] (71 percent).

TABLE 2.3. IRRI Lines Named as Varieties in the Philippines

Variety	Year released	Line name	Parents
IR8	1966	IR8-288-3	Peta
			Dee-geo-woo-gen
IR5	1967	IR5-47-2	Peta
			Tangkai Rotan
IR20	1969	IR532-E576	IR262-24-3
			TKM6
IR22	1969	IR579-160-2	IR8
			TADUKAN
IR24	1971	IR661-1-140-3	IR8
			IR127-2-2
IR26	1973	IR1541-102-7	IR24
			TKM6
IR28	1974	IR2061-214-3-8-2	IR833-6-2-1-1
			IR1561-149-1//IR24*4/*O. Nivara*
IR29	1974	IR2061-464-4-14-1	IR833-6-2-1-1
			IR1561-149-1//IR24*4/*O. Nivara*
IR30	1974	IR2153-159-1-4	IR1541-102-6-3
			IR20*4/*O. Nivara*
IR32	1975	IR2070-747-6-3-2	IR20*2/*O. Nivara*
			CR94-13
IR34	1975	IR2061-213-2-17	IR833-6-2-1-1
			IR1561-149-1//IR24*4/*O. Nivara*
IR36	1976	IR2071-625-1-252	IR1561-228-1-2/IR1737
			CR94-13
IR38	1976	IR2070-423-2-5-6	IR20*2/*O. Nivara*
			CR94-13
IR40	1977	IR2070-414-3-9	IR20*2/*O. Nivara*
			CR94-13
IR42	1977	IR2071-586-5-6-3	IR1561-228-1-2/IR1737
			CR94-13
IR43	1978	IR1529-430-3	IR305-3-17-1-3
			IR661-1-140-3
IR44	1978	IR2863-38-1-2	IR1529-680-3/CR94-13
			IR480-5-9-3
IR45	1978	IR2035-242-1	IR1416-128-5/IR1364-37-3-1

TABLE 2.3 (continued)

Variety	Year released	Line name	Parents
			IR1824-1
IR46	1978	IR2058-78-1-3-2	IR1416-131-5/IR1364-37-3-1
			IR1366-120-3-1/IR1539-111
IR48	1979	IR4570-83-3-3	IR1702-74-3-2/IR1721-11-6-8-3
			IR2055-481-2
IR50	1979	IR9224-117-2-3-3-2	IR2153-14-1-6-2/IR28
			IR36
IR52	1980	IR5853-118-5	NAM SA-GUI 19/IR2071-88
			IR2061-214-3-6-20
IR54	1980	IR5853-162-1-2-3	NAM SA-GUI 19/IR2071-88
			IR2061-214-3-6-20
IR56	1982	IR13429-109-2-2-1	IR4432-53-33/PTB33
			IR36
IR58	1983	IR9752-71-3-2	IR28/KWANG-CHANG-AI
			IR36
IR60	1983	IR13429-299-2-1-3	IR4432-53-33/PTB33
			IR36
IR62	1984	IR13525-43-2-3-1-3-2	PTB33/IR30
			IR36
IR64	1985	IR18348-36-3-3	IR5657-33-2-1
			IR2061-465-1-5-5
IR65	1985	IR21015-196-3-1-3	BATATAIS/IR36
			IR52
IR66	1987	IR32307-107-3-2-2	IR13240-108-2-2-3
			IR9129-209-2-2-2-1
IR68	1988	IR28224-3-2-3-2	IR19660-73-4/IR2415-90-4-3-2
			IR54
IR70	1988	IR28228-12-3-1-1-2	IR19660-73-4/IR54
			IR9828-36-3
IR72	1988	IR35366-90-3-2-1-2	IR19661-9-2-3/IR15795-199-3-3
			IR9129-209-2-2-2-1
IR74	1988	IR32453-20-3-2-2	IR19661-131-1-2
			IR15795-199-3-3
PSBRC2	1991	IR32809-26-3-3	IR4215-301-2-2-6/BG90-2
			IR19661-131-1-2

Variety	Year released	Line name	Parents
PSBRC4	1991	IR41985-111-3-2-2	IR4547-4-1-2/IR1905-81-3-1
			IR25621-94-3-2
PSBRc10	1992	IR50404-57-2-2-3	IR33021-39-2-2
			IR32429-47-3-2-2
PSBRc18	1994	IR51672-62-2-1-1-2-3	IR24594-204-1-3-2-6-2
			R28222-9-2-2-2-2
PSBRc20	1994	IR57301-195-3-3	IR35293-125-3-2-3/IR32429-47-3-2-2
			PSBRc4
PSBRc28	1995	IR56381-139-2-2	IR28239-94-2-3-6-2
			IR64
PSBRc30	1995	IR58099-41-2-3	IR72
			IR24632-34-2
PSBRc52	1997	IR59682-132-1-1-2	IR48613-54-3-3-1
			IR28239-94-2-3-6-2
PSBRc54	1997	IR60819-34-2-1	IR72
			IR48525-100-1-2
PSBRc64	1997	IR59552-21-3-2-2	PSBRc2
			IR39292-142-3-2-3
PSBRc80	2000	IR62141-114-3-2-2-2	IR50401-77-2-1-3
			IR42068-22-3-3-1-3
PSBRc82	2000	IR64683-87-2-2-3-3	IR47761-27-1-3-6
			PSBRc28
PSBRc84	2000	IR65185-3B-8-3-2	CSR10
			TCCP266-B-B-B-10-3-1
PSBRc86	2000	IR65195-3B-8-2-3	IR10198-66-2
			TCCP266-B-B-B-10-3-1
PSBRc88	2000	IR52713-2B-8-2B-1-2	IR64
			IR 4630-22-2-5-1-3/IR9764-45-2-2

Total rice production has more than doubled since 1966. The extent of increase in production for major rice-growing countries since the introduction of HYVs is given in Table 2.4 (Khush, 1999). In most countries, rice production grew faster than population. Indonesia, Philippines, and India, which were major rice-importing countries, became self-sufficient in rice production in the late 1970s. This also helped the increase in cereal consumption and calorie intake per capita. During 1965 to 1990, the daily calorie supply in relation to requirement improved from 81 to 120 percent for In-

TABLE 2.4. Total Area, Coverage of High-Yielding Varieties, and Increase in Rice Production in Selected Countries of Asia

Country	Total area planted (M. hectares)		Area planted to HYV	Production (M. tons)		Increase in production
	1966	1996	%	1966	1996	%
Bangladesh	9.1	10.3	46	14.3	28.0	96
China	31.3	31.4	100	98.5	190.1	93
India	35.2	42.7	75	45.6	120.0	163
Indonesia	7.7	11.3	77	13.6	51.2	276
Myanmar	4.5	6.5	58	6.6	20.9	217
Pakistan	1.4	2.3	41	2.0	5.6	180
Philippines	3.1	4.0	94	4.1	11.3	176
Sri Lanka	0.5	0.8	94	1.0	2.2	120
Thailand	7.3	9.2	13	13.5	21.8	61
Vietnam	4.7	7.3	80	8.5	26.3	209

donesia, 86 to 110 percent for China, 82 to 99 percent for Philippines, and 89 to 94 percent for India (UNDP, 1994).

The increase in per capita availability of rice and a decline in the cost of production per ton of output contributed to a decline in real prices both in the international and domestic markets of rice and wheat. The unit cost of production is about 20 to 30 percent lower for high-yielding varieties compared to traditional varieties of rice (Yap, 1991), and the price of rice adjusted for inflation was 40 percent lower compared with the level in the mid-1960s (Figure 2.4). The decline in food prices has benefited the urban poor and the rural landless who are not directly involved in food production but who spend more than half of their income on food grains. As net consumers of grain, the small and marginal farmers, who are the dominant rice producers in most Asian countries, have also benefited from the downward trend in real prices of rice (Khush, 1999).

The diffusion of high-yielding varieties has also contributed to the growth of income of the rural landless workers (Hossain, 1988; Hayami et al., 1978). High-yielding varieties require more labor per unit of land because of more intensive care in agricultural operations and harvesting of larger output. Labor requirement has also increased from higher intensity of cropping made possible through reduction of time in crop growth. As farm income increases, better-off farm households substitute leisure for family labor and hire more landless workers to do the job. Marketing of a larger volume of produce and increased demand for nonfarm goods and services from larger farm incomes generated additional employment in rural trade, transport, and construction activities. The economic miracle underway in

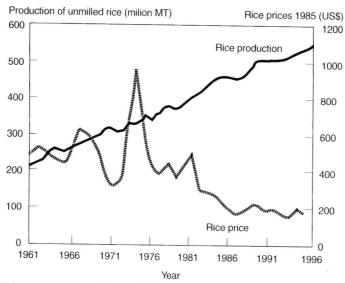

FIGURE 2.4. Trends in world rice production and price, 1961-1996.

many Asian countries was triggered by the growth in agricultural income and its equitable distribution, which helped expansion of the domestic market for nonfarm goods and services.

In sharp contrast to rich countries, where more environmental problems have been urban and industrial, the critical environmental problems in most of the low-income developing countries remain rural, agricultural, and poverty based. More than half of the world's very poor live on land that is environmentally fragile and rely on natural resources over which they have little control. Land-hungry farmers resort to cultivating unsuitable areas, such as erosion-prone hillsides, semiarid areas where soil degradation is rapid, and tropical forests, where crop yields on cleared fields drop sharply after just a few years (Khush, 1999). The widespread adoption of high-yielding varieties has helped most Asian countries meet their growing food needs from productive lands, thereby reducing the pressure to open up more fragile lands. Had 1961 yields still prevailed today, three times more land in China and two times more land in India would be needed to equal 1992 cereal production. If Asia attempted to produce today the 1990 harvest at the yield/ha levels of the 1960s, most of the forests, woodlands, pastures, and rangelands would have disappeared; and mountainsides would have eroded, with disastrous consequences for the upper watershed and productive lowlands (Khush, 1999). In addition, extinction of wildlife habitats and destruction of biodiversity would have been inevitable (Khush, 1999). As an example, to produce the 1999 world rice production of 587 million tons at the yield levels of 1965 would have required an extra 130 million ha of land.

Availability of rice varieties with multiple resistance to diseases and insects reduced the need for application of agrochemicals and facilitated the adoption of integrated pest-management practices. Reduced insecticide use helps: (1) enhance environmental quality, (2) improve human health of farming communities, (3) make safer food available, and (4) protect fauna and flora.

FUTURE CHALLENGES

Rice production has more than doubled since 1966. This is a remarkable achievement in the history of agriculture: that rapidly growing demand was met through the use of fertilizer-responsive improved varieties. Annual rice production expanded at the rate of 2.8 percent between 1975 and 1985. However, the growth in rice harvest slowed down to 1.2 percent between 1990 and 1999. Signs of slower growth were evident in the late 1980s as the growth in grain production fell below that of population, dropping the harvest per person from an all-time high of 342 kg in 1984 to 335 kg in 1990. By 1996, the harvest per person had fallen to 321 kg (Brown, 1997). The world grain stocks similarly declined. For a minimal level of food security, food grain stocks for 70 days of world consumption are needed. Whenever stocks fall below the 60-day supply, prices become volatile. The carryover stocks reached a low of 52 days' consumption in 1996, causing the world price of wheat and maize to more than double and the price of rice to increase by 50 percent.

For the world's affluent, who spend only a small share of their income for food, even a doubling of world grain price would not have a major immediate effect. But for the 1.3 billion people in the world who live on a dollar a day or less, such a rise could be life threatening. Heads of household unable to buy enough food for their families would hold their governments responsible and would likely take to the streets in protest. The resulting political instability in third-world cities would affect the earnings of multinational corporations, the performances of stock markets, the earnings of pension funds, and the stability of the international monetary system. In short, it could disrupt economic progress (Brown, 1997). These gloomy forecasts are reminiscent of the 1960s.

With 1.4 percent annual rate of growth, the world is continuing to add 80 million people a year, 90 percent of which occurs in the developing countries of Asia, Africa, and Latin America. Feeding the additional mouths would require an expansion in the world grain production of 26 million tons per year. Moreover, due to rising living standards, food habits are changing in many countries, particularly in Asia, and people are consuming more meat, eggs, and milk. This is driving the demand for grain up at a rapid rate. A kilogram of beef produced in the feedlot, a kilogram of pork, and a kilogram of poultry require, respectively, 7, 4, and >2 kg of grain (Brown, 1997).

More than a billion people in developing countries live below the poverty level and have poor access to food. As the poverty-alleviation programs in developing countries start to make an impact, the purchasing power of poor people will increase and so will the demand for grain. Based on population projections and improved consumption patterns in developing countries, food grain production will have to increase almost 400 percent in Africa, 200 percent in Latin America, and 50 percent in Asia. Overall, world food grain production must increase by 50 percent (Sinha et al., 1989).

The challenge for increasing rice production is daunting as the population of rice consumers is increasing at the rapid rate of 2 percent annually. The number of rice eaters is expected to increase by 40 percent and rice requirement by 50 percent during the next 25 years. This increase is unlikely to be met by increasing the area planted with rice. In fact, some of the best lands are being taken out of agricultural production due to urbanization, industrialization, and infrastructure needs, such as roads and parking lots. Several other factors contributing to shrinking croplands are soil erosion and depletion of aquifers.

Other constraints that are likely to affect rice production are: limited availability of irrigation and changing labor patterns. Water is an indispensable input for crop production. Water resource development has been the key to increasing grain production between 1950 and 1990. Among the activities involving exploitation of natural resources, irrigation is by far the most important. Worldwide, about 253 million ha of cropland are irrigated, and China and India alone account for 100 million ha of irrigated land (Fredericksen et al., 1993). As population increases and economic development intensifies, satisfying the need for drinking water, sanitation, and industrial activities must be accorded a high priority in allocating water resources.

Economic growth brings dramatic changes in the employment structure. In most developing countries, the rate of growth in the working-age population has been around 2 to 3 percent (World Bank, 1995). A faster rate of economic growth in nonagricultural sectors, particularly in Asia, has caused agricultural labor to seek more remunerative employment opportunities in the nonagricultural sector. This trend is likely to intensify in the next century, which would lead to labor shortages in the agricultural sector.

Rice is grown under diverse environments that are classified into five major categories for convenience (Khush, 1984; 1997). *Irrigated rice* accounts for about 55 percent of the total area. Another 5 percent is favorable *rainfed lowland* where improved varieties and technology have been adopted. About 80 percent of world rice production comes from 60 percent of these favorable rice-growing areas. Generally, rainfed lowland is regarded unfavorable, but because limited irrigation is becoming available in certain areas that were previously considered "rainfed unfavorable," these same areas can now be classified as favorable. However, very few varieties and/or new technologies have been developed for the unfavorable *(rainfed lowland, upland, deepwater,* and *tidal wetland)* environments that constitute 40 percent of the area planted to rice. The future challenges for rice improvement are twofold:

1. Develop varieties for the irrigated areas with higher yield potential and greater yield stability, and
2. Develop improved varieties with tolerance to abiotic stresses and with higher yield potential for unfavorable environments.

Increasing the Yield Potential of Irrigated Rice

Conventional hybridization and selection is a time-tested strategy for developing high-yielding crop cultivars. This approach has been the basis of variety development at IRRI. In the late 1960s and early 1970s, yield of IR8 was in the range of 9 to10 tons/ha under favorable, irrigated conditions at IRRI in the Philippines. However, the yield of IR8 now, under similar conditions at IRRI, ranges between 7 and 8 tons/ha, whereas the recently bred cultivars, such as IR72 and PSBRc 52 (IR59682-132-1-1-2), yield between 9 and 10 tons/ha. This suggests that yield potential of indica-inbred cultivars has not changed during the past 30 years. However, regression of yield against year of release of IRRI-bred cultivars indicates an annual gain of about 1 percent increase in yield per year based on the present yield of IR8 (Peng et al., 1999). It appears that due to new biotic or abiotic constraints or changes in soil biology, IR8 cannot attain the same yield potential now as it did in the late 1960s. The variety improvement during the past 30 years has resulted in cultivars resistant to these constraints. Without these improvements, the yield of high-yielding cultivars would have eroded by 25 to 30 percent.

The yield potential of indica-inbred cultivars in the tropics is 9 to 10 tons/ha during the dry season. Plant physiologists have suggested that physical environment in the tropics is not a limiting factor to increasing rice yield. Maximum yield potential was estimated to be 15.9 tons/ha (Yoshida, 1981).

In the past, quantum jumps in yield potential of crop plants have generally resulted from the modification of plant types. The modification in plant architecture allowed the yield potential of rice to be doubled in the 1960s.

To further enhance the genetic potential of rice from its present level of 10 tons/ha to 12 tons/ha, a new plant type was conceptualized at IRRI in 1989.

New Plant Type (NPT) for Increased Yield Potential

Yield is a function of total dry matter (biomass) and harvest index. Modern high-yielding rice varieties have a harvest index of 0.5 and a total biomass of about 20 tons/ha under optimal conditions. By raising biomass to about 22 tons/ha and harvest index to 0.55, it should be possible to obtain a yield of over 12 tons/ha. Breeding work for the new plant type was initiated in 1989 when about 2,000 entries from IRRI's genetic resources center were grown to identify donors possessing desirable traits for the NPT project. The suggested modifications to HYV's plant architecture included a reduction in tiller number from about 25 to about 20 (a decrease of about five tillers), an increase in the number of grains per panicle, deeper root system, thicker and dark green leaves, and straw stiffness (see Figure 2.5) (Khush, 1995a). Donors for the target traits were identified in the "bulu" or javanica germplasm mainly from Indonesia. This germplasm is now referred to as tropical japonica. First pedigree nursery from this material was grown in 1991. Since then, more than 2,000 crosses have been made and more than 100,000 pedigree lines evaluated. To date, more than 500 NPT lines have been evaluated in observational yield trials. As envisioned, the NPT lines possess large panicles, few unproductive tillers, and lodging resistance. However, the first prototype lines had low grain yield because of low bio-

FIGURE 2.5. Sketches of different plant types of rice. Left: tall conventional plant type. Center: improved high yielding, high tillering plant type. Right: proposed low tillering ideotype with higher harvest index and higher yield potential.

mass production and poor grain filling. The next logical step was to focus on increasing grain filling and biomass production. Since poor grain filling is inherited, only parents with good grain-filling characteristics were used in subsequent crosses (Khush, 1996). Many improved NPT lines now have normal grain filling and yield levels better than those of high-yielding indica varieties.

Further refinements in the original NPT design were made. However, results of comparative studies between NPT and high-yielding indicas indicated that NPT lines did not have sufficient biomass. An increase in tillering capacity was needed to increase biomass production and to compensate for any yield loss when loss of tillers is caused by insect damage, especially stem borers. Moreover, in the tropical japonica germplasm, which was used for developing NPT lines, there were no donors for resistance to brown planthopper and tungro. Improved NPT lines were, therefore, crossed with high-yielding indicas to increase slightly the tillering ability and to incorporate resistance to diseases and insects. Currently, a number of lines developed from crosses involving NPTs and elite indica lines are being field tested. Many of these lines have better tillering capacity (10 to 12) and preliminary results indicate their yield superiority over the improved indica varieties. More important, like improved indicas, their grains also are of superior quality. These lines have also inherited improved disease and insect resistance of indica germplasm.

Increasing Yield Potential Through Exploitation of Heterosis

Another approach for increasing the yield potential of rice in the tropics is the exploitation of hybrid vigor or heterosis through hybrid rice breeding. Hybrid rices have a yield advantage of 15 to 20 percent over conventional high-yielding varieties and have been grown in China since 1976. However, the Chinese hybrids were not adapted to the tropical environment. IRRI initiated work on hybrid rice in 1978. In the past five years, hybrid rices developed at IRRI have also shown a yield advantage of about 15 percent compared to the best inbred cultivars when grown in farmers' fields (Virmani, 1996). At IRRI and PhilRice farms, an increase in yield of about 9 percent compared with the best indica inbred cultivars was observed during 1996-1997. The higher yield of indica/indica hybrids is due to increased total biomass and higher spikelet number and grain weight (Ponnuthurai et al., 1984).

Almost all rice hybrids developed at IRRI and those grown in China have come from crosses between indica lines. However, greater diversity between parents is expected to enhance heterosis. As expected, hybrids developed from crosses between indica and japonica lines had greater heterosis for yield as compared to indica/indica hybrids (Yuan et al., 1989). However, indica/temperate japonica hybrids are partially sterile (Zhu et al., 1997). Initial results show that hybrids between indica and tropical japonica lines also are more heterotic than indica/indica hybrids. The utilization of NPT tropical japonica inbreds with wide compatibility genes overcomes the problem of intersubspecific sterility. Thus, the NPT development has complemented the hybrid-breeding program at IRRI and should lead to a further increase in yield potential of irrigated rice.

Developing Improved Germplasm for Unfavorable Environments

Forty percent of rice land is not suitable for growing varieties developed for irrigated lowland areas. These areas are characterized by variable water regimes, occurrence of drought, submergence, waterlogging, and soil toxicities and deficiencies. Development of improved varieties for these niches has been understandably very slow in the past. Hence, very little improvement in productivity in these environments has been made. The reasons for the slow progress follow.

- Adaptabilty of the genotypes to each of the specific growing conditions, which are very heterogeneous, is very important. Hence, it is essential to retain, as much as possible, the adaptability traits of the locally adapted varieties during the breeding process.
- Productivity and adaptability traits are sometimes negatively correlated. Hence, it is very difficult to combine various component traits.
- Inheritance of adaptability traits, such as drought and submergence tolerance, elongation ability, and tolerance to mineral stresses, is not sufficiently understood.
- Selection for local adaptability can be done only by evaluating segregating materials in target environments. On the other hand, research institutes and experimental stations are often located in more favorable rice-growing areas. Therefore, breeders are not able to pay adequate attention to the variety-development programs for such areas.
- The available screening procedures for selecting for tolerance to various stresses do not always give consistent results.
- The generation advancement during the breeding process is slower because of the complexity of the evaluation procedures and inherent photoperiod sensitivity in some populations.

Strategies adopted and some accomplishments in the development of improved germplasm for these unfavorable environments follow.

- Ideal plant types for each of the unfavorable environments have been conceptualized (IRRI, 1989).
- Evaluation of breeding materials under target environments in different countries is being done through a consortium approach involving key sites through shuttle breeding.

- Donors for specific abiotic stresses have been identified.
- The rapid generation advance (RGA) technique is being used for selected crosses involving photoperiod-sensitive parents.
- Recombinant-inbred and doubled-haploid lines are now available from crosses involving stress-tolerant parents.
- Inheritance of stress traits is now being investigated and some of the stress-tolerance traits, such as drought, submergence tolerance, and salinity, have been tagged with molecular markers (Gregorio, 1997; Mackill, 1999).

These strategies have facilitated the development of improved varieties for unfavorable environments. Recently, some IRRI-bred improved varieties for the various unfavorable environments have been released in the Philippines. For example, PSBRc 48 and PSBRc 50 were released in 1995, and PSBRc 84, PSBRc 86, and PSBRc 88 were released in 2000, for highly saline areas. Three additional varieties, viz. PSBRc 60, PSBRc 68, and PSBRc 70, were recommended for rain-fed lowland areas in 1997.

Widening the Gene Pool of O. sativa

Gene pools can be widened through hybridization of crop cultivars with wild species and weedy races, as well as through intraspecific crosses between diverse germplasm groups. Experimental evidence indicates that wide crosses can be used for improving many traits, including yield. Lawrence and Frey (1975) reported that 25 percent of the lines from BC_2 to BC_4 segregants from the *Avena sativa* × *Avena sterilis* matings had significantly higher grain yield than the recurrent parent. Nine lines with 10 to 29 percent higher yield were very similar to the recurrent parent.

Recently, QTLs for higher yield potential were discovered from exotic germplasm and tagged with molecular markers. The tagged QTLs were transferred to elite lines in tomato and rice (Tanksley and McCouch, 1997). A cross involving *O. rufipogon,* a wild species of rice, and an elite hybrid variety of cultivated rice was studied. In a BC_2 testcross population, transgressive segregants outperforming the original hybrid variety were observed (Xiao et al., 1996). At IRRI, a similar strategy to enhance yield of HYVs is being followed.

Several biotechnological approaches for increasing rice yield potential are being investigated. These include introduction of cloned novel genes through transformation and use of molecular-marker technology. Starch biosynthesis plays a pivotal role in plant metabolism, both as a transient storage metabolite of leaf tissue and as an important energy and carbon reserve for sink organs, such as seed, roots, tubers, and fruit. Several enzymatic steps are involved in starch biosynthesis in plants. ADP-glucose pyrophosphorylase (ADPGPP) is a critical enzyme in regulating starch biosynthesis in plant tissues. Even in storage organs with high levels of ADPGPP, its activity is still limiting. It should be possible to enhance starch production in storage tissues by regulating the expression of the gene encoding this enzyme (Kishore, 1994). Starch levels and dry matter accumulation were enhanced in potato tubers of plants transformed with the $glgC^{16}$ gene from *Escherichia coli* encoding ADPGPP (Stark et al., 1992). The transformed potato plants had tubers with higher dry matter and starch content under both growth chamber and field conditions. The $glgC^{16}$ gene has also been transferred into rice and the transformed progenies are being evaluated.

Full yield potential of crops is not realized primarily because of the toll taken by diseases and insects. It is estimated that diseases and insects cause yield losses of up to 25 percent annually in cereal crops. Similarly, crop yields are reduced and fluctuate greatly due to abiotic stresses, such as drought, excess water, mineral deficiencies and toxicities, and abnormal temperatures. Genetic improvement of crops to withstand biotic and abiotic stresses can impart yield stability. Host-plant resistance is the logical approach for minimizing crop losses from pest attacks. As mentioned earlier, numerous rice varieties with multiple resistance have been developed. Recent breakthroughs in cellular and molecular biology have opened new vistas for developing crop cultivars with durable resistance. It is now possible to introduce into crop cultivars novel genes for pest resistance from unrelated plants, animals, and microorganisms. Protocols for transformation of most of the important food crops have now been developed (Uchimiya et al., 1989), and novel genes are being successfully incorporated into crop species. The delta endoprotein produced by the bacterium *Bacillus thuringiensis* (Bt) is lethal to caterpillars of lepidopteran insects. When the cloned *Bt* gene, after proper modifications, is transferred into crop plants through genetic engineering, plants produce their own biocides. The *Bt* gene has been incorporated into maize, cotton, and rice. The *Bt*-containing maize strains were planted to millions of hectares in the United States in 1997 and transgenic plants exhibited high levels of resistance to maize insects. The *Bt*-containing rice plants showed high levels of resistance to rice stem borers (Ghareyazie et al., 1997). Currently, genes for resistance to sheath blight, e.g., genes for chitinases and glucanases, are being incorporated into rice.

The progress in developing crop cultivars for tolerance to abiotic stresses has been slow because of the lack of knowledge of mechanisms of tolerance, poor understanding of inheritance of tolerance, low heritability, and a lack of efficient techniques for screening germplasm and breeding materials. Genetic engineering techniques hold great promise in developing crop cultivars with higher levels of tolerance to abiotic stresses. Glycinebetaine, an osmoprotectant, is widely distributed in plants and allows cells to adjust their cytoplasm to maintain appropriate water content. This solute also protects proteins from salt-induced dissociation of their respective subunits. However, rice does not accumulate glycinebetaine. Therefore, it is sensitive to salt stress. Sakamoto and colleagues (1998) isolated the *codA* gene conditioning choline oxidase, which converts choline to glycinebetaine, from soil bacterium *Anthrobacter globiformis* and introduced it into rice. The transgenic rice plants accumulated glycinebetaine and were tolerant to salt stress.

CONCLUSIONS

Major increases in rice production have occurred as a result of widescale adoption of improved rice varieties developed at IRRI and by national rice improvement programs. These varieties are high yielding and possess good grain quality, shorter growth duration, multiple resistance to diseases and insects, and tolerance to some of the abiotic stresses. Improved varieties are now planted on 70 percent of the world's rice land. Rice production increased from 257 million tons in 1966 to 587 million tons in 1999, and most of the major rice-growing countries became self-sufficient. The price of rice adjusted for inflation is now 40 percent lower than it was in the early 1960s. Availability of pest-resistant varieties reduced the need for application of pesticides and facilitated the adoption of integrated pest-management practices.

The population of rice consumers is increasing at the rate of 1.8 percent each year. But the rate of growth in rice production has slowed down. It is estimated that rice production must increase by 50 percent by 2025 to meet demand for rice. This increase must occur in spite of decreasing water supplies, less labor, and reduced use of pesticides. Thus, we need varieties with higher yield potential and higher yield stability as well as better technologies for pest, soil, water, and crop management. Various strategies for increasing the yield potential include: ideotype breeding, heterosis breeding, wide hybridization, and genetic engineering. Varieties with location specificity and tolerance to abiotic stresses are being developed for the variable rainfed environments.

REFERENCES

Brown, L. 1997. The agricultural link. How environmental deterioration could disrupt economic progress. World Watch Paper 136, World Watch Institute, Washington, DC.

Fredericksen, H.D., Berkoff, J., and Barber, W. 1993. *Water resources management in Asia.* Vol. 1. Main Report. World Bank Technical Paper, No. 212. World Bank, Washington, DC.

Ghareyazie, B., Alinia, F., Menguito, C.A., Rubia, L., de Palma, J.M., Liwanag, E.A., Cohen, M.B., Khush, G.S., and Bennett, J. 1997. Enhanced resistance to two stem borers in an aromatic rice containing a synthetic *cry1*A(b) gene. *Molecular Breeding* 3(5):401-414.

Gregorio, G.B. 1997. Tagging salinity tolerance genes in rice using amplified fragment length polymorphism (AFLP). Doctoral thesis, University of Philippines, Los Baños.

Hayami, Y., Kikuchi, M., Moya, P., Bambo, L., and Marciano, E. 1978. *Anatomy of peasant economy: A rice village in the Philippines.* International Rice Research Institute, Los Baños, Philippines.

Hossain, M. 1988. *Nature and impact of green revolution in Bangladesh.* Research Report, No. 67. International Food Policy Research Institute, Washington, DC.

IRRI. 1982. *IR36, The world's most popular rice.* International Rice Research Institute, Los Baños, Philippines.

IRRI. 1989. *IRRI towards 2000 and beyond.* International Rice Research Institute, Los Baños, Philippines.

Khush, G.S. 1977. Disease and insect resistance in rice. *Advances in Agronomy.* 29:265-341.

Khush, G.S. 1984. Terminology for rice growing environments. In *Terminology of rice growing environments* (pp. 5-10). International Rice Research Institute, Manila, Philippines.

Khush, G.S. 1987a. Development of rice varieties suitable for double cropping. In *Tropical Agriculture Research,* Series 20 (pp. 235-246). Tropical Agricultural Research Center, Ministry of Agriculture, Forestry and Fisheries, Japan.

Khush, G.S. 1987b. Rice breeding: Past, present, and future. *Journal of Genetics* 66(3): 195-216.

Khush, G.S. 1992. Selecting rice for simply inherited resistances. In H.T. Stalker and J.P. Murphy (Eds.), *Plant Breeding in the 1990s* (pp. 303-322). C.A.B. International, Wallingford, U.K.

Khush, G.S. 1995a. Breaking the yield frontier of rice. *GeoJournal* 35(3):329-332.

Khush, G.S. 1995b. Modern varieties—their real contribution to food supply and equity. *GeoJournal* 35(3):275-284.

Khush, G.S. 1996. Prospects of and approaches to increasing the genetic yield potential. In R.E. Evenson, R.W. Herdt, and M. Hossain (Eds.), *Rice research in Asia: Progress and priorities* (pp. 59-71). International Rice Research Institute, Manila, Philippines.

Khush, G.S. 1997. Origin, dispersal, cultivation and variation of rice. *Plant Molecular Biology* 35(3):25-34.

Khush, G.S. 1999. Green revolution: Preparing for the 21st century. *Genome* 42(4):646-655.

Kishore, G.M. 1994. Starch biosynthesis in plants: Identification of ADP glucose pyrophosphorylase as a rate-limiting step. In K.G. Cassman (Ed.), *Breaking the yield barrier* (pp. 117-119). International Rice Research Institute, Manila, Philippines.

Lawrence, P.L. and Frey, K.J. 1975. Backcross variability for grain yield in oat species crosses, *Avena sativa* L. × *A. sterilis* L. *Euphytica* 24(1):77-85.

Mackill, D. 1999. Genome analysis and breeding. In K. Shimamoto (Ed.), *Molecular biology of rice* (pp. 17-41). Springer-Verlag, New York.

Peng, S., Cassman, K.G., Virmani, S.S., Sheehy, J., and Khush, G.S. 1999. Yield potential trends of tropical rice since the release of IR8 and the challenge of increasing rice yield potential. *Crop Science* 39(6):1552-1559.

Ponnuthurai, S., Virmani, S.S., and Vergara, B.S. 1984. Comparative studies on the growth and grain yield of some F_1 rice (*O. sativa* L.) hybrids. *Philippine Journal of Crop Science* 9(3):183-193.

Sakamoto, A., Murata, A., and Murata, N. 1998. Metabolic engineering of rice leading to biosynthesis of glycinebetaine and tolerance to salt and cold. *Plant Molecular Biology* 38(6):1011-1019.

Sinha, S.K., Rao, N.H., and Swaminathan, M.S. 1989. Food security in the changing global climate. In *Climate and Food Security* (p. 579-597). International Rice Research Institute, Manila, Philippines and the American Association for the Advancement of Science, Washington, DC.

Stark, D.M., Timmerman, K.P., Barry, G.F., Preiss, J., and Kishore, G.M. 1992. Regulation of the amount of starch in plant tissues by ADP glucose pyrophosphorylase. *Science* 285(5080):287-292.

Tanksley, S.D. and McCouch, S.R. 1997. Seed banks and molecular maps: Unlocking genetic potential from the wild. *Science* 277(5329):1063-1066.

Uchimiya, H., Handa, T., and Brar, D.S. 1989. Transgenic plants. *Journal Biotechnology* 12(1):1-20.

UNDP (United Nations Development Program). 1994. *Human development report.* Oxford University Press, Oxford.

Virmani, S.S. 1996. Hybrid rice. *Advances in Agronomy* 57:377-462.

World Bank. 1995. *World Development Report.* Oxford University Press, Oxford.

Xiao, J., Grandillo, S., Ahn, S.N., McCouch, S.R., and Tanksley, S.D. 1996. Genes from wild rice improve yield. *Nature* 384(6606):223-224.

Yap, C.L. 1991. *A comparison of the cost of producing rice in selected countries.* FAO Economic and Social Development Paper 101. FAO, Rome.

Yoshida, S. 1981. *Fundamentals of rice crop science.* International Rice Research Institute, Manila, Philippines.

Yuan, L.P., Virmani, S.S., and Mao, C.X. 1989. Hybrid rice—achievements and future outlook. In *Progress in irrigated rice research* (pp. 219-235). International Rice Research Institute, Manila, Philippines.

Zhu, Q., Zhang, Z., Yang, J., Cao, X., Lang, Y., and Wang, Z. 1997. Source-sink characteristics related to the yield in intersubspecific hybrid rice. *Scientia Agricultura Sinica* (In Chinese) 30(3):52-55.

Chapter 3

Tropical Maize: Innovative Approaches for Sustainable Productivity and Production Increases

Ripusudan L. Paliwal
Margaret E. Smith

INTRODUCTION

Maize Classification and Origin

Maize, *Zea mays* L., is one of the oldest food grains; its origin dates back approximately 8,000 to 10,000 years. It belongs to the grass family Poaceae (Gramineae) and tribe Maydeae. Maize is the only cultivated plant in this tribe. Teosinte is the closest wild relative of maize. Earlier taxonomists had classified *Zea* and *Euchlaena* (to which teosinte belonged) as two separate genera. Now, based on cross compatibility in hybridization of these two groups of plants and cytogenetic studies, it is generally agreed that maize and teosinte belong to the same genus, *Zea* (Reeves and Mangelsdorf, 1942).

Tripsacum is the other new world genus of the tribe Maydeae and has at least 13 species, all of which are wild. Some species have limited commercial value, primarily as forage crops (e.g., *Tripsacum dactyloides* L., commonly called gamagrass). *Zea* and *Tripsacum* are called "new world" Maydeae because their center of origin is in the Americas. Both teosinte and *Tripsacum* are important as potential sources of desirable traits for improvement of maize.

The tribe Maydeae has five genera of Asiatic origin: *Coix, Sclerachne, Polytoca, Chionachne,* and *Trilobachne.* These are generally referred to as oriental Maydeae. Only *Coix lacryma-jobi* L., or Job's tears, is of some economic significance in South and Southeast Asia and is used for forage and for popping the seeds as a snack food.

Molecular genetic evidence indicates that maize is an ancient tetraploid, but it is a functional diploid plant with a basic set of 10 chromosomes. Other annual species of *Zea* are also diploid. One perennial species, *Zea diploperennis*, is also diploid as the name implies. *Zea perennis*, a perennial species, is tetraploid with $2n = 40$. The other wild relative, *Tripsacum*, has a basic chromosome number of 18. *Tripsacum dactyloides* is diploid with $2n = 36$. In the oriental Maydeae, the genus *Coix* has the lowest basic chromosome number ($n = 5$). Both *Tripsacum* and *Coix* have species with higher levels of ploidy and numbers of chromosomes ranging from 36 to more than 90 (Paliwal, 2000b).

The debate on the origin of maize still continues, focused more on how than where. It is generally agreed that maize originated in Mesoamerica, consisting of the midaltitude regions of Mexico and Guatemala, where maize and teosinte have coexisted since ancient times and where both of these species have a wide genetic diversity (Weatherwax, 1955; Iltis, 1983; Galinat, 1988; Wilkes, 1989). Fossil pollen and archeological maize cobs found in caves in Mexico strongly support the hypothesis that maize originated in this region.

The debate on how maize originated is more intense. Nonetheless, most maize researchers agree that maize's wild relative teosinte was linked with the origin of maize. Wilkes and Goodman (1995) have summarized the major theories to explain the teosinte-maize link and the origin of maize into three groups: (a) that maize is the domesticated form of its wild ancestor teosinte, (b) that an ancestral form of maize (archeological evidence for which has never been found) gave rise to both maize and the annual teosintes, and (c) that maize is the product of an ancient hybridization between teosinte and another unknown grass species.

Whether maize originated from teosinte or teosinte and maize originated separately, an undisputed fact is that teosinte germplasm has introgressed quite extensively into maize during its evolution and domestication in Mexico (Wilkes, 1977; Kato-Y, 1997). The available evidence indicates that the origin of maize involved mutations at several major loci in ancestral teosinte. Later, these key alleles moved into favorable genetic backgrounds and came under the effect of more numerous minor loci (Galinat, 1988; Doebley, 1994). The transformation of a weedy grass into a highly productive plant with an ear full of edible grains in such a short span of time, whether through natural selection or under the watchful eyes of the early farmer-breeders, is certainly a remarkable feat of crop improvement.

The spread of maize from its center of origin to various parts of the world has been as remarkable as its evolution into a cultivated and productive food plant. It quickly spread from its center of origin in Mesoamerica to as far as Canada in the north and Chile in the south. With the arrival of Europeans in the Americas in the late fifteenth century, maize began to move around the world; in less than 300 years, it had become an important food crop in many countries in temperate, subtropical, and tropical regions of the world (Dowswell et al., 1996; Paliwal, 2000b).

Cultivated maize is a fully domesticated plant. Maize and humans have lived and evolved together since ancient times. It does not grow in the wild, cannot survive in nature, and is completely dependent on human husbandry. Modern maize is one of the most productive species of food plants. It is a C4 plant with a high rate of photosynthetic activity, and has the highest potential for carbohydrate production per unit area per day among cereal crops. It was the first major cereal to undergo rapid and widespread technological transformation in its cultivation, as evidenced by the well-documented story of hybrid maize in the United States and later in Europe. It is the top-ranking cereal in grain yield per hectare (FAOSTAT, 1998). The success of science-based technology in maize cultivation stimulated a more general agricultural revolution in many parts of the world.

Global Maize Picture

The global maize area in 1997 was about 140 million hectares and total production was a little over 560 million metric tons. The average global productivity was four metric tons per hectare (CIMMYT, 1999b). Today maize is the second most important cereal grain in total production after wheat, with milled rice occupying third place (FAOSTAT, 1998). Maize is of great economic significance worldwide as human food, animal feed, and a source of many industrial products. The diversity of environments under which maize is grown is unmatched by that of any other crop. Having originated in the tropics, it is now latitudinally grown up to 58 degrees north in Canada and Russia and up to 40 degrees south in Chile and Argentina. Most of the maize crop is grown at moderate altitudes, but it is also grown below sea level in the Caspian plain and up to an altitude of 3800 meters in the Andean mountains. The crop continues to expand to new areas and environments.

Maize has many and diverse uses. It is the only cereal that can be used as food at various stages of the development of the plant. Young maize ear shoots (baby corn) are harvested as soon as the plant flowers and used as vegetables. Tender green ears of sweet corn are a delightful delicacy and are consumed in various ways. Green ears of normal field maize are also used on a large scale for roasting and boiling and consumed as food in several countries. Dry maize grain provides the starchy staple in much of Latin America, Africa, and parts of Asia. It is consumed in many ways, including being boiled, toasted, as porridge, and in soups and breads of various sorts. The maize plant, which is still green when ears are harvested as baby ears or green ears, makes good forage. Large acreages of maize are grown globally for green fodder harvested at various stages of plant development or for ensilage at the dough stage. Maize fits very well in intensive cropping schemes, in subsistence as well as commercial farming (Dowswell et al., 1996; Smith and Paliwal, 1997).

Though maize is a tropical plant by origin, its yield potential is low in typical tropical environments that have high day and night temperatures. Its yield potential is better expressed in temperate and subtropical environments where day temperatures are high but nights are cooler. There is a wide variation in maize productivity in various regions of the world. Maize yields in temperate environments are more than four times greater than in tropical environments. Statistics on maize area, productivity, and production in various regions of the world differing in maize productivity are shown in Table 3.1. Even in the developing countries, maize yields are almost three times higher in temperate environments than in tropical environments.

Maize in the Tropics—Increasing Importance

The demand for and production of maize in developing countries is rising at a higher rate than in developed countries. A graph prepared by CIMMYT (1999b) compares the rate of growth of maize production during the period 1961 to 1997 (see Figure 3.1). Another emerging trend in developing countries is increasing the use of maize as animal feed. With improved economies and rising per capita income in the tropical developing countries and the consequent improvement in purchasing power, substantially more people will be able to afford and consume animal proteins. This will lead to an even greater increase in demand for maize (Byerlee and Saad, 1993; Pingali and Heisey, 1996). It is estimated that by the year 2020, developing countries will demand 55 percent of global maize production compared to their present 45 percent share (Rosegrant et al., 1995; Donald N. Duvick, personal communication, e-mail, June 2001). The earlier estimates of increased demand for maize for animal feed based on high economic growth

TABLE 3.1. Maize Area, Productivity, Production, and Use in Various Regions of the World (1995-1997 Average)

Region	Developed countries	Eastern Europe	Developing countries	
Climate	Temperate	Temperate	Temperate	Tropical
Area (million ha)	33.8	9.2	31.5	65.0
Yield (tons/ha)	7.8	3.7	4.9	1.7
Production (million tons)	263.4	33.8	155.0	110.0
Utilization as:				
Food (%)	4	7	12	50
Feed (%)	76	79	75	40

Source: Adapted from *Maize Facts and Trends* (CIMMYT, 1999b).

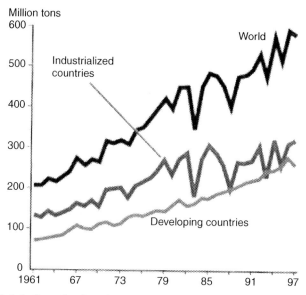

FIGURE 3.1. Growth of maize production: 1961 to 1997 (*Source:* CIMMYT, 1999b).

in developing countries, particularly Asia, have now been scaled down, taking into consideration the slowing down of those economies (CIMMYT, 1999b). In this scenario, a greater percentage of maize will be available for direct consumption as food.

Most of the increased maize production to meet the emerging demand in developing countries will need to come from increased productivity. Increases in area under maize will have a comparatively limited impact, though maize is expected to enter more and more into intensive crop rotations as well as expand to some new environments. A great need and considerable scope exists for improving the potential productivity of maize in tropical environments. New improved maize types with high yield potential in truly tropical and marginal environments will have to be developed with more imaginative conventional breeding and also newer techniques involving the use of molecular genetics, genetic engineering, and gene transfers.

TROPICAL MAIZE-GROWING ENVIRONMENTS

Broad Classification of Maize-Growing Environments

The classification of maize environments is based first on major climatic regions corresponding to the latitudes where maize is grown. The countries/regions falling between the equator and 27° North and South latitude constitute the tropical environment, and maize grown in this belt is classified as tropical maize. The tropical environment is subdivided into three categories based on altitude. These are lowland tropical (sea level to 1200 meters above sea level [masl]), midaltitude (1200 to 1800 masl) and highland (>1800 masl), and the maize genotypes specific to these environments are classified as lowland tropical, midaltitude, and highland germplasm. The countries/regions falling between 27° to 36° North and South latitude are classified as subtropical. A range of genotypes from tropical to subtropical (the latter having been derived from combinations of tropical and temperate germplasm) differing in altitudinal adaptation is grown in this belt. The temperate maize growing environments (above 36° North and South latitude), on the other hand, need and grow only one type of germplasm—temperate.

CIMMYT (1988) made the first attempt to classify maize growing environments in developing countries. This classification took into consideration ecological adaptation, major biotic and abiotic factors affecting maize productivity in various ecological regions, and the maturity, grain color, and type preferred by farmers in these environments (Pham and Edmeades, 1987). Pollak and Pham (1989) refined the classification of maize environments in sub-Saharan Africa by applying multivariate statistical techniques to long-term agroclimatic data available from regional agricultural research stations. Pollak and Corbett (1993) used spatial agroclimatic data sets and geographical information system (GIS) technology to classify maize environments of Mexico into 10 groups. Dowswell and colleauges (1996) have described maize environments of the third world and the germplasm requirements for these various environments. Corbett (1998) has refined and provided the map of maize-growing environments of Kenya using GIS applications. Ecological classification for delineating different maize-growing environments is shown in Table 3.2. Very recently, Hartkamp and colleagues (1999) suggested a numerical scale for ecological classification of maize production environments. They used four daylength groups (<11 h, 11-12.5 h, 12.5-13.4 h, and >13.4 h) for expressing latitude groups. Within various latitude groups, mean temperature regimes (>24°, >18-<24°, >18-<22°, and <18°) were used corresponding to elevation. This classification seems to be less subjective, but its wider applicability needs to be tested.

Complexity of Maize-Growing Environments in the Tropics

Smith and Paliwal (1997) and Paliwal (2000b) emphasized the diversity of maize-cropping environments in developing countries. This diversity is dictated by various climatic and geographical factors, biotic and abiotic stresses, maturity, and grain type preferences. It is also due to the large maize farming spectrum—from intensively managed monocultures grown by commercial farmers with high levels of mechanization, fertilizer, water,

TABLE 3.2. Classification of Maize-Growing Environments

Environment	Latitude (° NS)	Altitude (masl)	Temperature (° C)	
			Average	TMx-TMn *
Lowland Tropical	<27	<1200	>25	<12
Subtropical	27-36	<1200	<25	>12
Midaltitude	<30	1200-1800	<23	>15
Tropical Highlands	<30	>1800	<18	>18
Temperate	>36	<1000	<24	>18

Source: Based on discussions with maize scientists at CIMMYT and National Maize Programs.

Note: *Difference between average maximum temperature (Tmx) and average minimum temperature (Tmn) during growing season.

and pesticide inputs, to subsistence farming on plots that are half a hectare or less in size, where the crop is grown in multicrop mixtures using only hand labor and naturally available fertility and water resources. Therefore, maize-growing environments in the tropics cannot be classified simply according to abiotic and biotic variables.

The classification of maize-growing environments in the tropics should include the following parameters: ecological adaptation, growing seasons, moisture source (rainfed or irrigated) and amount, maturity, grain type preference, biotic stresses, abiotic stresses, soil fertility, nature of farming (subsistence or commercial), cultivar type commonly used (open-pollinated variety or hybrid), cropping pattern (solo, mixed, crop rotation), maize area, average yield, achievable yield, maize uses, and major productivity and production constraints. All these traits and parameters directly influence the type of germplasm that could be successful and acceptable to the farmers in a specific environment. Paliwal (2000a) has described the maize-growing environments of individual countries in the tropics that grow maize on 100,000 or more hectares, using ecological adaptation as shown in Table 3.2 and the parameters given above.

CHALLENGES OF MAIZE IMPROVEMENT FOR THE TROPICS

Current Status of Maize Improvement in the Tropics

The overall low productivity of maize in the tropics is due to several factors, which include shorter daylengths, shorter growing period, low radiation intensity due to cloud cover, higher night temperatures, and more severe biotic and abiotic stresses. The low productivity is also a reflection of the intensity of maize breeding efforts devoted to the maize crop in the tropics in general and for difficult and marginal environments in particular. The need is to significantly increase investment in research on tropical maize, particularly in research targeted to marginal and unfavorable environments.

In temperate environments, nearly all the maize area is planted to hybrids developed from inbred lines crossed in various ways (double, three-way, or single crosses). In contrast, farmers in tropical environments use many seed types, making tropical maize germplasm needs more complex. Open-pollinated categories of maize include landraces, farmer varieties, improved open-pollinated varieties (OPVs), composites (populations developed by combining mostly OPVs), and synthetics (populations developed by combining inbred lines). Hybrids in the tropics may be produced using various inbred line, inbred × non-inbred, and exclusively noninbred parent combinations.

Progress from improvement of maize in the tropics has been reviewed by several authors in recent years (Paliwal and Sprague, 1981; Brandolini and Salamini, 1985; Paterniani, 1990; Pandey and Gardner, 1992; Eberhart et al., 1995; Dowswell et al., 1996). Only some of the most significant achievements in tropical maize improvement are mentioned in this paper.

The wide diversity of maize races is very well documented and reasonably well conserved in gene banks. Several of these also are conserved in situ by farmers. Some excellent maize populations have been developed, e.g., ETO, Tuxpeño and its derivatives like La Posta and Pop. 49, Suwan-1, Amarillo Dentado, Kitale Synthetic, and Katumani Synthetic (details about these populations are found in Dowswell et al., 1996). These populations show wide adaptability and have been the sources of several outstanding varieties and hybrids. Harvest index in typical tropical germplasm was low (about 0.3 to 0.35). Through recurrent selection, the plant height has been shortened to a more manageable level and the harvest index has gone up to about 0.47 in the improved populations. Fair to good levels of resistance are available in the improved germplasm for most of the important tropical diseases. However, insect resistance levels need to be improved.

Contrary to general belief, the rate of gain per year in population improvement and development of open-pollinated varieties has been as good as that achieved in development of hybrid maize (Coors, 1999; Duvick, 1992, 1999). Some improved OPVs perform as well as hybrids in better environments and are planted across a much larger area than any single hybrid, e.g., BR-106 in Brazil, ICA-156 in Colombia, INIAP-526 in Ecuador, Marginal 28 in Peru, Suwan-1 in Thailand, and Lakshmi in India (Paliwal, personal observation). It is important to maintain the diversity of improved germplasm in tropical maize, including the range from OPVs to various kinds of hybrid combinations, to meet the needs of diverse tropical maize-growing environments.

Almost all the maize improvement work in the tropics has been done and targeted to favorable environments with high productivity potential. A combination of superior cultivars and production technology is used in such regimes and is supported by requisite infrastructure for inputs, marketing, and utilization of maize. However, this situation occurs only in some favorable high-yielding tropical environments. Where it does occur, this has led to a shift of maize from subsistence farming to commercial farming, e.g., the Nile Valley in Egypt, the highlands of Kitale in Kenya, and winter-season maize in northeastern and southern India.

Missing Links in Tropical Maize Improvement

The extent of favorable and high-yielding maize environments in the tropics is low compared to less favorable and marginal environments. As expected, statistics on this vary somewhat depending on the source and author (CIMMYT, 1988; Dowswell et al., 1996; Heisey and Edmeades, 1999; Paliwal, 2000a). About 60 to 70 percent of maize in the tropics suffers from one or more stresses. Yet maize-improvement efforts in the tropics are generally planned and operated based on the technologies used by researchers in more favorable and uniform maize environments. Such approaches generally do not address the real problems of the diverse and difficult tropical environments. Adaptability of germplasm to location-specific farmers' conditions and requirements, germplasm for marginal environments, economic stability, and other similar factors that are important from the farmers' point of view are generally not considered in setting research agendas. These traits are complex and difficult to analyze and handle. Under such circumstances, it is no wonder that after more than half a century of maize research work in the tropics, 60 percent of tropical maize area is still planted by farmers with their own saved seed, including their own varieties and landraces (CIMMYT, 1999b).

Pandey and Gardner (1992) reported the results of a survey of 48 maize scientists from Africa, Asia, Latin America, and the Middle East on the traits they considered important for maize improvement. Yield was reported as the most important trait across all four continents. The overriding concern with an improved cultivar is its yield, but yield potential must occur in combination with other traits that will add to its adaptability, stability, and superior economic performance under farmers' conditions in the target area. Moreover, yield calculated on a per hectare basis is not appropriate for the tropics where maize is grown in diverse environments differing in the growth cycle of the crop and in the cropping patterns in which maize is included along with other crops in space (mixed cropping) or time (sequential or relay cropping). The crop growth cycle can range from less than two months (as in the case of maize for green ear shoots) or three months (in the case of early varieties) to nearly 14 months (in the case of extra late varieties in some highland areas). Maize yield calculated on a per unit area/day basis is more appropriate under these conditions.

Need for Targeting Maize Research and Improvement to Specific Environments

Simmonds (1984) emphasized the need for decentralized breeding and selection in the target environments. Miranda (1985) suggested that breeding methodologies for tropical regions should be viewed with respect to the situations in the target areas, i.e., developed areas (favorable high-yielding environments), undeveloped areas (unfavorable low-yielding environments), and intermediate areas (marginal environments). The importance of local specificity and breeding in target environments and the need to efficiently harness genotype × environment interactions to achieve sustainable productivity gains across the spectrum of diverse maize-growing environments in the tropics have been emphasized by several workers in recent years (Bramel-Cox et al., 1991; Smith and Zobel, 1991; Smith and Paliwal, 1997; Ceccarelli, 1994, 1997).

Most breeders feel more comfortable in making selections at high-yield potential levels (high productivity levels) attained by adding fertilizer and other key inputs. Ceccarelli (1994, 1997) strongly advocates the importance of selection in target areas for low-fertility environments in particular, and selection for adaptability and stability in low-productivity/marginal environments. In barley, spectacular yield increases in varieties specifically adapted to very marginal conditions are possible if the breeding work is done on locally adapted germplasm and selections are made in the target environment.

Smith and Zobel (1991) added a point of caution, emphasizing that there was no direct experimental evidence to indicate that decentralized breeding in specific target environments will be more useful and productive and that the added efforts and expenses involved would be justified compared with a single, centralized breeding effort with well-chosen objectives. Ceccarelli (1997) also suggests that specific adaptation is not a very popular concept among breeders and they prefer to attain wide adaptation from their breeding efforts. The seed industry, which is likely to play a greater role in crop-breeding research in tropical environments, is also interested in producing widely adapted varieties in the interest of greater seed sale of each variety. Significantly added resources would be required for research focus and efforts to be expanded to address specific target environments. The required level of additional investment in maize breeding toward such efforts seems unlikely.

An alternative approach would be to clearly define and delineate various maize-growing environments in each country. These specific environments could then serve as the basis for wider testing of varieties and other technologies for the farming conditions of each target environment. This would help in acquiring a reliable level of confidence and assurance of the usefulness of technologies for subsistence farmers in their target environments.

Participatory Plant Breeding

Maize germplasm, being heterozygous and also heterogeneous, carries with it considerable genetic diversity when moved from one place to another or exchanged among farmers. Farmer selection pressure results in the development of more adaptable and improved germplasm for that specific environment. Such farmers' varieties often possess unique combinations of traits for adaptation to the environment and cropping system in which they are grown and the uses to which they are put (Brush, 1995). This has led to recent emphasis on "participatory" plant breeding, where farmers can participate in the process of selection of new and improved varieties (Hardon, 1995). Duvick (1996b) emphasizes that during the past half century, maize breeding methodologies have not changed. Maize breeding, like any other crop breeding, still remains in the domain of art and experience. Therefore, it seems logical that farmers, having the advantages of knowledge of their farming environment and their varietal requirements, could do a better job of selection under difficult farming situations.

How successful participatory plant breeding might be in development of really superior germplasm for marginal and unfavorable maize environments is still a subject of exploration. Ceccarelli and Grando (1999) argue that if decentralized selection is accepted as an essential methodology for successful breeding for marginal environments, this would inevitably lead to the adoption of participatory plant breeding as a tactical necessity. They classify selection and breeding into four types as follows.

1. *Centralized nonparticipatory selection*—selection and breeding is done by professional breeders at the research station.
2. *Decentralized nonparticipatory selection*—breeders use farmers' fields for selection, but without much involvement of the farmer in the selection process.
3. *Centralized participatory selection and breeding*—farmers are involved in the selection process at the research station.

4. *Decentralized participatory selection and breeding*—selection is done by farmers in their own fields.

Ceccarelli and colleagues (2000) have reported preliminary findings from their experiments on barley in which various systems of participatory and nonparticipatory breeding were compared. They found that farmers were slightly more efficient than professional breeders in identifying the highest yielding entries in their own fields. The professional breeder was more efficient than the farmers in selecting on the research station located in a high rainfall area, but less efficient than the farmers on another research station located in a low rainfall area. To be efficient and successful as a breeder, one must know the environment well. Since maize breeders have been largely working in favorable environments and know these environments well, they have been quite successful in breeding superior maize cultivars for these environments. There is no reason to believe that professional breeders will not be as successful in unfavorable environments if they get to know these environments well. Breeders should always emphasize the importance of keeping in constant touch with farmers in the target area to understand fully the traits that are of importance to them and working with farmers for improvement of those traits.

NEW APPROACHES AND RESEARCH THAT COMBINE CONVENTIONAL BREEDING AND MOLECULAR AND BIOTECHNOLOGICAL TOOLS

Genetic Yield Increases

No evidence exists of plateauing of genetic yield potential in temperate maize or in tropical maize (Duvick, 1996a). In fact, higher genetic gains can be expected to be harnessed in tropical maize for a longer time. Conventional plant breeding is expected to remain the dominant approach for development of superior cultivars, though biotechnology tools will start playing an increasing role in practical plant breeding (Duvick, 1996a, 1996b; Smith and Paliwal, 1997; Paliwal, 2000b). Some authors strongly emphasize the need to increase use of biotechnology tools for accelerating the rate of gains in tropical agriculture (Van Montagu, 1998; Serageldin, 1999).

Concerns have been expressed about the possibility and feasibility of maintaining essential rates of genetic gain in productivity. Evans (1998) argues that so far the cereal yield increases that have kept pace with population growth have been largely due to a rise in harvest index associated with dwarfing in small grains, and due to better adaptation to climate and modern agronomy in other crops like maize. He warns that such productivity advances may not be repeated because further increases in the harvest index are likely to be limited and breeders, to date, have been unable to increase the maximum rates of photosynthesis and crop growth.

There has been considerable research and effort to improve the adaptation of maize to cooler environments but not so to the warmer or hot environments. Al-Khatib and Paulsen (1999), commenting on the effects of high temperatures on photosynthetic processes in cereals, state that high temperatures limit the production of crops in many regions of the world. Yet the physiological basis of tolerance to high-temperature stress is not known. The C4 pathway of millets (and by implication maize) by itself is probably inadequate to accord full and satisfactory adaptation to warm habitats.

Increased Tolerance and Adaptation of Tropical Maize to Heat Stress

It has often been suggested that the productivity of tropical maize in the main rainy season can be significantly improved if the crop can be sown well before the onset of the rains. This will enable the plants to get well established before the rains start and overcome the problems of weeds and moisture stresses. This is good logic, but farmers are unable to adopt this technology because suitable germplasm with heat tolerance and insect resistance, two factors that affect the early sown crop adversely, is not available. The adverse effect of heat stress on reproductive development in maize has been amply quantified (Dale, 1983; Stamp et al., 1983; Hall, 1992; Commuri et al., 1996). In a recent study, Cantarero and colleagues (1999) discussed the adverse effect of increased night temperature at silking on kernel set in maize. Increased night temperature hastened the crop development rate. Kernel set was adversely affected because of the increased developmental rate that occurs with higher temperatures. An insufficient photosynthate supply seemed to be a major cause of greater kernel abortion at the tip of the ear under warmer nights.

Genetic variability for heat tolerance has been well exploited in cowpea, cotton, and tomato, and germplasm with good levels of heat tolerance has been developed (Hall, 1992; Ismail and Hall, 1998). Screening of germplasm for heat tolerance in tropical maize could possibly provide desirable quantitative trait loci (QTLs) for heat tolerance. Another approach could be a search for heat tolerance QTLs in maize's wild relatives that tolerate heat effectively.

Heat-shock proteins and cold-shock proteins induce physiological changes in plants that enable them to better tolerate subsequent temperature extremes. Jaglo-Ottosen and colleagues (1998) have shown that overexpression of a single transcription factor, CBF1, in *Arabidopsis* can induce the expression of a suite of cold-regulated (COR) genes even in the absence of the normal stimulus of low temperatures. Such transgenic plants are more resistant to the damaging effects of low temperatures than are unmodified plants. This work has opened the possibility of using the transcriptional activator by turning on the whole battery of COR genes to enhance the freeze tolerance of crops. It may be possible to find similar systems that will work for activation of heat-shock proteins. This approach has acquired considerable importance in view of potential global warming that could have an adverse effect on tropical maize and other crops.

Identification and Use of Adaptability/Stability QTLs from Farmers' Maize Varieties with a Long History of Successful Cultivation

In the preceding paragraphs, the needs for site-specific germplasm improvement, selection in target areas (particularly for low-fertility environments), and selection for adaptability and stability in low-productivity/marginal environments have been discussed. Ceccarelli (1994, 1997) emphasizes that spectacular yield increases in barley varieties specifically adapted to low-input conditions are possible if the breeding work is done on locally adapted germplasm and effective selection and improvement are done in the target environment. As mentioned earlier, this may not be a cost-effective approach. However, considerable merit exists in Ceccarelli's suggestion regarding the use of locally adapted germplasm for the development of stable and adaptable cultivars with increased productivity for marginal environments. Identification and use of adaptability/stability QTLs from farmers' varieties with a long history of successful cultivation in the marginal environments and the incorporation of such QTLs into appropriate germplasm with higher yield potential may be a good strategy.

Nitrogen-Efficient Maize

The reaction and acceptance of improved maize technology by farmers in the tropics is highly variable. In general, maize farmers fall into three major categories: (1) resource-poor farmers or farmers in resource-poor environments who are not in a position to try and use any risky technology, (2) middle-level farmers who like to try new technology and are prepared to take modest risks, and (3) resource-rich commercial farmers who readily adopt new, improved, and expensive technology. Resource-rich farmers are committed to commercial cultivation of hybrid maize with use of high inputs. Resource-poor farmers are reluctant to use improved germplasm since they believe it needs high inputs that are quite expensive. Many middle-level farmers shift between hybrid maize and OPVs depending on their perception of risk involved and the returns on applied inputs. It is generally expected that improved germplasm, whether OPV or hybrid, will give better returns for the inputs applied by the farmer. Unless the return of grain for each kilogram of fertilizer applied to the maize crop is substantial, most farmers will not adopt improved OPV or hybrid maize on a large scale or on a permanent basis.

In the case of wheat, farmers generally have taken to improved varieties and are sticking to them, as they believe they get a higher return of grain for each kilogram of applied nitrogen. CIMMYT (1998) has published a graph that illustrates this point very well (see Figure 3.2). The improved 1980 series of wheat cultivars requires less area and total nitrogen to produce a specified amount of grain (5 tons) than tall wheat cultivars, or the wheat varieties of the 1960s and 1970s. In maize, we have not come across similar studies and graphs. Certainly improved cultivars need less area to produce a specified amount of grain, but the same cannot be said for fertilizer use. In the United States and Argentina, cultivars from different eras of maize breeding have been compared (Carlone and Russell, 1987; Duvick, 1984, 1997). The newer generation hybrids gave higher yields because of their response to higher plant densities under high levels of fertilizer use. No clear indication exists that maize breeding under high fertility levels has developed hybrids with a greater responsiveness to increased nitrogen levels. We have not come across similar comparative reports on the performance of improved cultivars and nitrogen-utilization efficiency in tropical maize.

Sinclair (1998) claims that increases in crop yields in this century have been largely due to selection of plants that could respond to applied nitrogen

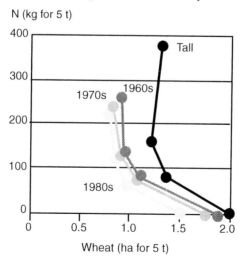

FIGURE 3.2. Yield and nitrogen requirement of tall and improved series of wheat (*Source:* CIMMYT, 1998).

by accumulating the nutrient and using it in the production of grain. At a constant crop mass, increase in harvest index is not possible without an increase in crop nitrogen accumulation, because grain is more nitrogen dense than other plant parts. Germplasm that can efficiently allocate accumulated nitrogen to the grain has an essential trait to allow subsequent increase in the harvest index. Simply adding nitrogen to existing varieties that may not partition it efficiently to grain would be a waste of this natural resource. In fact, it will result in a decreased harvest index, as vegetative plant growth would be stimulated by the added nitrogen rather than grain development and growth (Sinclair, 1998).

Lafitte and colleagues (1997) examined several landraces of maize and found that some races are more efficient in the use of soil nitrogen. The work initiated to identify nitrogen-efficient maize genotypes under low nitrogen environments seems to be a step in the right direction. Lafitte and Edmeades (1994) and Bänziger and colleagues (1997) report that improvement in yield for low-nitrogen environments can be achieved faster following selection for yield under low nitrogen conditions. The advantage of selection under low nitrogen became statistically significant when relative yield reduction under low-nitrogen environments was more than 43 percent. As we have noted, such situations occur quite frequently in farmers' maize fields in the tropics.

Bänziger and colleagues (1999) recently reported that it is possible to select germplasm that possesses simultaneous tolerance for two stress factors—drought and low nitrogen. They found that selection for tolerance to midseason drought stress also increased maize yields in four lowland tropical populations when grown across a wide range of nitrogen levels including low to severe nitrogen stress. These populations showed a gain in biomass production and nitrogen accumulation and a more efficient use of leaf nitrogen for grain production across different nitrogen levels. These results need to be confirmed and repeated by more studies with more maize populations and different environments. Development of more nitrogen-efficient populations will encourage even small farmers in marginal environments to use superior germplasm with increased productivity.

Value-Added Maize Germplasm

Recently, maize breeding efforts have included more emphasis on specialty end-use traits to increase the yield of particular grain components, improve varietal adaptation to postharvest conditions and uses, and reduce environmental impacts of maize production. Breeding is increasingly focused on varieties that can provide value-added products that retain their identity through the product stream, rather than simply on better yielding varieties that will produce standard yellow dent grain (or any other standard type of grain). A recent report estimates that about 5 percent of U.S. maize plantings in 1999 (roughly 1.6 million hectares) were dedicated to high-oil and other value-enhanced varieties (U.S. Grains Council, 1999). Similar trends will likely occur in tropical maize improvement. To date, efforts to breed for improved end-use traits have focused on oil, protein quality, phytic acid, and micronutrient content. Work on storability, particularly under less-than-ideal conditions, is also needed. Breeding efforts on high quality-protein maize have been extensively discussed and reviewed in other publications (e.g., Bjarnason and Vasal, 1992; Magnavaca et al., 1993; Vasal, 1994; Bittel et al., 1996), so these will not be addressed here. Brief discussions follow concerning some of the other value-added traits emphasized in recent maize breeding efforts.

High-Oil Maize

The longest term and perhaps the first effort to improve grain quality in maize was initiated in 1896 at the University of Illinois. Selection was made in the variety 'Burr's White' for both high and low oil and high and low protein in the grain (Dudley and Lambert, 1992). This study continues to provide unique results concerning response to continuous generations of selection for altered grain quality and remains of great interest scientifically. However, in recent years, high-oil hybrids have become a subject of considerable attention commercially as well.

Most maize cultivars have about 4 percent oil in the kernel (dry weight basis). High-oil maize is an attractive option for livestock feed because oil provides over twice the calories per unit weight compared to carbohydrates, and, thus, high-oil maize is a more energy-dense feed than normal maize. Feeding trials have shown, in most cases, that feeding efficiency and rate of weight gain are greater for livestock fed high-oil maize than for comparison groups fed normal maize (Alexander, 1988). These benefits may apply where maize is consumed as human food also, providing a more energy-dense starchy staple in the diet. However, storability concerns related to elevated oil content would need to be investigated.

The oil content of the Illinois high-oil selection group has reached as high as 20 percent in recent generations, but this material was not simultaneously selected for yield and thus its yield potential is low. Even where both yield and oil content were breeding objectives, commercial use of high-oil hybrids had been limited due to reduced grain yield potential relative to normal hybrids (Lambert, 1994). However, varieties have recently been released that combine the yield potential of normal hybrids with elevated oil quality from high-oil parents in what is called the TOPCROSS grain production system (Edge, 1997). This system uses a mixture of two types in the commercial seed (called TC BLEND seed corn): 90 to 92 percent of the seed is a male sterile version of an elite, high-yielding, normal grain hybrid; the remaining 8 to 10 percent of the seed is a high-oil pollinator. The xenia effect from the pollinator causes the grain harvested from the elite hybrid to be higher in oil content than it would have been if pollinated by a normal oil content parent.

Evaluations have shown that such seed blends result in grain oil ranging from 6.4 to 7.9 percent, compared with 4.0 to 4.4 percent for the same hybrids without the high-oil pollinator (Thomison et al., 1997). Yields for the high-oil mixtures ranged from 6.7 to 8.4 t/ha, while their normal counterparts yielded 7.3 to 9.1 t/ha, suggesting a slight reduction in grain yield for high-oil TC BLENDS compared to normal grain hybrids. However, certain individual TC BLENDS showed grain yields comparable to or only very slightly reduced in comparison to their normal-grain hybrid counterparts, suggesting that choosing the right hybrid can result in higher oil grain with little or no yield sacrifice. Efforts are now underway to combine high oil with other value-added traits (high protein, increased levels of key amino acids) to provide highly tailored feed grain hybrids that would be sold through identity-preserved channels at premium prices (Bowditch Group, 1999). If grain storability concerns are manageable, seed blends designed to produce high-oil grain may have an important role in both food and feed production.

Low-Phytic-Acid Maize

Phosphorus in maize grain is stored primarily as phytic acid, which accounts for 65 to 85 percent of total phosphorus in the grain but is not efficiently utilized by nonruminant animals (Lott, 1984; Raboy, 1997). Most of the phytic acid from dietary grain fed to poultry and swine is excreted and contributes to environmental contamination associated with animal manure. To provide livestock with a balanced ration, inorganic phosphorus is typically added to corn- and soybean-based feeds (Cromwell and Coffey, 1991). This adds cost to the feed and does not reduce the phosphorus load in manure. More recently, a synthetic version of the enzyme phytase, which results in breakdown of phytic acid when both phytase and phytic acid are consumed, has been added to feeds (Gibson and Ullah, 1990; Jongbloed and Lenis, 1992). This frees up more of the phosphorus bound in phytic acid for utilization by the animal, reducing the need for phosphorus supplementation in feed, the phosphorus excreted in manure, and the potential for environmental contamination.

A breeding effort at the USDA station in Aberdeen, Idaho, targeted identification of maize mutants with low phytic acid but normal total phosphorus content to genetically achieve what had been achieved by adding phytase to feeds (Raboy et al., 1997). Two mutations were identified, designated *low phytic acid 1-1* and *low phytic acid 2-1* and representing two distinct genetic loci (Raboy and Gerbasi, 1996). Both mutants result in significantly reduced phytic acid content in grain (50 to 65 percent of normal) but maintain the level of total grain phosphorus, meaning that most of the grain phosphorus is not bound as phytic acid (Raboy et al., 1997).

Further work with the *low phytic acid 1-1* mutant has indicated that a yield reduction may be associated with this mutation, resulting either from a pleiotropic effect or possibly from linkage drag (since the original flint source in which the mutation was isolated is not elite germplasm). Raboy and colleagues (1997) speculate on possible physiological causes that might explain a yield reduction associated with low phytic acid content. Feeding trials with chickens showed available phosphorus in the low-phytic-acid hybrids ranged from 70 to 90 percent (compared with 30 to 48 percent in counterpart normal hybrids) and reductions in phosphorus content in manure ranged from 9 to 40 percent (Raboy et al., 1997).

Research is underway to assess the potential benefit of low-phytic-acid grain relative to its effects on mineral nutrition and other dietary processes in humans. To date, a study comparing tortillas prepared with normal and low-phytic-acid maize showed greater iron retention after consuming the low-phytic-acid-based tortillas compared to the normal grain-based tortillas (Mendoza et al., 1998). Although low-phytic-acid maize might enhance micronutrient nutrition in humans, concerns related to the potential beneficial roles of phytic acid in human diets (including antioxidant and anticarcinogenic roles) must be addressed (Graf and Eaton, 1993; Harland and Morris, 1995). Additionally, research is needed to investigate the potential for low-phytic-acid grain to be more susceptible to grain molds in storage, due to its higher available phosphorus level, and to assess seedling vigor and yield of low-phytic-acid genotypes when planted in phosphorus-deficient soils.

Micronutrient Content in Maize Grain

Mineral and vitamin deficiencies (often called "micronutrient malnutrition") have been recognized as major human nutritional and health problems in the past decade (Bouis, 1996). Deficiencies of vitamin A, iron, and iodine are now recognized as global and prevalent problems (Combs et al., 1996). Iron deficiency affects over 2.1 billion people worldwide. Preschool children in the tropics and reproductive-age women are the most highly affected groups, but by no means the only groups affected (Combs et al., 1996). Zinc deficiency is also common among women, infants, and children in many parts of the world.

Several maize breeding efforts have been initiated to explore the possibility of improving iron and zinc content and bioavailability in maize. Results from these efforts are preliminary at this point. The major share of various micronutrients, including iron and zinc, is concentrated in the aleurone layer of maize kernels. Typically the aleurone layer is only one cell thick, but the multiple aleurone *(mal)* mutant in maize results in multiple cell layers in the aleurone, thus potentially increasing micronutrient content of kernels (Smith, 1996; Welch et al., 1993). Lines carrying the *mal* allele also have higher lysine content in the kernels. The opaque-2 *(o2)* allele has long been known to increase the fraction of lysine in maize kernel protein, as well as total protein content in the kernels. In certain genetic backgrounds, *o2* also increases content of various micronutrients in the grain (Welch et al., 1993). Since lysine is known to increase the bioavailability of iron and zinc to humans, either *mal* or *o2* alone might increase not only the content of iron and zinc, but also the dietary availability of these critical micronutrients. Furthermore, both alleles in combination could have an even greater effect on content and bioavailability of iron and zinc than either allele alone.

Selection for increased levels of iron and zinc within maize breeding populations is underway at CIMMYT's maize breeding program in Harare, Zimbabwe (M. Bänziger, personal communication, January 2000, in Harare, Zimbabwe). Over 1,400 maize germplasm sources have been screened for iron and zinc content in the grain. Initial results revealed dramatic differences in content, with the best materials exceeding the mean by up to 50 percent. However, with refinement of the screening methodology to include a sodium hypochloride rinse to remove any possible soil contamination of grain samples, these differences appear to have dropped to less than 30 percent. Goals of future work are to confirm screening methodology and ensure that soil contamination is not a problem, verify the actual levels of micronutrients in high iron and zinc selections, continue divergent selection to determine the potential for improvement in iron and zinc content, and assess the benefits to human nutritional status that might result from high-iron and high-zinc maize grain (M. Bänziger and J. Long, personal communication, January 2000, in Harare, Zimbabwe).

Grain and Seed Storability

Little work has been done to date on breeding for improved maize grain storability under tropical conditions. As noted above, there are grain storability concerns associated with several of the grain types that might be selected for their improved human nutritional qualities (e.g., high-oil, low-phytic acid, and high-quality protein maize). Studies are needed to determine how these altered grain types affect their storability.

Resistance to stored grain insects is another important concern for tropical maize breeding that has not been widely addressed. Some progress has been made in recent years toward understanding the mechanisms of resistance to stored grain pests in maize and screening and selection for resistance. Arnason and colleagues (1997), Classen and colleagues (1990), Horber (1989), Serratos and colleagues (1997), and Wright and colleagues (1989) have described methodology for culturing of stored grain insects, screening for resistance, and biochemical and genetic mechanisms providing resistance to pests in storage. Considerable variability has been observed in various maize populations for resistance to storage pests. Phenolic acid content in the grain is shown to be strongly and positively correlated with resistance to storage pests. These are linked with hardness of grain, which may be related to the mechanical contributions of the phenolic dimers to the cell wall strength. Phenolic acid amines with toxic effects on insects have also been detected in the aleurone layer of the grain (Arnason et al., 1997). Hopefully, in the near future, efforts to breed maize genotypes for improved levels of resistance to stored grain insects will be successful.

Seed Storage

Basic studies have identified clear genetic differences in maintenance of maize seed viability under stress conditions (high humidity and temperature). Bernal-Lugo (personal communication, November 1998, e-mail) showed that the inbred Pa33 maintained very good viability despite storage at 30°C and 75 percent relative humidity for four months, while the inbreds Mo17 and RD4503 lost viability very quickly after only a month at these conditions. This suggests that there may be unexploited potential to breed for improved seed storability for farmers who may be saving and storing their own seed under tropical conditions.

MOVING THE BENEFITS OF MAIZE
GENETIC IMPROVEMENT TO FARMERS

Use of Improved Seeds in the Tropics

Seed is the end product of a plant breeding program. A program can be deemed successful *only* when the end product is available to and used by farmers. Unfortunately, this principle is not fully appreciated or practiced by many maize research programs in developing countries. As evidence, one need only cite the small percentage of the tropical maize crop that is planted using seed of improved varieties or hybrids. Less than 40 percent of the area under maize in tropical environments is planted with improved seed (CIMMYT, 1999b).

This nonacceptance/nonuse of improved seeds by tropical maize farmers could be because improved seeds are not suited to their needs or because they are not readily available to them. The first point has been discussed at length in the preceding sections of this chapter. In this section, we will cover the inadequacies of the seed production and supply system.

Types of Maize Seed

In temperate environments, nearly all the maize area is planted with hybrids developed from inbred lines crossed in various ways (double, three-way, or single crosses). In contrast, farmers in tropical environments use many seed types, making tropical maize seed systems more complex. Open-pollinated categories of maize include landraces, farmer varieties, improved open-pollinated varieties (OPVs), composites (populations developed by combining mostly OPVs), and synthetics (populations developed by combining inbred lines). Hybrids in the tropics may be produced using various inbred lines, inbred × non-inbred, and exclusively noninbred parent combinations (Smith and Paliwal, 1997; Paliwal, 2000b).

Of the improved seed types, OPVs are easier to develop and maintain, their seed production is relatively simpler and less expensive, and thus OPV seed is less expensive than that of hybrids. In addition, farmers can save and replant OPV seed for two to three seasons/years. This reduces their yearly dependence on organized seed systems, whose performance, more often than not, may be erratic in many developing-country situations. The management practices and input requirements for the cultivation of improved OPVs can be similar to those of landraces and farmers' own varieties. Adoption and farmer-to-farmer seed movement of OPVs is easier than that of hybrids and can result in larger area coverage. This explains why OPVs are preferred and needed for many maize-growing environments in the tropics, in spite of a higher yield potential offered by well-adapted hybrids.

Statistics reported on the use of improved seeds in developing countries in the tropics present a paradox. About 80 percent of the area in tropical environments is planted to OPVs (unimproved and improved), of which only 20 percent is planted with improved OPVs. The area planted with hybrid seeds is about the same (19 percent) as the area planted with improved seeds of OPVs. The hybrid seed production and distribution system, which is largely private sector, is much better organized and more efficient than the OPV seed production and distribution system, which is almost exclusively in the public sector. The coverage by hybrid seeds is limited to irrigated and well-watered environments, which represent more productive land but a small percentage of the area under maize in the tropics. OPVs are used in the rainfed and less-assured environments, which represent a much larger percentage of the tropical maize area.

The hybrid seed production technology developed for temperate maize has generally been transferred and adapted for production of tropical maize hybrid seed. The technology for OPV seed production is less developed. In most cases, OPV seed is multiplied with only as much care as is given to a good field of commercial maize grain. In hybrid maize seed production, more than one parent is involved, and these parents have to be crossed each year to produce fresh hybrid seed. In the case of OPVs, only one parent is involved in seed production and the process is called seed "multiplication," as in self-pollinated crops. Seed production of maize OPVs needs more care and attention than simple seed multiplication, though, because maize is cross-pollinated (Pandey, 1998; CIMMYT, 1999a; Paliwal, 2000b). The true potential for increasing productivity through OPVs is not realized in the absence of a sound seed production system.

The extent of use of improved and quality seed of both OPVs and hybrids will have to be increased to realize faster and higher gains in tropical maize productivity. It is imperative that the entire tropical seed industry be revitalized. Policies and incentives must be set up so that seed production of nonhybrid cultivars can be as efficient as hybrid seed production in the private sector.

Public and Private Sector Seed Systems

The private sector plays an increasing role in maize seed production and distribution in most countries in the tropics. The importance of both public and private organizations in seed production and distribution is generally recognized. On the other hand, it is also emphasized that the private sector has a distinct advantage in organizing efficient seed production and distribution systems. On the whole, the increasing presence of the private-sector seed system in tropical environments is a positive development. There are, however, some matters of considerable concern. One is the declared or undeclared division of responsibilities between the two sectors. The private sector is largely involved in the production of hybrid seeds only for favorable maize production environments. The public sector is left with the responsibility of producing seeds of OPVs, which are largely used in marginal to unfavorable environments.

This dichotomy is not a correct approach. Good quality seed production is required for both improved OPVs and hybrid seeds. The impression should not be allowed to continue that seed production of OPVs can be or should be entrusted to public sector undertakings, leaving the private sector largely responsible for producing superior hybrid seed. Under this construct, the public sector would meet maize seed demands for a comparatively larger area (i.e., that sown to OPVs) until hybrid seed is more widely adopted—probably not for some time to come. If the public-sector seed system is weakened further without alternative effective arrangements, maize productivity in the tropics will not increase at the rate required to meet increasing demands.

Weakening of the Public Maize Research System

There is another, even more serious, long-term problem in the present scenario—the weakening of public sector maize research in most developing countries. It is a cumulative effect of the increasing presence and publicity of the private sector and the impression that the public sector has failed to deliver research results to farmers. The latter is largely attributable to the failure of the seed production and supply system, rather than a lack of good germplasm from public-sector breeding efforts. A weakened public sector, however, will seriously hamper the emergence of an indigenous private sector, which often depends on publicly developed experimental varieties and hybrids. (Such was the case of the early maize seed industry in countries such as the United States.) Lack of a strong public sector will also harm poor farmers in marginal and unfavorable areas, who are less likely to be served by the private sector. The downward trend of public sector research and seed production investments must be reversed in developing countries.

CONCLUSIONS: TROPICAL MAIZE IMPROVEMENT NEEDS

From what has been said in this chapter, we emphasize the following shifts in paradigms needed for tropical maize research and improvement:

1. A better understanding, definition, and delineation of maize-growing environments
2. Increased emphasis on maize research for marginal environments
3. Increased emphasis on stress-tolerant germplasm with stable performance across time and space
4. Decentralization of maize research and emphasis on improvement for specific target environments
5. Selection and testing under unfavorable/marginal environments for which technology is targeted
6. Identification and use of appropriate sites for multilocation testing in the target environments
7. Development of suitable germplasm for mixed, relay, and multiple cropping systems
8. Emphasis on development of diversified types of improved germplasm (OPVs and various types of hybrids) for various farming conditions
9. Emphasis on value-added traits

10. Close interaction with farmers and their involvement in research planning and execution
11. Urgent revitalization of public-sector maize research and seed production systems

REFERENCES

Alexander, D.E. 1988. High oil corn: Breeding and nutritional properties. In D.B. Wilkinson (Ed.), *Proceedings of the Forty-Third Annual Corn and Sorghum Research Conference* (pp. 97-105). ASTA, Washingon, DC.

Al-Khatib, K. and Paulsen, G.M. 1999. High-temperature effects on photosynthetic process in temperate and tropical cereals. *Crop Sci.* 39(1):119-125.

Arnason, J.T., Conilh de Beyssac, B., Philogene, B.J.R., Bergvinson, D., Serratos, J.A., and Mihm, J.A. 1997. Mechanisms of resistance in maize to the maize weevil and the larger grain borer. In J.A. Mihm (Ed.), *Insect resistant maize: Recent advances and utilization. Proceedings of an International Symposium held at the International Maize and Wheat Improvement Center* (pp. 91-95.) CIMMYT, Mexico DF.

Bänziger, M., Betran, F.J., and Lafitte, H.R. 1997. Efficiency of high nitrogen selection environments for improving maize for low-nitrogen target environments. *Crop Sci.* 37(4):1103-1109.

Bänziger, M., Edmeades, G.O., and Lafitte, H.R. 1999. Selection for drought tolerance increases maize yields across a range of nitrogen levels. *Crop Sci.* 39(4):1035-1040.

Bittel, D.C., Shaver, J.M., Somers, D.A., and Gengenbach, B.G. 1996. Lysine accumulation in maize cell cultures transformed with a lysine insensitive form of maize dihydrodipicolinate synthase. *Theoretical and Applied Genetics* 92(1):70-77.

Bjarnason, M. and Vasal, S.K. 1992. Breeding of quality protein maize (QPM). *Plant Breeding Reviews* 9:181-216.

Bouis, H. 1996. *Plant breeding strategies for improving human mineral and vitamin nutrition.* Micronutrients and Agriculture No. 1, Federation of American Scientists, Washington, DC.

Bowditch Group 1999. *Electronic AgBiotech Newsletter* 163, January 20, 1999. Online <http://www.bowditchgroup.com/index2.htm>.

Bramel-Cox, P.J., Barker, T., Zavala-García, F., and Eastin, J.D. 1991. Selection and testing environments for improved performance under reduced-input conditions. In D.A. Sleper, T.C. Barker, and P.J. Bramel-Cox (Eds.), *Plant breeding and sustainable agriculture: Considerations for objectives and methods* (pp. 29-56). CSSA Special Publication No. 18, Madison, WI.

Brandolini, A. and Salamini, F. (Eds.). 1985. *Breeding strategies for maize production improvement in the tropics.* FAO and Instituto Agronomico Per L'Oltremare, Firenze, Italy.

Brush, S.B. 1995. In situ conservation of landraces in centers of crop diversity. *Crop Sci.* 35(2):346-354.

Byerlee, D. and Saad, L. 1993. *CIMMYT's economic environment to 2000 and beyond: A revised forecast.* CIMMYT, Mexico DF.

Cantarero, M.G., Cirilo, A.G., and Andrade, F.H. 1999. Night temperature at silking affects kernel set in maize. *Crop Sci.* 39(3):703-710.

Carlone, M.R. and Russell, W.A. 1987. Response to plant densities and nitrogen levels for four maize cultivars from different eras of breeding. *Crop Sci.* 27(3):465-470.

Ceccarelli, S. 1994. Specific adaptation and breeding for marginal conditions. *Euphytica* 77(3):205-219.

Ceccarelli, S. 1997. Adaptation to low/high input cultivation. In P.M.A. Tigerstedt (Ed.), *Adaptation in Plant Breeding* (pp. 225-236). Kluwer Academic Publishers, Dordrecht, The Netherlands.

Ceccarelli, S. and Grando, S. 1999. Decentralized-participatory plant breeding. *Plant Breeding News* Edition 103, July 28, 1999. PBN-L@mailserv.fao.org.

Ceccarelli, S., Grando, S., and Michael, M. 2000. A methodological study on participatory barley breeding. I. Selection phase. *Euphytica* 111(2):91-104.

CIMMYT. 1988. *Maize production regions in developing countries.* CIMMYT Maize Program, Mexico DF.

CIMMYT. 1998. *CIMMYT in 1997-98: Change for the better.* CIMMYT, Mexico DF.

CIMMYT. 1999a. *Development, maintenance and seed production of open-pollinated varieties* Second edition. CIMMYT Maize Program, Mexico DF.

CIMMYT. 1999b. *World maize facts and trends 1997/98.* CIMMYT, Mexico DF.

Classen, D., Arnason, J.T., Serratos, J.A., Lambert, J.D.H., Nozzolillo, C., and Philogene, B.J.R. 1990. Correlation of phenolic acid content of maize to resistance to *Sitophilus zeamais*, the maize weevil, in CIMMYT's collections. *J. Chem. Ecol.* 16(2):301-315.

Combs, G.F., Welch, R.M., Duxbury, J.M., Uphoff, N.T., and Nesheim, M.C. 1996. *Food-based approaches to preventing micronutrient malnutrition: An international research agenda.* CIIFAD, Ithaca, NY.

Commuri, P.D., Jones, R.J., Koch, K.E., and Hannah, C.L. 1996. Maize kernel development under heat stress: Effects on sugar metabolism and starch biosynthesis. *Agronomy Abstracts* (p. 111). November 3-8.

Coors, J.G. 1999. Selection methodology and heterosis. In J.G. Coors and S. Pandey (Eds.), *Genetics and exploitation of heterosis in crops* (pp. 225-245). ASA, CSSA, and SSSA, Madison, WI.

Corbett, J.D. 1998. Classifying maize production zones in Kenya through multivariate cluster analysis. In R.M. Hassan (Ed.), *Maize technology development and transfer: A GIS application for research planning in Kenya* (pp. 15-25). CAB International (Oxford), CIMMYT (Mexico) and KARI (Kenya), Oxford University Press, Cambridge.

Cromwell, G.L. and Coffey, R.D. 1991. Phosphorus—A key essential nutrient, yet a possible major pollutant—Its central role in animal nutrition. In T.P. Lyons

(Ed.), *Biotechnology in the Feed Industry* (pp. 133-145). Alltech Technical Publications, Nicholasville, KY.

Dale, R.F. 1983. Temperature perturbations in the midwestern and southeastern United States important for corn production. In C.D. Raper Jr. and P.J. Kramer (Eds.), *Crop reactions to water and temperature stresses in humid temperate climates* (pp. 21-32). Westview Press, Boulder, CO.

Doebley, J. 1994. Genetics and the morphological evolution of maize. In M. Freeling and V. Walbot (Eds.), *The maize handbook* (pp. 66-77). Springer-Verlag, New York.

Dowswell, C.D., Paliwal, R.L., and Cantrell, R.P. 1996. *Maize in the third world.* Westview Press, Boulder, CO.

Dudley, J.W. and Lambert, R.J. 1992. Ninety generations of selection for oil and protein in maize. *Maydica* 37(2):81-87.

Duvick, D.N. 1984. Genetic contributions to yield gains of U.S. hybrid maize, 1930 to 1980. In W. Fehr (Ed.), *Genetic contributions to yield gains in four major crop plants* (pp. 15-47). CSSA Special Publication No. 7, Madison, WI.

Duvick, D.N. 1992. Genetic contribution to advances in yield of U.S. maize. *Maydica* 37(1):69-79.

Duvick, D.N. 1996a. Crop improvement—Emerging trends in maize. In R.B. Singh, V.L. Chopra, and A. Varma (Eds.), *Crop Productivity and Sustainability: Proceedings of the Second International Crop Science Congress* (pp. 127-138). Oxford and IBH Publishing Co., New Delhi, India.

Duvick, D.N. 1996b. Plant breeding, an evolutionary concept. *Crop Sci.* 36(3):539-548.

Duvick, D.N. 1997. What is yield? In G.O. Edmeades, M. Banziger, H.R. Michelson, and C. B. Pena-Valdivia (Eds.), *Developing drought and low-N tolerant maize. Proceedings of a Symposium, March 25-29, 1996* (pp. 332-335). CIMMYT, Mexico DF.

Duvick, D.N. 1999. Heterosis: Feeding people and protecting natural resources. In J.G. Coors and S. Pandey (Eds.), *Genetics and exploitation of heterosis in crops* (pp. 19-29). ASA, CSSA, and SSSA, Madison, WI.

Eberhart, S.A., Salhuana, W., Sevilla, R., and Taba S. 1995. Principles for tropical maize breeding. *Maydica* 40(4):339-355.

Edge, M. 1997. Seed management issues for "TopCross High Oil Corn." In J.E. Cortes (Ed.), *Proceedings of the Nineteenth Annual Seed Technology Conference* (pp. 49-55). February 18, 1997. Seed Science Center, Iowa State University, Ames, IA.

Evans, L.T. 1998. Greater crop production: Whence and whither. In J.C. Waterlow, Leslie Fowden, D.G. Armstrong, and Ralph Riley (Eds.), *Feeding a world population of more than eight billion people: A challenge to science* (pp. 89-97). Oxford University Press, New York, Oxford.

FAOSTAT 1998. *FAO production statistics 1998.* FAO, Rome.

Galinat, W.C. 1988. The origin of corn. In G.F. Sprague and J.W. Dudley (Eds.), *Corn and corn improvement* (pp.1-31). American Society of Agronomy. Madison, WI.

Gibson, D.M. and Ullah, A.B.J. 1990. Phytases and their action on phytic acid. In D.J. Morre, Wendy F. Boss, and Frank A. Loewus (Eds.), *Inositol metabolism in plants* (pp. 77-92). Wiley-Liss, New York.

Graf, E. and Eaton, J.W. 1993. Suppression of colonic cancer by dietary phytic acid. *Nutr. Cancer* 19(1):11-19.

Hall, A.E. 1992. Breeding for heat tolerance. *Plant Breed. Rev.* 10:129-168.

Hardon, J. 1995. (Rapporteur). Participatory plant breeding. Workshop sponsored by IDRC, IPGRI, FAO, and CGN at Wageningen, the Netherlands on June 26-29, 1995. *Issues in Genetics Resources* No. 3, October 1995. IPGRI, Rome, Italy.

Harland, B.R. and Morris, E.R. 1995. Phytate: A good or a bad food component. *Nutr. Res.* 15(5):733-754.

Hartkamp, A.D., White, J.W., and Aguilar, A.R. 1999. Global maize production environments. A GIS-based approach. *Agronomy Abstracts* (p.99). October 31-November 4.

Heisey, P.W. and Edmeades, G.O. 1999. Maize production in drought stressed environments: Technical options and research resource allocation. Part I of CIMMYT 1999. *World maize facts and trends 1997/98.* CIMMYT, Mexico DF.

Horber, E. 1989. Methods to detect and evaluate resistance in maize to grain insects in the field and in storage. In CIMMYT 1989. *Towards insect resistant maize for the third world: Proceedings of the International Symposium on Methodologies for Developing Host Plant Resistance to Maize Insects* (pp. 140-150). CIMMYT, Mexico DF.

Iltis, H.H. 1983. From teosinte to maize: The catastrophic sexual transmutation. *Science* 222(4626):886-894.

Ismail, A.M. and Hall, A.E. 1998. Positive and potential negative effects of heat-tolerance genes in cowpea. *Crop Sci.* 38(2):381-390.

Jaglo-Ottosen, K.R., Gilmour, S.J., Zarka, D.G., Schabenberger, O., and Thomashow, M.F. 1998. Arabidopsis CBF1 overexpression induces COR genes and enhances freezing tolerance. *Science* 280(5360):104-106.

Jongbloed, A.W. and Lenis, N.P. 1992. Alteration of nutrition as a means to reduce environmental pollution by pigs. *Livestock Production Sci.* 31(1-2):75-94.

Kato-Y, T.A. 1997. Review of introgression between maize and teosinte. In T.A. Serratos, M.C. Wilcox, and F. Castillo-Gonzalez (Eds.), *Gene flow among maize landraces, improved maize varieties, and teosinte: Implications for transgenic maize* (pp. 44-53). CIMMYT, Mexico DF.

Lafitte, H.R. and Edmeades, G.O. 1994. Improvement for tolerance to low soil nitrogen in tropical maize. II. Grain yield, biomass production, and N accumulation. *Field Crops Res.* 39(1):15-25.

Lafitte, H.R. Edmeades, G.O., and Taba, S. 1997. Adaptive strategies identified among tropical maize land races for nitrogen-limited environments. *Field Crops Res.* 49(2/3):187-204.

Lambert, R.J. 1994. High oil hybrids. In Hallauer, A.R. (Ed.), *Specialty corns* (pp. 123-145). CRC Press, Boca Raton, FL.

Lott, J.N.A. 1984. Accumulation of seed reserves of phosphorus and other minerals. In Murray, D.R. (Ed.), *Seed physiology* (pp. 139-166). Academic Press, New York.

Magnavaca, R., Larkins, B.A., Schaffert, R.E., and Lopes, M.A. 1993. Improving protein quality of maize and sorghum. In D.R. Bruxton, R. Shibles, R.A. Forsberg, B.L. Blad, K.H. Asay, G.M. Paulsen, and R.F. Wilson (Eds.), *International Crop Science I* (pp. 649-653). Crop Science Society of America, Madison, WI.

Mendoza, C., Viteri, F., Lonnerdal, B., Young, K., Raboy, V., and Brown, K.H. 1998. Effect of genetically modified, low-phytate maize on iron absorption from tortillas. *American Journal of Clinical Nutrition* 68(5):1123-1127.

Miranda, J.B. 1985. Breeding methodologies for tropical maize. In A. Brandolini and F. Salamini (Eds.), *Breeding strategies for maize production improvement in the tropics* (pp. 177-206). FAO and Instituto Agronomico Per L'Oltremare, Firenze, Italy.

Paliwal, R.L. 2000a. *Individual country maize growing environments in the tropics.* (Unpublished).

Paliwal, R.L. 2000b. *Tropical maize improvement and production.* FAO Plant Protection and Production Series No. 28, Rome.

Paliwal, R.L. and Sprague, E.W. 1981. *Improving adaptation and yield dependability in maize in the developing world.* CIMMYT, Mexico DF.

Pandey, S. 1998. Varietal development: Conventional plant breeding. In M.L. Morris (Ed.), *Maize seed industries in developing countries* (pp. 57-76). Lynne Rienner Publishers, Boulder, CO.

Pandey, S. and Gardner, C.O. 1992. Recurrent selection for population, variety, and hybrid improvement in tropical maize. *Advances in Agronomy* 48:1-87.

Paterniani, E. 1990. Maize breeding in the tropics. *Crit. Rev. Plant Sci.* 9(2):125-154.

Pham, H.N. and Edmeades, G.O. 1987. Delineating maize production environments in developing countries. In *CIMMYT research highlights 1986* (pp. 3-11). CIMMYT, Mexico DF.

Pingali, P.L. and Heisey, P.W. 1996. Cereal crop productivity in developing countries: Past trends and future prospects. *Conference on Global Agricultural Science Policy for the Twenty-First Century* (pp. 51-94). Melbourne, Australia.

Pollak, L.M. and Corbett, J.D. 1993. Using GIS datasets to classify maize-growing regions in México and Central America. *Agron. J.* 85(6):1133-1139.

Pollak, L.M. and Pham, H.N. 1989. Classification of maize testing locations in sub-Saharan Africa by using agroclimatic data. *Maydica* 34(1):43-51.

Raboy, V. 1997. Accumulation and storage of phosphate and minerals. In B.A. Larkins and I.K. Vasil (Eds.), *Cellular and molecular biology of plant seed development* (pp. 441-477). Kluwer Academic Publishers, Dordrecht, Netherlands.

Raboy, V. and Gerbasi, P. 1996. Genetics of *myo*-inositol phosphate synthesis and accumulation. In B.B. Biswas and S. Biswas (Eds.), *Subcellular biochemistry, Vol. 26: Myo-Inositol phosphates, phosphoinositides, and signal transduction* (pp. 257-285). Plenum Press, New York.

Raboy, V., Young, K.A., and Ertl, D.S. 1997. Breeding corn for improved nutritional value and reduced environmental impact. In *Proceedings of the Fifty-second Annual Corn and Sorghum Research Conference* (pp. 271-282). ASTA Publ. No. 52, Washington, DC.

Reeves, R.G. and Mangelsdorf, P.C. 1942. A proposed taxonomic change in the tribe Maydeae. *Am. J. Bot.* 29(8):815-817.

Rosegrant, M.W., Agcaoili-Sombilla, M., and Perez, N.D. 1995. *Global food production to 2020: Implications for investment.* IFPRI—Food, Agriculture and Environment Discussion Paper No. 5. IFPRI, Washington, DC.

Serageldin, I. 1999. Biotechnology and food security in the 21st century. *Science* 285(5426): 387-389.

Serratos, J.A., Blanco-Labra, A., Arnason, J.T., and Mihm, J.A. 1997. Genetics of maize grain resistance to maize weevil. In J.A. Mihm (Ed.), *Insect resistant maize: Recent advances and utilization. Proceedings of an International Symposium held at the International Maize and Wheat Improvement Center* (pp. 132-138). CIMMYT, Mexico DF.

Simmonds, N.W. 1984. *Principles of crop improvement.* Longman, London and New York.

Sinclair, T.R. 1998. Historical changes in harvest index and crop nitrogen accumulation. *Crop Sci.* 38(3):638-643.

Smith, M.E. 1996. The potential of the opaque-2 and multiple aleurone genes for improving the nutritional value of maize for human consumption. *Micronutrients and Agriculture* 1:11-12.

Smith, M.E. and Paliwal, R.L. 1997. Contributions of genetic resources and biotechnology to sustainable productivity increases in maize. In K. Watanabe and E. Pehu (Eds.), *Plant biotechnology and plant genetic resources for sustainability and productivity* (pp. 133-144). R.G. Landes Company and Academic Press Inc., Austin, TX.

Smith, M.E. and Zobel, R.W. 1991. Plant genetic interactions in alternative cropping systems: Considerations for breeding methods. In D.A. Sleper, T.C. Barker, and P.J. Bramel-Cox (Eds.), *Plant breeding and sustainable agriculture: Considerations for objectives and methods* (pp. 57-82). CSSA Special Publication No. 18, Madison, WI.

Stamp, P., Geisler, G., and Thiraporn, R. 1983. Adaptation to suboptimal and supraoptimal temperatures of inbred maize lines differing in origin with regard to seedling development and photosynthetic traits. *Physiol. Plant.* 58(1):62-68.

Thomison, P., Geyer, A., Lotz, L., and Siegrist, H. 1997. Using TC Blends? In high oil corn production. In *Report of the Fifty-Second Annual Corn and Sorghum Research Conference* (pp. 283-298). ASTA Publ. No. 52, Washington, DC.

U.S. Grains Council 1999. *Annual report.* USGC, Washington, DC.

Van Montagu, M. 1998. How and when will plant biotechnology help? In J.C. Waterlow, L. Fowden, D.G. Armstrong, and R. Riley (Eds.), *Feeding a world population of more than eight billion people: A challenge to science* (pp. 98-106). Oxford University Press, New York, Oxford.

Vasal, S.K. 1994. High quality protein corn. In A.R. Hallauer (Ed.), *Specialty corns* (pp. 79-121). CRC Press Inc., Boca Raton, FL.

Weatherwax, P. 1955. History and origin of corn. I. Early history of corn and theories as to its origin. In G.F. Sprague (Ed.), *Corn and corn improvement* First edition (pp. 1-16). Academic Press, New York.

Welch, R.M., Smith, M.E., Van Campen, D.R., and Schaefer, S.C. 1993. Improving the mineral reserves and protein quality of maize (*Zea mays* L.) kernels using unique genes. In N.J. Barrow (Ed.), *Plant nutrition: From genetic engineering to field practice* (pp. 235-238). Kluwer Academic Publishers.

Wilkes, H.G. 1977. Hybridization of maize and teosinte in México and Guatemala and the improvement of maize. *Econ. Bot.* 31(3):254-293.

Wilkes, H.G. 1989. Maize: Domestication, racial evolution and spread. In D.R. Harris and G.C. Hillman (Eds.), *Forage and farming* (pp. 440-454). Unwin Hyman, London.

Wilkes, H.G. and Goodman, M.M. 1995. Mystery and missing links: The origin of maize. In S. Taba (Ed.), *Maize genetic resources* (pp. 1-6). CIMMYT, Mexico DF.

Wright, V.F., Mills, R.B., and Willcutts, B.J. 1989. Methods for culturing stored-grain insects. In CIMMYT, *Towards insect resistant maize for the third world. Proceedings of the International Symposium on Methodologies for Developing Host Plant Resistance to Maize Insects* (pp. 74-83). CIMMYT, Mexico DF.

Chapter 4

Maize Improvement
for Resistance to Aflatoxins:
Progress and Challenges

Manjit S. Kang
Orlando J. Moreno

INTRODUCTION

Aflatoxins are specific mycotoxins produced, as secondary metabolites, by some strains of *Aspergillus flavus*, most strains of *A. parasiticus*, *A. nomius*, and *A. tamarii* (Gourama and Bullerman, 1995; Payne, 1998). However, only *A. flavus* and *A. parasiticus* are agriculturally important. Aflatoxins cause toxicities in animals and humans. They are capable of causing liver cancer, mutations, and deformities in developing embryos. There are several compounds designated as aflatoxin, but important ones are B1, B2, B2a, G2, G2a, M1, M2, P1, Q1, and R0 (Ong, 1975).

Aflatoxin B1 is the most potent member of the family and has demonstrated an extremely high carcinogenicity toward some species of animals (Wogan et al., 1974). Aflatoxins B1 and B2 are produced by *A. flavus*, whereas four toxins (B1, B2, G1, and G2) are produced by *A. parasiticus* (CAST, 1989). Aflatoxins cause chromosomal aberrations in animal as well as plant cells and induce gene mutations in several organisms (Ong, 1975).

The Food and Agriculture Organization (FAO) estimates that 25 percent of the world's food crops are affected by aflatoxins. Losses to livestock and poultry producers from aflatoxin-contaminated feeds include death and more subtle effects, such as immune suppression, reduced growth rates, and loss in feed efficiency. Thus, aflatoxins in food and feeds pose serious health hazards to animals and humans. Eradication of mycotoxins in food and feeds is, therefore, of prime importance (Haumann, 1995; Robens and Dorner, 1997).

In the United States, attention was focused on the aflatoxin problem in 1977 when heavy contamination of field maize (*Zea mays* L.) occurred throughout the Southeast. Preharvest aflatoxin contamination of the crop is chronic primarily in the southeastern United States and is not considered a threat in other areas of the United States. Nevertheless, more than 21 states have reported the problem (Wilson and Payne, 1994). The U.S. Food and Drug Administration has established an aflatoxin limit of 20 parts per billion (ppb) (20 µg/g) for foods and for most feeds and feed ingredients. This acceptance level is equivalent to about one ounce in 3,125 tons (Hardin, 1998) or one infected kernel in 652 bushels of grain (Seedburo, 1995). Products in international commerce may be subject to tolerance levels commonly ranging between 5 and 50 ppb. The EU has tolerance levels as low as 2 ppb (Hansen, 1993). On June 19, 1993, the World Health Organization's International Agency for Research on Cancer designated aflatoxin as a Group 1 category carcinogen because of its carcinogenic effect on humans (Hansen, 1993). The U.S. Food and Drug Administration considers the occurrence of aflatoxin in food as a potential threat to the food supply, the public, and world markets for exported commodities. Regulatory concerns mandate that maize meet the established tolerance levels, but it is extremely difficult to produce maize free of these exceedingly small amounts of aflatoxin. Thus, it is necessary to develop methods that will prevent aflatoxin production.

Economic Impact

The impact of aflatoxin contamination is pronounced on the agricultural economy in drought-stricken years, with estimated losses ranging in the hundreds of millions dollars. For farmers, losses can be in the form of yield, nonmarketable grain, restricted markets, increased transportation costs, discounts, increased costs of drying and selling, and inability to obtain loans on stored grain (Nichols, 1983). In 1980, maize growers in the southeastern United States lost an estimated $97 million (Nichols, 1983). In addition, direct losses to grain merchandisers totaled $14 million.

In many countries, maize is a primary staple for humans, and many people are being exposed to aflatoxin well above accepted standards (e.g., mean 37 ppb in Benin to 292 ppb in Nigeria) (Cardwell et al., 1997). The concern, therefore, is worldwide for two reasons: first, because U.S. maize is shipped routinely to importing countries, and second, because maize consumed in numerous countries throughout the world has been reported as contaminated (Wood, 1989).

Preharvest Contamination

Prior to the 1970s, aflatoxin contamination was believed to be a storage problem. However, it is well established now that aflatoxin contamination of maize grain occurs both at preharvest stage and during storage (Anderson et al., 1975; Lillehoj et al., 1980). Infection by *A. flavus* and subsequent aflatoxin production have been recognized as a problem not only in the southeastern United States since the mid-1970s but occasionally also in the Corn Belt (Zuber et al., 1976; Lillehoj et al., 1980; Kilman, 1989).

If contamination is prevented before storage, the problem can be managed. Host-plant resistance is one of the best options to reduce preharvest aflatoxin contamination. Because of the sporadic nature of natural fungal infection and variability of aflatoxin levels, however, consistency and accuracy in field experimentation have been difficult to achieve.

CONTAMINATION PROCESS

Because of the exigent need for information, specific research areas of aflatoxin in maize have previously been reviewed (Shotwell and Hurburgh, 1991; Gorman and Kang, 1991; Payne, 1992; Widstrom, 1996; Moreno and Kang, 1999; Brown et al., 1999). Because improving maize for resistance to aflatoxin is the most attractive strategy, all available information on genetic mechanisms that control kernel infection and aflatoxin production, as well as on the biochemical pathway(s) of aflatoxin production, needs to be disseminated widely throughout the world.

Since two living organisms (maize and fungus) are involved in the grain contamination problem, a better understanding and application of interorganismal genetic principles to this complex disease are warranted.

Kernel Infection and Aflatoxin Production

Morphologically, *A. flavus* and *A. parasiticus* are distant relatives of *A. nidulans* (Brown et al., 1996; Bhatnagar, 1997). *A. flavus* and *A. parasiticus* produce aflatoxin and have no sexual stage (Bhatnagar, 1997). It is interesting that *A. nidulans* produces no aflatoxins but has a sexual stage.

Angle et al. (1989) found that *A. flavus* overwintered in the soil as conidia and mycelia. It is not clear how *A. flavus* survives in the soil. Boller and Schroeder (1974) reported that moistened conidia of *A. flavus* lose their viability in 21 to 60 days. Wicklow and colleagues (1984) demonstrated that *A. flavus* can produce sporogenic sclerotia and that these sclerotia were present in naturally infected maize kernels. Colonization of the kernel surfaces by *A. flavus* is extremely important in the epidemiology of this disease. Marsh and Payne (1984) and Smart and colleagues (1990) found that the fungus colonized the surface of kernels and the glume tissue surrounding the kernel. From this tissue, the fungus enters into intact seeds in one of two ways. It grows on the surfaces of the rachis and spikelet and invades at the junction of the bracts and rachillas. At this stage, the cells are large, thinwalled, and highly vacuolated. Smart and colleagues (1990) also found that the fungus could grow up through the rachis (cob) into the spikelet through the continuous air space in these tissues. Hyphae were common in all embryonic tissue, especially in the scutellum. *A. flavus* was not commonly found in the endosperm (Jones et al., 1980). Payne and colleagues (1988) detected little or no invasion of seed by *A. flavus* before the seed reached physiological maturity (32 percent moisture). Marsh and Payne (1984) found that colonization and infection were highest following inoculation of yellow-brown silks with *A. flavus* conidia.

The kernel development stage at which the fungus invades is important. Rambo and colleagues (1974) and Zuber and Lillehoj (1979) reported that kernels were most susceptible to infection in the late-milk to the early-dough stage. Marsh and Payne (1984) followed the colonization of maize silks in a controlled environment chamber using a tan color mutant of *A. flavus.* Yellow-brown and turgid silks best supported the fungal growth. Failure to grow on unpollinated silks was likely caused by an inhibitor, but it could also be caused by some nutrient deficiency.

Factors Favoring Infection by Aspergillus

The incidence and severity of aflatoxin contamination are highly dependent on environmental conditions. As will be discussed later, most of the environmental stresses act in concert with one another, which exacerbates the aflatoxin contamination problem.

Aflatoxin-producing fungi can grow at temperatures as high as 48°C as well as at water potentials as low as –35 MPa (Klitch et al., 1994). Plant stress favors colonization of maize kernels and infection by *Aspergillus.* High aflatoxin levels are often associated with abiotic stresses, such as drought, heat, nitrogen deficiency, and tillage operations, as well as with biotic stresses, such as insects and weeds (Darrah and Barry, 1991). Factors significantly related to high toxin levels were insect damage, cropping system, prolonged field drying, and storing practices (Cardwell et al., 1997).

A flow chart illustrates the various biotic and abiotic interactions at various stages of maize production and handling (see Figure 4.1). It also shows certain FDA guidelines for the disposal of aflatoxin-contaminated grain.

Abiotic Stresses

Drought. Soil water deficit, especially coupled with excessively high temperatures, is the most common yield-limiting factor in maize production areas around the world. There is strong evidence for drought stress per se to be a contributor to elevated aflatoxin levels. Payne and colleagues (1986) showed that irrigation or subsoiling was associated with less aflatoxin contamination. If irrigation is not an option, growers may consider alternative control measures/cultural practices.

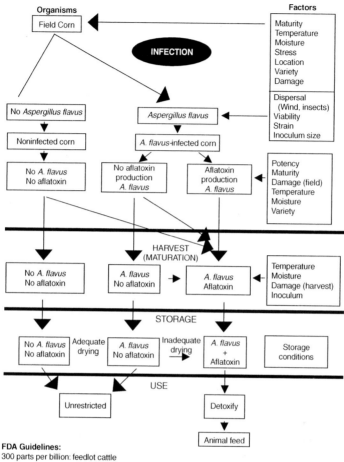

FIGURE 4.1. An outline of the *Aspergillus flavus*-corn (maize) infection process and aflatoxin production as affected by environmental and other factors during pre- and postharvest stages. (Modified from an unpublished version prepared by the late Dr. Marcus S. Zuber, University of Missouri-Columbia.)

Despite the common occurrence of drought and the advances in breeding methodology over the years, breeding for drought tolerance has received little or no emphasis. A problem is the difficulty to select for a trait when conditions necessary for its expression cannot be generated at will.

Temperature. Aflatoxin found in preharvest grain pointed to a higher incidence of aflatoxin contamination of maize grown in warmer regions of the United States (Anderson et al., 1975). The most serious aflatoxin problems have occurred in years with above-average temperatures and below-average rainfall. In 1983, temperatures in July and August were reportedly 2 to 3°C higher than normal in Iowa, which led to a high incidence of aflatoxin contamination. In addition, the areas of the state with the highest incidence of aflatoxin received six inches (15 cm) less than normal rainfall.

The temperature requirement for optimal aflatoxin production by *A. flavus* and *A. parasiticus* is 30°C; whereas it is 37°C for sterigmatocystin (a precursor to aflatoxin) production by *A. nidulans* (Bhatnagar, 1997). The importance of temperature and humidity was corroborated by Widstrom and colleagues (1990), who indicated that early planting in the Deep South increased the risk of aflatoxin contamination as compared with planting aimed at changing the time period for grain filling. No control strategy was completely effective for commercial hybrids when environmental conditions were extremely favorable for growth of the fungus (Payne, 1992).

Nutrients. Plant stress caused by limitations of nutrients reduces yield. Effects of nitrogen stress were investigated by Payne and colleagues (1989). The experiment consisted of Pioneer brand hybrid 3320, silk inoculation or kernel wounding, and nitrogen levels of 0, 56, 112, 168, and 224 kg/ha. Optimum yield was achieved at or just above the 168 kg/ha nitrogen level. Kernels from plants receiving no nitrogen had 28 percent more aflatoxin as compared with those plants receiving an optimum nitrogen dose. The correlation of yield and aflatoxin level was -0.99 ($P = 0.01$) for ears inoculated by kernel wounding, -0.04 for silk-inoculated ears (three environments), and -0.94 ($P = 0.05$) for one environment in which noninoculated ears were sampled.

Mineralization of nitrogen on highly organic soils, as compared with sandy soils, tended to reduce aflatoxin contamination (Jones, 1987). Availability of nitrogen is reduced with drought stress, which compounds effects of low moisture per se. Kang (1996) pointed out that impact of aflatoxin contamination could be minimized by reducing stress on the plant; for instance, use of optimal fertilization and irrigation, lime application, crop rotation, weed control, use of adapted hybrids, etc.

Biotic Stresses

Insects. The role of insects in the infection and contamination processes has been well established (Widstrom, 1979; McMillian, 1983,1987; Barry, 1987). Taubenhaus (1920) implicated insects invading maize ears as contributors to infection by fungi. He did not suggest *A. flavus* to be the cause of feeding problems in animals, however, probably because of previous reports indicating that *Aspergillus* spp. other than *A. flavus* were responsible for the disease symptoms (Mayo, 1891; Dalrymple, 1893).

In general, insect damage to the ear was consistently associated with increased sporulation by *A. flavus* on the ear and increased aflatoxin contamination of the grain. Differences among hybrids for aflatoxin contamination have been regarded as a function of husk coverage and plant resistance to insects (Zuber et al., 1983; McMillian et al., 1985). Insects have been shown to enhance *A. flavus* infection and aflatoxin production (Widstrom et al., 1975). The European corn borer (*Ostrinia nubilalis* Hubner) contributed more to the contamination process than either the corn earworm (*Helicoverpa zea* Boddie) or the fall armyworm (*Spodoptera frugiperda* J. E. Smith).

Development of resistance to insect feeding should be effective in reducing fungal infection and aflatoxin accumulation in the grain (Williams et al., 1997). Transgenic maize hybrids that express insecticide proteins from *Bacillus thuringiensis* could reduce fungal infection. Williams and colleagues (1997) reported that the southwestern corn borer *(Diatraea grandiosella)* population was significantly reduced in *Bt* hybrids. Damage from the fall armyworm and the corn earworm was not affected in such hybrids, however.

Maize weevil (*Sitophilus zeamais* Motschulsky) is an effective vector of *A. flavus* spores and is capable of increasing aflatoxin levels in kernels by as much as 100 times in the presence of the fungus (Rodriguez et al., 1983; Barry et al., 1985). Weevil activity in stored maize grain enhanced *A. flavus* growth (Dix and All, 1987). Zhang and colleagues (1997) found a strong, negative genetic relationship ($r = -0.84^*$) between general combining ability (GCA) effects for nonpreference for grain by the maize weevil and the GCA effects for percentage kernel infection, which implied that the higher the resistance to kernel infection by *A. flavus*, the higher the nonpreference for grain by the maize weevil. Another vector of the fungus is the sap beetle *(Carpophilus freemani)* (Nitidulidae) (Lussenhop and Wicklow, 1990; Rodriguez-del-Bosque et al., 1998).

Weeds. Weed competition during the reproductive stages (silking, pollen shedding, pollination) can be especially problematic. A good program of weed control is necessary for successful maize production (Lillehoj, 1983; Widstrom, 1996). Eliminating weeds aids in preventing water stress on the crop and yield losses in dryland-production systems.

REDUCING AFLATOXIN CONTAMINATION IN MAIZE

Detoxification

Contaminated maize can be effectively detoxified and made acceptable as feed-grade grains. Several techniques, such as physical separation of af-

---

* denotes significant at $P < 0.01$

latoxin-contaminated seed, solvent extraction of aflatoxin, and chemical in-activation of the toxin, have been developed (Lillehoj and Wall, 1987; McKenzie et al., 1998).

Electronic kernel sorters have been developed that, after examining each kernel, reject discolored ones, causing a reduction in aflatoxin (Dickens and Whitaker, 1975). A new type of grain cleaner (The SV4-C separator) has been developed by Camas International, Pocatello, Idaho. The machine, based on fluidized technology, is considered the first of its kind. It reduces levels of aflatoxin contamination to less than 20 ppb, the legal cutoff for food, while removing only about 7 percent of the grain (House, 1992).

Heat treatments have been used to detoxify aflatoxin (Rehana et al., 1979; Seenappa and Nyagahungu, 1982). These are steaming under pressure, dry roasting, and other cooking techniques (Coomes et al., 1966; Schroeder et al., 1985). These techniques can reduce aflatoxin levels but do not elimi-nate the toxin.

Ultraviolet radiation has been employed to degrade aflatoxin M1 in milk (Yousef and Marth, 1985). Gamma-ray treatment to remove aflatoxin pro-duced mixed results. Feuell (1966) observed no reduction in toxicity, whereas Ogbadu and Bassir (1979) detected decreased toxicities of contam-inated feeds after gamma radiation.

Chemical procedures, such as solvent extractions of aflatoxin from con-taminated commodities, in situ chemical inactivation of aflatoxin, and oxi-dation and reduction agents for degradation and destruction of aflatoxin, have been suggested (Feuell, 1969; Goldblatt, 1971; Detroy et al., 1971). Chemical inactivation of aflatoxin is a promising approach for control of toxic commodities (Lillehoj and Wall, 1987). Many oxidation and reduction agents reduce aflatoxin (Goldblatt, 1971). They usually partially eliminate one or two kinds of aflatoxins from commodities, and, in most cases, they cause some loss of desirable components in the commodity.

Aqueous or gaseous ammonia with or without elevated temperature and pressure appears to be a most efficient approach to decontaminating com-modities containing aflatoxin (Lillehoj and Wall, 1987; Bothast, 1991). The ammoniation procedure can reduce aflatoxin levels from 100 ppb to 10 ppb in small lots, but its effectiveness is reduced in larger lots (about 40 tons) (Brekke et al., 1978). The feed efficiency of ammonia-treated maize grain has not been affected negatively.

Electrochemically produced ozone (O_3) has been shown to provide pro-tection against AFB_1 in young turkey poults, without altering their perfor-mance (McKenzie et al., 1998). Thus, electrochemically generated O_3 might be an effective way to remediate, at a minimal cost, bulk quantities of maize with little or no destruction of important nutrients (McKenzie et al., 1998).

Breeding for Resistance

Although many control measures, such as insecticides, biological controls, and chemical and physical detoxification, have been tried to reduce or eliminate aflatoxin contamination in maize, none seems to be economically feasible (Lillehoj and Wall, 1987). Host-plant resistance is a most logical and economical way to solve the aflatoxin problem in maize (Zuber et al., 1978; Gorman and Kang, 1991; Widstrom, 1996). However, combining aflatoxin resistance with other pertinent traits into maize lines continues to be a challenge for breeders. In some cases, research funding is a problem because breeding is a long-term enterprise.

Inoculation Techniques

For successfully breeding for resistance to aflatoxin accumulation in maize, an effective inoculation technique is required. There is no consensus on the most effective inoculation method to use for field screening. Artificial kernel wounding was previously thought to be necessary to obtain sufficiently high aflatoxin levels for differentiation of genotypes. Rambo et al. (1974) compared silk inoculation with two kernel-injury methods: (1) injection of spores into kernels with a syringe and needle, and (2) insertion of a spore-impregnated cotton swab into a hole in the ear. Kernel infection was observed only with the injury methods. Calvert and colleauges (1978) reported that injury-inducing techniques, pinboard and razor blades mounted in a holder, produced higher levels of aflatoxin than did hypodermic syringe and needle. Thick-pericarp genotypes had lower aflatoxin levels than did thin-pericarp genotypes (Calvert et al., 1978; Gorman and Kang, 1991). Wounding techniques were used to simulate insect damage, but they compromised resistance offered by the aleurone or pericarp layers. Wallin (1986), using whole kernels and decapped kernels, suggested that pericarp and/or aleurone layers contributed to resistance.

To determine genotype resistance to natural fungal invasion, noninjury techniques were needed for screening. Some noninjury methods for genotype screening purposes have shown promise. Jones and colleagues (1980) observed significant infection by *Aspergillus* with the silk inoculation procedure; inoculations were most effective one and two weeks after midsilk. Scott and Zummo (1988) compared the pinbar method to two noninjury techniques. The pinbar method gave highest infection, but the two noninjury methods also effectively differentiated between hybrids. Wounding procedures are more stringent than noninjury ones to select for resistance to *A. flavus,* since morphological barriers (e.g., pericarp) to the fungus are breached. Most of the artificial inoculation techniques for differentiating genotypes for aflatoxin contamination have given relatively large coefficients of variation (CV $\geq$ 44 percent), which makes them unreliable for genetic studies and for effectively screening maize inbred lines, hybrids, and segregating populations. High CVs have been common in most studies on aflatoxin or percent kernel infection. Zhang and colleauges (1998) found that toothpick-under-husk (TUH) inoculation method, which involves placement of a toothpick laden with *A. flavus* on the surface of kernels following an incision in the middle of husks of an ear, was reliable for evaluating resistance to aflatoxin. They reported a substantially reduced CV of 20.6 percent.

Characteristics of Resistant Genotypes

It is helpful for breeders to know characteristics of resistant genotypes to facilitate selection. No single most important characteristic can be employed as an effective selection criterion to reduce preharvest aflatoxin contamination.

Kang and Lillehoj (1988) reported the results of two leafy-gene *(Lfy)* synthetics, developed by Cornnuts (a product of Nabisco), and two nonleafy *(lfy lfy)* counterparts that were inoculated with *A. parasiticus* (three replications). The *Lfy* gene caused about 50 percent reduction in aflatoxin (AF) B_1, AFB_2, AFG_1, and AFG_2.

Leafy-gene containing genotypes have been shown to allow low levels of aflatoxin accumulation in other studies as well (Gorman et al., 1992a,b). The *Lfy* gene materials are characterized by extra leaves above the ear and have been reported to impart tolerance to drought and insects (Shaver, 1983).

Recently, Guo and colleauges (1995) found that wax and cutin layers of maize kernel pericarp played a role in resistance to aflatoxin accumulation in MAS:gk and some other genotypes. Russin and colleagues (1997) indicated that kernel wax was an important factor in maize kernel resistance to *A. flavus*. Noninjury inoculation methods, such as toothpick-under-husk and bag inoculation (Zhang et al., 1998), may be sufficiently reliable to evaluate the effectiveness of a preformed chemical within the seed, assuming that a normal dose of inoculum is applied; a massive dose could compromise the effectiveness of such a compound.

Brown and colleagues (1995) suggested the existence of high-lipid content in the seed embryo of an aflatoxin-resistant maize genotype, GT-MAS:gk. Huang and colleagues (1997) reported the presence of two inhibitors in seed extracts of the resistant inbred Tex6. A 28 kDa protein inhibits fungal growth, whereas a 100 kDa protein inhibits aflatoxin biosynthesis. Guo and colleagues (1998) indicated that kernel proteins were important in two genotypes (GT-MAS:gk and MP420) resistant to *A. flavus* infection and aflatoxin contamination. These compounds could serve as chemical markers (or the encoding genes as molecular markers) to screen for aflatoxin resistance in maize. In addition, it would augment research on protein structure and function and on genetic regulation of protein expression, which may have long-term significance in engineering aflatoxin resistance into maize. The α-amylase produced by *A. flavus* has a role in aflatoxin production in diseased maize kernels. Action of fungal α-amylase on kernel starch results in the production of fermentable sugars leading to an induction of aflatoxin biosynthesis (Fakhoury and Woloshuk, 1997). Woloshuk and colleagues (1997) suggested that inhibiting the action of the amylase produced by *A. flavus* might be a viable strategy to control aflatoxin production. They disrupted the α-amylase gene *amyA* in an aflatoxin-producing strain of *A. flavus*. Whether amylase inhibitors in maize have an effect on amylase from *A. flavus* is a critical question to be answered.

Genetic Nature of Resistance

Though initial investigations on the potential of host-plant resistance for directly controlling *Aspergillus* spp. infection and aflatoxin contamination were not optimistic (LaPrade and Manwiller, 1976; Zuber, 1977), a pattern of resistance began to emerge. The resistance mechanisms appeared to be complex and responded to insect damage and environmental conditions (Lillehoj et al., 1976; LaPrade and Manwiller, 1977). Genetic differences existed in popcorn (McMillian et al., 1982), dent maize hybrids (Scott et al., 1991; Scott and Zummo, 1990a; Widstrom et al., 1978), and other noninbred germplasm (Kang et al., 1990; Wallin et al., 1991).

Hybrid response is difficult to predict for any location (Wilson et al., 1989; Scott et al., 1991) because differences in environmental conditions, planting date, harvest dates, inoculation procedure, and insect injury greatly affect aflatoxin accumulation. Testing of genotypes across years and/or locations is deemed necessary because large genotype × environment interactions are often encountered in aflatoxin and kernel infection studies (McMillian et al., 1982; Zuber et al., 1983; Zhang et al., 1997). Gardner and colleagues (1987) suggested that for aflatoxin field studies, eight replications were necessary from the standpoint of efficiency, cost effectiveness, and reduction in standard error. Genotype × environment interactions have important implications in resource allocation/management and developing breeding strategies (Kang and Gauch, 1996; Magari and Kang, 1997; Kang, 1998).

Hybrids with good husk coverage and insect resistance showed reduced aflatoxin levels in the grain. It is well established now that the infection and contamination processes are under genetic control (Widstrom et al., 1978; Zuber et al., 1978; and Zuber and Lillehoj, 1979; Gorman et al., 1992a; Zhang et al., 1997). Zuber and colleagues (1978) conducted the first genetic study in the United States on maize resistance to AFB_1 production using an eight-inbred diallel. Resistance to aflatoxin accumulation in maize kernels appeared to be quantitative in nature; general combining ability (GCA) was more important than specific combining ability (SCA), and reciprocal mean squares were not significant. Later, Widstrom and colleagues (1984) and Darrah and colleagues (1987) reported similar results as Zuber and colleagues (1978) did. Not all studies conclude that GCA was more important (Gardner et al., 1987; Gorman et al., 1992a; Campbell and White, 1995), however.

White and colleagues (1997) reported inbred line Tex6 to be a source of resistance, which was derived from a white maize population (PI 401762) from the southern United States. Generation mean analyses indicated additive gene action to be of primary importance for resistance to aflatoxin production (White et al., 1997; Hamblin and White, 2000).

All inheritance studies are in agreement that resistance to aflatoxin accumulation in maize kernels is a quantitative trait and is highly influenced by environmental factors. This makes traditional field identification of sources of resistance difficult. Restriction fragment length polymorphisms (RFLPs) and other molecular marker techniques offer the possibility of locating quantitative trait loci (QTL) through linkage with known molecular markers, although the expression of the trait is extremely environment-dependent. Marker-assisted selection, when perfected, should help develop resistant germplasm incorporating QTL from different chromosomal regions that have so far been identified (e.g., 1L, 2L, 4S) (White et al., 1997), which would be very useful in enhancing germplasm. To develop commercial hybrids with resistance to aflatoxin and/or *A. flavus,* both parents must be resistant because of the polygenic control, where each gene contributes a small effect. Environment × QTL interactions for aflatoxin accumulation need to be investigated to better understand the genetic nature of resistance and to achieve some degree of success in the selection process.

Since mature maize ears serve as substrate for fungal growth, resistance should be evaluated at the mature-ear stage. Greenhouse screening is not practical. Furthermore, currently there is no way to visually evaluate parental lines or hybrids for resistance. Each genotype must be screened for field aflatoxin contamination.

There is no general agreement on whether percent kernel infection (PKI) or aflatoxin per se is a better approach to study inheritance of resistance/susceptibility. For example, Tucker and colleagues (1986) reported a Pearson correlation coefficient of $r = 0.88$ ($P \leq 0.01$) between kernel infection and aflatoxin contamination when 260 kernels were plated from each of four genotypes. Windham and Williams (1998) suggested, however, that measuring aflatoxin concentration is more definitive than kernel infection to select for resistance, and they indicated that field plots should be harvested about 63 days after midsilk.

From an economic standpoint, PKI determination is much cheaper and simpler than aflatoxin quantification. General combining ability (GCA) effects for maize weevil (*Sitophilus Zeamais* Motschulsky) nonpreference for grain were reported by Kang and colleagues (1995) for the 10 inbred lines that Zhang and colleagues (1997) employed for PKI evaluation. Interestingly, a strong, negative genetic relationship ($r = -0.84*$) between the GCA effects for maize weevil nonpreference for grain and the GCA effects for PKI was detected. This suggested that the higher the resistance to kernel infection by *A. flavus,* the higher the nonpreference for grain by maize weevil in these 10 inbred lines. Maize weevils' nonpreference/preference for grain might be conditioned by the same or similar genes that condition resistance/susceptibility to kernel infection by *A. flavus.* Thus, selection for low PKI could not only reduce aflatoxin contamination but also provide, as correlated response, much needed resistance to weevil feeding damage.

Sources of Resistance

Several important sources of resistance have been identified as a result of screening and genetic studies (Scott and Zummo, 1988; Kang et al., 1990; Campbell et al., 1993), which should be useful as resistant germplasm for breeding purposes (McMillian et al., 1993; Scott and Zummo, 1990b, 1992). The chances for finding and developing other sources of resistance are excellent and should, in the future, provide maize with a defense against invasion by *A. flavus* and subsequent aflatoxin production (Widstrom, 1996). Other resistant germplasms also have been reported (White et al., 1997; Windham and Williams, 1998; Hamblin and White, 2000).

Systemic Acquired Resistance and Pathogenesis-Related Proteins

Plants respond to fungal attack by complex defense mechanisms (Dixon and Harrison, 1990). Dicots have been have been extensively studied in regard to β-1,3-glucanase that is triggered in response to fungal invasion, but only limited information is available about induction of β-1,3-glucanase in monocots. Chitinases and β-1,3-glucanases serve protective functions in maize tissue challenged with *A. flavus* (Wu et al., 1994). Resistant maize genotypes were shown to have higher β-1,3-glucanase activity before and after treatment with fungus than susceptible ones (Lozovaya et al., 1998). Genotypes with enhanced levels of β-1,3-glucanase activity could be more resistant to aflatoxin accumulation. Genetic manipulation aimed at overexpression of this enzyme to increase resistance to *A. flavus* might be a viable strategy.
Understanding plant-pathogen interactions is the key to solving the aflatoxin problem. Signal transduction pathways that lead to changes in gene

---

* denotes significant at $P < 0.01$

expression in response to pathogens, such as *Aspergillus flavus*, as well as other plant stresses, need to be explored. Plant responses are generally common to most stresses (A.-H.-Mackerness and Thomas, 1999). A recurring theme in plant responses to diverse environmental stresses is the production of reactive oxygen species (ROS) (Smallwood et al., 1999). Studies have shown that ROS are involved in UV-B signal pathways leading to the down-regulation of photosynthetic genes and up-regulation of the acidic-type pathogenesis-related (PR) genes (A.-H.-Mackerness et al., 1998). The PR genes such as β-1,3-glucanase (PR2) and chitinase (PR3) (see Table 4.1) are important genetic defense mechanisms of resistance in response to stresses (Schraudner et al., 1992).

TABLE 4.1. Natural Compounds That Inhibit *Aspergillus flavus* Growth

Compound	Effect	Commodity	References
Ajoene and diallyl sulfide	Inhibit fungal growth	Garlic	Appleton and Tansey, 1975; Mabrouk and El-Shayeb, 1981; Yoshida et al., 1987; Kang et al., 1994
Onion extract	Inhibits fungal activity	Onion	Sharma et al., 1979; Sharma et al., 1981
Welsh onion extract	Inhibits growth of and aflatoxin production by *A. flavus*	Welsh onion (*Allium fistulosum* L.)	Fan and Chen, 1999
Acetosyringone, syringaldehyde, sinapinic acid	Inhibit aflatoxin production	Tree nuts, peanuts, cotton, and maize	Hua et al., 1997
α-linolenic acid	Inhibits *A. flavus* spore germination	Maize (embryo)	Norton, 1997
α-Tocol (tocopherol)	Inhibits aflatoxin formation	Maize (embryo)	Norton, 1997
Anacardic acids	Inhibit aflatoxin biosynthesis	Pistachio (hulls)	Campbell et al., 1997
Anthocyanins	Reduce growth of *A. flavus* and/or the production of aflatoxin	Maize	Davis et al., 1997
β-carotene	Inhibits aflatoxin biosynthesis	Many plant species	Wicklow et al., 1997
Chitinase	Inhibits fungal growth	Maize	Moore et al., 1997; Payne, 1997
Naphthoquinones	Inhibit germination of *A. flavus* conidia	Walnut (hulls)	Campbell et al., 1997; Mahoney et al., 1997

Compound	Effect	Commodity	References
Zeamatin and a ribosome inactivating protein (RIP)	Inhibit fungal growth	Maize	Brown, Chen, et al., 1997; Guo et al., 1997; Payne, 1997
n-decyl aldehyde	Inhibit aflatoxin biosynthesis	Maize	Wright et al., 1997
n-decyl aldehyde, octanal, hexanal	Reduce growth of *A. parasiticus*	Maize	Wright et al., 1997
3-methyl-1butanol, nonanol, camphene, and linonene	Inhibit fungal growth	Maize	Wright et al., 1997
13S-hydroxperoxy-9z, 11E-octadecadienoic acid	Suppresses growth of *A. flavus* and inhibits aflatoxin production	Soybean	Burow et al., 1997
14 kDa trypsin protein	Inhibits fungal growth	Maize	Chen et al., 1997; Chen et al., 1998

Systemic acquired resistance (SAR) is another mechanism that has been shown to be important in combating stresses in plants (Smallwood et al., 1999; Kang, 1998). The strategy is to enhance resistance by the development of chemicals that induce an inherent resistance response in plants (Gatz, 1997). The SAR is a long-lasting disease resistance that can provide a significant level of protection against pathogens (Durner et al., 1997). The SAR is generally accompanied by the accumulation of endogenous salicylic acid (SA) (Ernst et al., 1999). Exogenous application of SA or structural analogs such as aspirin, 2,6-dichloroisonicisotinic acid, and benzothiadiazole induces resistance against pathogens and the activation of SAR (Ernst et al., 1999). These are signal molecules involved in local (hypersensitive response) and systemic reactions of plants (Moeder et al., 1999). Induction of PR1 and glucanase and chitinase are mediated by SA (Sharma and Davis, 1997).

INTERRUPTING/REDUCING TOXIN PRODUCTION ABILITY OF THE FUNGUS

Chemical and Biological Interference in Toxin Production

Imposition of chemicals or conditions on the fungus to inhibit aflatoxin production, and the genetic manipulation of fungus or host plant to interrupt the capability of the fungus to produce aflatoxins, should constitute long-term solutions. The initial efforts in this area involved the use of chemicals

(Davis and Diener, 1967; Rao and Harein, 1972; Shroeder et al., 1974; Bothast et al., 1976) and atmospheric gases (Landers et al., 1967; Wilson and Jay, 1975; Wilson et al., 1975). Chemicals within the host plant were also suggested as potential inhibitors of aflatoxin production (Nagarajan and Bhat, 1972).

A number of chemicals and biological compounds that inhibit or suppress fungal growth and/or aflatoxin production have been evaluated. Most of them have been tested on maize under storage conditions and, therefore, may not be useful as control agents on field maize. Compounds isolated from maize itself have also been reported as having limited activity against *A. flavus* and aflatoxin production. β-Ionone affected morphology and sporulation of *Aspergilli* (Wilson et al., 1981) while other volatiles also showed some activity against these fungi (Gueldner et al., 1985). Salt-extracted and base-soluble proteins from maize kernels showed sporadic fungicidal effects on *A. flavus*, but the effects were not always pronounced in the resistant variety (Neucere and Zeringue, 1987; Neucere and Godshall, 1991; Neucere, 1992). Kang and colleagues (1994) evaluated garlic *(Allium sativa)*. When garlic concentration (1 g of garlic in 2 mL H_2O = 100 percent) was 5 percent or above, growth of both *A. flavus* and *A. parasiticus* was inhibited. Kang and colleagues (1995) also reported that inhibiting ability of cultivated garlic extract was higher than that of wild garlic (a weed) extract. Inconsistent results were characteristic of evaluations of the activity of chitinase from mature kernels (Neucere et al., 1991). Brown and colleagues (1993) suggested that resistance to aflatoxin contamination was related to metabolic activities in the living maize embryo. Antifungal compounds in plants include ribosome inactivating proteins, lectins, certain low molecular weight polypeptides, cell-surface glycoproteins, hydrolases, and certain basic proteins (McCormick et al., 1988). Some of the known, naturally occurring compounds that inhibit fungal growth and/or aflatoxin production are presented in Table 4.1.

Atoxigenic-Toxigenic Fungal Competition

Another approach to controlling aflatoxin contamination is the use of nontoxigenic species of *Aspergillus* or other fungi as competitors to limit growth of toxigenic strains of *Aspergillus*. Calvert and colleagues (1978) first compared aflatoxin production of *A. flavus* and *A. parasiticus* as competitors, and Wicklow and colleagues (1980) followed with tests of aflatoxin production by *Aspergillus* spp. in competition with other nontoxigenic species. Additionally, several isolates of the *A. flavus* group have been shown to vary in both aflatoxin-producing ability and aggressiveness in competing for infection sites (Zummo and Scott, 1994).

Exploitation of competition between fungal species appears to be an attractive approach to reduce aflatoxin contamination. *Fusarium moniliforme* is highly effective in reducing infection and contamination by *A. flavus* (Zummo and Scott, 1992; McAlpin et al., 1998). *Aspergillus flavus* and *A. parasiticus* are both aggressive colonizers of maize, but *A. flavus* is more persistent than *A. parasiticus* (Zummo and Scott, 1990, 1994). Brown and colleagues (1991) suggested the use of competitive atoxigenic strains of *A. flavus* to reduce aflatoxin production on maize grain. Cotty and Bhatnagar (1994) found variation among atoxigenic strains of *A. flavus* in their ability to prevent production of aflatoxin and biosynthetic pathway enzymes. Kale and colleagues (1996) characterized, by use of electron microscopy, six nonaflatoxigenic variants of *A. parasiticus* (*sec* mutants). The loss of aflatoxigenic capabilities in the *sec* variant was caused by alterations in the conidial morphology of the fungus, suggesting that the regulation of aflatoxin synthesis and conidiogenesis might be interrelated. None of the chemical and biological methods tested thus far can be recommended as fully practical or economical (Bilgrami and Misra, 1981).

Genetic Manipulation of the Fungus

The aflatoxin biosynthetic pathway is one of the best characterized pathways of secondary metabolism in fungi (Brown et al., 1999). Work on the biosynthetic pathway(s) of aflatoxins began in the 1970s (Detroy and Hesseltine, 1970; Heathcote et al., 1976). Papa (1976, 1977, 1979, 1980) pioneered the development of a linkage map of the fungus and determination of the inheritance of *A. flavus* genes in the pathway(s) and their interrelationships. Many researchers used *A. parasiticus* as a model to work out the relationship among intermediates (Bhatnagar et al., 1987; Hsieh et al., 1973; Hseih, Yao, Fitzell, and Reece, 1976; Hseih, Yao, Fitzell, and Singh, 1976; Lee et al., 1976). The aflatoxin biosynthetic pathway has been elucidated mainly by use of *A. flavus* and *A. parasiticus*. These two fungi, coupled with *A. nidulans* whose genetics is well understood, have advanced this area of research.

New knowledge on aflatoxin biosynthesis has resulted in new strategies for identifying genes and pathways for different aflatoxins (Bennett et al., 1994; Bhatnagar et al., 1989, 1991). A corollary of the work on the genetics of *A. parasiticus* was the identification of a red-brown mutant (Wilson et al., 1986). This mutant produces norsolorinic acid (an orange intermediate) that can be monitored in the aleurone layer of maize kernels and can serve as an indicator of the effectiveness of inoculation and aflatoxin production (Keller et al., 1992).

One powerful tool for understanding the factors that affect growth and aflatoxin production by *A. flavus* is a gene reporter system (Du et al., 1997; Du and Payne, 1997; Flaherty et al., 1995, 1997). Experiments with laboratory and field infection of maize kernels by reporter strains of *A. flavus* allow quantitation of the amount of fungal invasion in kernels, a critical measurement in the assessment of resistance during maize varietal screening by breeders and agronomists (Robens and Dorner, 1997). Brown, Cleveland, and colleagues (1997) showed that a reporter construct containing the promoter of the *A. flavus* B-tubulin gene fused to the *E. coli* β-glucuronidase gene (GUS) reliably measured fungal growth in maize kernels. Brown-Jenco and colleagues (1997) reported the construction and analysis of a new reporter gene construct using the *omtA* gene promoter of the aflatoxin biosynthetic pathway.

The work on genetics of aflatoxin biosynthesis has led to the cloning of *apa-2*, a gene associated with regulation of the process (Chang et al., 1993). Studies using blocked mutants, metabolic inhibitors, and radioactively labeled precursors in bioconversion experiments have provided a relatively clear picture of the biochemical pathway involved in the synthesis of AFB$_1$. It is a complex pathway controlled by at least 19 gene products. Since 1992, 17 genes catalyzing 12 enzymatic steps have been cloned and characterized (Bhatnagar et al., 1992; Brown et al., 1998). Several genes directly involved in AFB$_1$ synthesis have been cloned and the functions of their gene products have been characterized. These genes include *nor1*, which is involved in conversion of norsolorinic acid to averantin (Chang et al., 1992; Trail et al., 1994); *ver-1A*, involved in the conversion of versicolorin A to demethylsterigmatocystin (Liang et al., 1996; Skory et al., 1992); *omtA*, involved in the conversion of sterigmatocystin to *O*-methyl-sterigmatocystin (Yu et al., 1993); and the *fast-1A* (Mahanti et al., 1996) and *pksA* genes (Chang et al., 1995; Trail et al., 1995) involved in the synthesis of the AFB$_1$ polyketide backbone.

The *aflR* gene, encoding a protein that is proposed to be a positive regulator of aflatoxins pathway functions, was identified in *A. flavus* (Payne et al., 1993) and *A. parasiticus* (Chang et al., 1993). At least 17 genes involved in the aflatoxin biosynthetic pathway are clustered within a 75-kb DNA fragment in the genome of *A. parasiticus* (Brown et al., 1998; Brown et al., 1999). Several additional transcripts have also been mapped to this gene cluster. A gene, *avnA* (previously named *ord-1*), corresponding to one of the two transcripts identified earlier between the *ver-1* and *omtA* genes on the gene cluster has been sequenced (Yu et al., 1997). The gene for a dehydrogenase *(adhA)* and a cytochrome P-450 coding gene *(avnA)* have been identified and characterized. Gene *aflJ* has been identified. The *aflJ* gene is involved in the regulation of the aflatoxin pathway along with the regulatory gene, *aflR*. They work in concert in regulating aflatoxin biosynthesis in *A. flavus* and *A. parasiticus*. The *aflJ* gene has not been identified on the *A. nidulans* sterigmatocystin gene cluster, suggesting different regulation of toxin synthesis (Bhatnagar, 1997). Sterigmatocystin (ST) biosynthetic pathway in *A. nidulans* is estimated to involve at least 15 enzymatic activities, whereas certain *A. flavus, A. parasiticus*, and *A. nomius* strains contain additional activities that convert ST to aflatoxin (AF). Sterigmatocystin is much less toxic to biological systems than AF and occurs in several genera of fungi. The AF is produced by only four species within the *Aspergillus* section *Flavi*. This has led to the speculation that the role of AF is more specific than that of ST (Brown et al., 1999). The AF biosynthesis appears to be more tightly regulated than ST biosynthesis (Brown et al., 1999).

New biological control strategies have been proposed based on these findings. A better understanding of the reactions in the biosynthesis of secondary metabolites, the cloning of genes associated with aflatoxin biosynthesis, and the functional characterization of these genes have been significant achievements toward understanding aflatoxin production (Keller et al., 1992) and how to control it prior to harvest (Widstrom, 1996). For a schematic depiction of the current aflatoxin biosynthetic pathway, please refer to Brown and colleagues (1999).

The AF pathway, because of the many genes (metabolic steps) involved in it, offers many mutational opportunities to block the aflatoxin production. Even if the block occurred at the penultimate step, just prior to AF production, the resultant chemical ST would be of much less risk to human and animal health. A summary of the various genes and their roles/possible roles is given by Moreno and Kang (1999).

CONCLUDING REMARKS

The most effective and economical control for aflatoxin contamination in maize is development and use of resistant hybrids. The aflatoxin contamination process is complicated because of the involvement of the biology/genetics of two organisms. Thus, breeding for resistance to this disease remains a difficult challenge for breeders. Several factors influence infection and aflatoxin production by *A. flavus* in commodities, but the biochemical basis of these effects is not completely elucidated.

Information gained through the enzymology and molecular biology of these fungi can be used to engineer a biological control agent, to identify compounds that block aflatoxin formation, and to identify compounds that inhibit the fungal activities. Successful control of aflatoxin contamination will likely involve a combination of many strategies. To manage the aflatoxin problem, it will be necessary to investigate all possible approaches. The approach should be an integrated one involving genetics, molecular biology, plant breeding, biochemistry and plant physiology, ecology, and plant pathology. Genetic manipulations of both the host plant and the fungus offer the best hope for achieving a satisfactory control of the aflatoxin problem with maize grain. A better knowledge of signal transduction in plants in response to biotic and abiotic stresses also should provide additional arsenal in combating maize diseases such as aflatoxin contamination by *Aspergillus flavus*.

REFERENCES

A.-H.-Mackerness, S., Surplus, S.L., Jordan, B.R., and Thomas, B. 1998. Effects of supplementary UV-B radiation on photosynthetic transcripts at different stages of development and light levels in pea: Role of ROS and antioxidant enzymes. *Photochem. Photobiol.* 68(1):88-96.

A.-H.-Mackerness, S. and Thomas, B. 1999. Effects of UV-B radiation on plants: Gene expression and signal transduction pathways. In M.F. Smallwood, C.M. Calvert, and D.J. Bowles (Eds.), *Plant responses to environmental stress* (pp. 17-24). BIOS Scientific, Oxford, U.K.

Anderson, H.W., Nehring, E.W., and Wichser, W.R. 1975. Aflatoxin contamination of corn in the field. *J. Agric. Food Chem.* 23(4):775-782.

Angle, J.S., Lindgren, R.L., and Gilbert-Effiong, D. 1989. Survival of *Aspergillus flavus* conidia in soil. In C.E. O'Rear and G.C. Llewellyn (Eds.), *Biodeterioration Research 2* (pp 245-250). Plenum Press, New York.

Appleton, J.A. and Tansey, M.R. 1975. Inhibition of growth of zoopathogenic fungi by garlic extract. *Mycologia* 67(4):882-885.

Barry, D. 1987. Insects of maize and their association with aflatoxin contamination. In M.S. Zuber, E.B. Lillehoj, and B.L. Renfro (Eds.), *Aflatoxin in maize* (pp. 201-211). CIMMYT, Mexico, DF.

Barry, D., Zuber, M.S., Lillehoj, E.B., Mcmillian, W.W., Adams, N.J., Kwolek, W.F., and Widstrom, N.W. 1985. Evaluation of two arthropod vectors as inoculators of developing maize ears with *Aspergillus flavus*. *Environ. Entomol.* 14(5):634-636.

Bennett, J.W., Bhatnagar, D., and Chang, P.K. 1994. The molecular genetics of aflatoxin biosynthesis. In K. A. Powell, A. Renwich, and J. F. Peberdy (Eds.), *The genus* Aspergillus. *From taxonomy and genetics to industrial applications* (pp. 51-58). FEMS Symposium. Plenum Press, New York.

Bhatnagar, D. 1997. Gene expression systems in aflatoxigenic fungi for monitoring aflatoxin elaboration in crops. In J. Robens and J. Dorner (Eds.), *Aflatoxin elimination workshop: A decade of research progress 1988-1997* (pp. 25-27). Food Safety and Health, USDA/ARS, Beltsville, MD.

Bhatnagar, D., Cleveland, T.E., and Lax, A.R. 1989. Comparison of the enzymatic composition of cell-free extracts of non-aflatoxigenic *Aspergillus parasiticus* with respect to late stages of aflatoxin biosynthesis. *Arch. Environ. Contam. Toxicol.* 18(3):434-438.

Bhatnagar, D., Ehrlich, K.C., and Cleveland, T.E. 1992. Oxidation-reduction reactions in biosynthesis of secondary metabolites. In D. Bhatnagar, E. B. Lillehoj, and D. K. Arora (Eds.), *Handbook of applied mycology*, Vol 5 (pp. 255-286). Marcel Dekker, New York.

Bhatnagar, D., McCormick, S.P., and Kingston, D.G.I. 1991. Enzymological evidence for separate pathways for aflatoxin B_1 and B_2 biosynthesis. *Biochemistry* 30(17):4343-4350.

Bhatnagar, D., McCormick, S.P., Lee, L.S., and Hill, R.A. 1987. Identification of O-methylstergmstocystin as an aflatoxin B_1 and G_1 precursor in *Aspergillus parasiticus*. *Appl. Environ. Microbiol.* 53(5):1028-1033.

Bilgrami, K. S., and Misra, R.S. 1981. Aflatoxin production by *Aspergillus flavus* in storage and standing maize crops. In K.S. Bilgrami, R.S. Misra, and P.C. Misra (Eds.), *Advancing frontiers of mycology and plant pathology* (pp. 67-78). Today and Tomorrow's Printers and Publishers, New Delhi, India.

Boller, R.A. and Schroeder, H.W. 1974. Production of aflatoxin by cultures derived from conidia stored in the laboratory. *Mycologia* 66(11):61-66.

Bothast, R.J. 1991. Processing of aflatoxin-contaminated corn. In O.L. Shotwell and C.R. Hurburgh Jr. (Eds.), *Aflatoxin in corn: New perspectives.* (pp. 369-376). Iowa State University, Ames, IA.

Bothast, R.J., Goulden, M.L., Shotwell, O.L., and Hesseltine, C.W. 1976. *Aspergillus flavus* and aflatoxin production in acid-treated maize. *J. Stored Prod. Res.* 12(3):177-183.

Brekke, O.L., Stringfellow, A.C., and Peplinski, A.J. 1978. Aflatoxin inactivation in corn by ammonia gas laboratory trials. *J. Agric. Food Chem.* 26(6):1383-1389.

Brown, D.W., Yu, J.-H., Kelkar, H.S., Fernandes, M., Nesbitt, J.C., Keller, N.P., Adams, T.H., and Leonard, T.J. 1996. Twenty-five coregulated transcripts define a sterigmatocystin gene cluster in *Aspergillus nidulans*. *Proc. Natl. Acad. Sci.* (U.S.) 93(4):1418-1422.

Brown, M.P., Brown-Jenco, C.S., and Payne, G.A. 1999. Genetic and molecular analysis of aflatoxin biosynthesis. *Fungal Genet. Biol.* 26(2):81-98.

Brown, R.L., Bhatnagar, D., Cleveland, T.E., and Cary, J.W. 1998. Recent advances in preharvest prevention of mycotoxin contamination. In K.K. Sinha and D. Bhatnagar (Eds.), *Mycotoxins in agriculture and food safety* (pp. 351-379). Marcel Dekker, Inc., New York, NY.

Brown, R.L., Chen, Z.-Y., Cleveland, T.E., and Russin, J.S. 1999. Advances in the development of host resistance in corn to aflatoxin contamination by *Aspergillus flavus*. *Phytopathology* 89(2):113-117.

Brown, R.L., Chen, Z.-Y., Lax, A.R., Cary, J.W., Cleveland, T.E., Russin, J.S., Guo, B.Z., Williams, W.P., Davis, G., Windham, G.L., and Payne, G.A. 1997. Determination of maize kernel biochemical resistance to aflatoxin elaboration: Mechanisms and biotechnological tools. In J. Robens and J. Dorner (Eds.), *Afla-*

toxin elimination workshop: A decade of research progress 1988-1997 (pp. 58-59). Food Safety and Health, USDA/ARS, Beltsville, MD.

Brown, R.L., Cleveland, T.E., Payne, G.A., Woloshuk, C.P., Campbell, K.W., and White, D.G. 1995. Determination of resistance to aflatoxin production in maize kernels and detection of fungal colonization using an *Aspergillus flavus* transformant expressing *Escherichia coli* β-glucoronidase. *Phytopathology* 85(9):983-989.

Brown, R.L., Cleveland, T.E., Payne, G.A., Woloshuk, C.P., and White, D.G. 1997. Growth of an *Aspergillus flavus* transformant expressing *Escherichia coli* β-glucuronidase in maize kernels resistant to aflatoxin production. *J. Food Prot.* 60(1):84-87.

Brown, R.L., Cotty, P.J., and Cleveland, T.E. 1991. Reduction in aflatoxin content of maize by atoxigenic strains of *Aspergillus flavus*. *J. Food Prot.* 54(8):623-626.

Brown, R.L., Cotty, P.J., Cleveland, T.E., and Widstrom, N.W. 1993. Living maize embryo influences accumulation of aflatoxin in maize kernels. *J. Food Prot.* 56(11):967-971.

Brown-Jenco, C.S., Brown, R.L., Bhatnagar, D., and Payne, G.A. 1997. Construction and analysis of an *Aspergillus flavus omtA* (P) GUS reporter construct. In J. Robens and J. Dorner (Eds.), *Aflatoxin elimination workshop: A decade of research progress 1988-1997.* (p. 73). Food Safety and Health, USDA/ARS, Beltsville, MD.

Burow, G.B., Nesbitt, T.C., Dunlap, J., and Keller, N.P. 1997. Seed lipoxygenase products modulate *Aspergillus* mycotoxin biosynthesis. *Mol. Plant. Interact.* 10(3):380-387.

Calvert, O.H., Lillehoj, E.B., Kwolek, W.F., and Zuber, M.S. 1978. Aflatoxin B_1 and G_1 production in developing *Zea mays* kernels from mixed inocula of *Aspergillus flavus* and *A. parasiticus. Phytopathology* 68(3):501-506.

Campbell, K.W., Hua, S., Light, D.M., Molyneux, R.J., Roitman, J., Merril, G., Mahoney, N., Goodman, N., Baker, J., and Mehelis, C. 1997. Role of natural products, semiochemicals and microbial agents in reducing insect infestations, *Aspergillus* infection and aflatoxigenesis in tree nuts. In J. Robens and J. Dorner (Eds.), *Aflatoxin elimination workshop: A decade of research progress 1988-1997* (p. 20). Food Safety and Health, USDA/ARS, Beltsville, MD.

Campbell, K.W. and White, D.G. 1995. Inheritance of resistance to *Aspergillus* ear rot and aflatoxin in corn genotypes. *Phytopathology* 85(8):886-896.

Campbell, K.W., White, D.G., and Toman, J. 1993. Sources of resistance in F_1 corn hybrids to ear rot caused by *Aspergillus flavus. Plant Dis.* 77(11):1169.

Cardwell, K.F., Udoh, J.M., and Hell, K. 1997. Assessment of risk of mycotoxic degradation of stored maize in Nigeria and Benin Republic, West Africa. In J. Robens and J. Dorner (Eds.), *Aflatoxin elimination workshop: A decade of research progress 1988-1997* (pp. 15). Food Safety and Health, USDA/ARS, Beltsville, MD.

CAST. 1989. In K.A. Nisi (Ed.), *Mycotoxins economic and health risks* (p. 1-90). Task Force Report No. 116. Council for Agricultural Science and Technology, Ames, IA.

Chang, P.-K., Cary, J.W., Yu, J., Bhatnagar, D., and Cleveland, T.E. 1995. The *Aspergillus parasiticus* polyketide synthase gene *pksA*, a homolog of *Aspergillus nidulans wA*, is required for aflatoxin B_1 biosynthesis. *Mol. Gen. Genet.* 248(3):270-277.

Chang, P.-K., Cary, J.W., Bhatnagar, D., Cleveland, T.E., Bennett, J.W., Linz, J.E., Woloshuk, C.P., and Payne, G.A. 1993. Cloning of the *Aspergillus parasiticus apa-2* gene associated with the regulation of aflatoxin biosynthesis. *Appl. Environ. Microbiol.* 59(10):3273-3279.

Chang, P.-K., Skory, C.D., and Linz, J.E. 1992. Cloning of a gene associated with aflatoxin B_1 biosynthesis in *Aspergillus parasiticus. Curr. Genet.* 21(3):231-233.

Chen, Z.-Y., Brown, R.L., Lax, A.R., Guo, B.Z., Cleveland, T.E., and Russin, J.S. 1997. A maize kernel trypsin inhibitor is associated with resistance to *Aspergillus flavus* infection. In J. Robens and J. Dorner (Eds.), *Aflatoxin elimination workshop: A decade of research progress 1988-1997* (p. 34). Food Safety and Health, USDA/ARS, Beltsville, MD.

Chen, Z.-Y., Brown, R.L., Lax, A.R., Guo, B.Z., Cleveland, T.E., and Russin, J.S. 1998. Resistance to *Aspergillus flavus* in corn kernels is associated with a 14-kDa protein. *Phytopathology* 88(4):276-281.

Coomes, T.J., Crowther, P.C., Feuell, A.J., and Francis, B.J. 1966. Experimental detoxification of groundnut meals containing aflatoxin. *Nature* 209(5021):406-407.

Cotty, P.J. and Bhatnagar, D.1994. Variability among atoxigenic *Aspergillus flavus* strains in ability to prevent aflatoxin contamination and production of aflatoxin biosynthetic pathway enzymes. *Appl. Environ. Microbiol.* 60(7):2248-2251.

Dalrymple, W.H. 1893. Report of the veterinarian. *La. Agric. Exp. Stn. Bull.* 22: 724-730.

Darrah, L.L. and Barry, D. 1991. Reduction of preharvest aflatoxin in corn. In G.A. Bray and D.H. Ryan (Eds.), *Mycotoxins, cancer and health* (pp. 288-310). Louisiana State University Press. Baton Rouge, LA.

Darrah, L.L., Lillehoj, E.B., Zuber, M.S., Scott, G.E., Thompson, D., West, D.R., Widstrom, N.W., and Fortnum, B.A. 1987. Inheritance of aflatoxin B_1 levels in maize kernels under modified natural inoculation with *Aspergillus flavus. Crop Sci.* 27(5):869-872.

Davis, G.L., Windham, G.L., and Williams, W.P. 1997. *Aspergillus flavus* growth and aflatoxin accumulation in 15 maize isolines containing genes for anthocyanin production. In J. Robens and J. Dorner (Eds.), *Aflatoxin elimination workshop: A decade of research progress 1988-1997* (p. 70). Food Safety and Health, USDA/ARS, Beltsville, MD.

Davis, N.D. and Diener, U.L. 1967. Inhibition of aflatoxin synthesis by p-aminobenzoic acid, potassium sulfite, and potassium fluoride. *Appl. Microbiol.* 15(6):1517-1518.

Detroy, E.W., Lillehoj, E.B., and Ciegler, A. 1971. Aflatoxin and related compounds. In A. Ciegler et al. (Eds.), *Microbial Toxins* (pp. 84-178). Academic Press, New York, NY.

Detroy, R.W., and Hesseltine, C.W. 1970. Secondary biosynthesis of aflatoxin B$_1$ in *Aspergillus parasiticus. Can. J. Microbiol.* 16(10):959-963.

Dickens, J.W. and Whitaker, T.B. 1975. Efficacy of electronic color sorting and hand picking to remove aflatoxin contaminated kernels from commercial lots of shelled peanuts. *Peanut Sci.* 2(2):45-50.

Dix, D.E. and All, J.N. 1987. Interactions between maize weevil (Coleoptera Curculionidae) infestations and infection by *Aspergillus flavus* and other fungi in corn. *J. Entomol. Sci.* 22(2):108-118.

Dixon, R.A. and Harrison, M. 1990. Activation, structure, and organization of genes involved in microbial defense of plants. *Adv. Genet.* 28:165-234.

Du, W., Flaherty, J.F., Huang, Z.-Y., and Payne, G.A. 1997. Expression of the green fluorescent protein in *Aspergillus flavus. Phytopathology* 87(6):S26.

Du, W. and Payne, G.A. 1997. Expression of the green fluorescent protein in *Aspergillus flavus* and its use as a marker to evaluate resistance in corn kernels. In J. Robens and J. Dorner (Eds.), *Aflatoxin elimination workshop: A decade of research progress 1988-1997* (p. 72). Food Safety and Health, USDA/ARS, Beltsville, MD.

Durner, J., Shah, J., and Klessig, D.F. 1997. Salicylic acid and disease resistance in plants. *Trends Plant Sci.* 2(7):266-274.

Ernst, D., Grimmig, B., Heidenreich, B., Schubert, R., and Sandermann Jr., H. 1999. Ozone-induced genes: Mechanisms and biotechnological applications. In M.F. Smallwood, C.M. Calvert, and D.J. Bowles (Eds.), *Plant responses to environmental stress* (pp. 33-41). BIOS Scientific, Oxford, U.K.

Fakhoury, A.M. and Woloshuk, C.P. 1997. Characterization of an alpha-amylase deficient mutant of *Aspergillus flavus*. In J. Robens and J. Dorner (Eds.), *Aflatoxin elimination workshop: A decade of research progress 1988-1997* (p. 37). Food Safety and Health, USDA/ARS, Beltsville, MD.

Fan, J.J. and Chen, J.H. 1999. Inhibition of aflatoxin-producing fungi by Welsh onion extracts. *J. Food Protect.* 62(4):414-417.

Feuell, A.J. 1966. Aflatoxin in groundnuts. IX. Problems of detoxification. *Tropical Science* 8(2):61-70.

Feuell, A.J. 1969. Types of mycotoxins in foods and feeds. In L. A. Goldblatt (Ed.), *Aflatoxin scientific background, control, and implications* (pp. 187-215). Academic Press. New York, NY.

Flaherty, J.F., Du, W.L., and Payne, G.A. 1997. Construction and expression of an AFLRGFP fusion protein. *Phytopathology* 87(6):S30.

Flaherty, J.F., Weaver, M.A., Payne, G.A., and Woloshuk, C.P. 1995. A beta-glucuronidase reporter gene construct for monitoring aflatoxin biosynthesis in *Aspergillus flavus. Appl. Environ. Microbiol.* 61(7):2482-2486.

Gardner, C.A.C., Darrah, L.L., Zuber, M.S., and Wallin, J.R. 1987. Genetic control of aflatoxin production in maize. *Plant Dis.* 71(5):426-429.

Gatz, C. 1997. Chemical control of gene expression. *Annu. Rev. Plant Physiol. Plant Mol. Biol.* 48:89-108.

Goldblatt, L.A. 1971. Control and removal of aflatoxin. *J. Am. Oil Chemists Society* 48(10):605-610.

Gorman, D.P. and Kang, M.S. 1991. Preharvest aflatoxin contamination in maize: Resistance and genetics. *Plant Breeding* 107(1):1-10.

Gorman, D.P., Kang, M.S., Cleveland, T.E., and Hutchinson, R.L. 1992a. Combining ability for resistance to field aflatoxin accumulation in maize grain. *Plant Breeding* 109(4):296-303.

Gorman, D.P., Kang, M.S., Cleveland, T.E., and Hutchinson, R.L.1992b. Field aflatoxin production by *Aspergillus flavus* and *A. parasiticus* on maize kernels. *Euphytica* 61(3):187-191.

Gourama H. and Bullerman, L.B. 1995. *Aspergillus flavus* and *Aspergillus parasiticus* Aflatoxigenic fungi of concern in foods and feeds: A review. *J. Food Prot.* 58(12):1395-1404.

Gueldner, R.C., Wilson, D.M., and Heidt, A.R. 1985. Volatile compounds inhibiting *Aspergillus flavus. J. Agric. Food Chem.* 33(3):411-413.

Guo, B.Z., Brown, R.L., Lax, A.R., Cleveland, T.E., Russin, J.S., and Widstrom, N.W. 1998. Protein profiles and antifungal activities of kernel extracts from corn genotypes resistant and susceptible to *Aspergillus flavus. J. Food Protect.* 61(1):98-102.

Guo, B.Z., Chen, Z.-Y., Brown, R.L., Lax, A.R., Cleveland, T.E., Russin, J.S., and Widstrom, N.W. 1997. Antifungal protein in corn kernels immunochemical localization and induction during germination. In J. Robens and J. Dorner (Eds.), *Aflatoxin elimination workshop: A decade of research progress 1988-1997* (p. 71). Food Safety and Health, USDA/ARS, Beltsville, MD.

Guo, B.Z., Russin, J.S., Cleveland, T.E., Brown, R.L., and Widstrom, N.W. 1995. Wax and cutin layers in maize kernels associated with resistance to aflatoxin production by *Aspergillus flavus. J. Food Protect.* 58(3):296-300.

Hamblin, A.M. and White, D.G. 2000. Inheritance of resistance to *Aspergillus* ear rot and aflatoxin production of corn from Tex6. *Phytopathology* 90(3):292-296.

Hansen, T.J. 1993. Quantitative testing for mycotoxins. *Am. Assoc. Cereal Chem.* 38(5):346-348.

Hardin, B. 1998. Testing for natural aflatoxin inhibitors. *Agric. Res.* 46(7):17.

Haumann, F. 1995. Eradicating mycotoxins in food and feeds. *Inform* 62:48-256.

Heathcote, J.G., Dutton, M.F., and Hibbert, J.R.1976. Biosynthesis of aflatoxins. Part II. *Chem. Ind.* 270-273.

House, C. 1992. New grain cleaner sorts by density weeds out aflatoxin. *Feedstuffs* 64(1):17.

Hsieh, D.P.H., Lin, M.L., and Yao, R.C. 1973. Conversion of sterigmatocystin to aflatoxin B_1 by *Aspergillus parasiticus. Biochem. Biophys. Res. Commun.* 52(3):992-997.

Hsieh, D.P.H., Yao, R.C., Fitzell, D.L., and Reece, C.A. 1976. Origin of the bisfuran ring structure in aflatoxin biosynthesis. *J. Am. Chem. Soc.* 98(6):1020-1021.

Hsieh, D.P.H., Yao, R.C., Fitzell, D.L., and Singh, R.1976. Biosynthesis of aflatoxins, conversion of norsolorinic acid and other hypothetical intermediates into aflatoxin B_1 by *Aspergillus parasiticus*. *J. Agric. Food Chem.* 24(6):1170-1174.

Hua, S.-S.T., Flores-Espiritu, M., Hong, A., Baker, J.L., and Grosjean, O.K. 1997. Repression of the Gus reporter gene of aflatoxin biosynthesis by plant signal molecules. In J. Robens and J. Dorner (Eds.), *Aflatoxin elimination workshop: A decade of research progress 1988-1997* (p. 33). Food Safety and Health, USDA/ARS, Beltsville, MD.

Huang, Z., White, D.G., and Payne, G.A. 1997. Corn seed proteins inhibitory to *Aspergillus flavus* and aflatoxin biosynthesis. *Phytopathology* 87(6):622-627.

Jones, R.K. 1987. The influence of cultural practices on minimizing the development of aflatoxin in field maize. In M.S. Zuber, E.B. Lillehoj, and B.L. Renfro (Eds.), *Aflatoxin in maize* (pp. 136-144). CIMMYT; Mexico, DF.

Jones, R.K., Payne, G.A., and Leonard, K.J. 1980. Factors influencing infection by *Aspergillus flavus* in silk-inoculated corn. *Plant Dis.* 64(9):859-863.

Kale, S.P., Cary, J.W., Bhatnagar, D., and Bennett, J.W. 1996. Characterization of experimentally induced, nonaflatoxigenic variant strain of *Aspergillus parasiticus*. *Appl. Environ. Microbiol.* 62(9):3399-3404.

Kang, M.S. 1996. Producers can minimize impact of aflatoxin on corn. *Delta Farm Press* 53(29):10, 22.

Kang, M.S. 1998. Using genotype-by-environment interaction for crop cultivar development. *Adv. Agronomy* 62:199-252.

Kang, M.S. and Gauch Jr., H.G. (Eds.) 1996. *Genotype-by-environment interaction.* CRC Press, Boca Raton, FL.

Kang, M.S. and Lillehoj, E.B. 1988. Aflatoxin resistance in maize. In *Report of projects for 1987* (p. 48). Department of Agronomy, Louisiana Agricultural Experiment Station, Baton Rouge, LA

Kang, M.S., Lillehoj, E.B., and Widstrom, N.W. 1990. Field aflatoxin contamination of maize genotypes of broad genetic base. *Euphytica* 51(1):19-23.

Kang, M.S., Zhang, Y., Magari, R., and Kondapi, N. 1994. Effect of garlic on *Aspergillus flavus* mycelial growth and corn kernel infection. *Abstracts Technical Papers* (p.3). So. Branch Am. Soc. Agron. No. 21.

Kang, M.S., Zhang, Y., and Myers, Jr., O. 1995. Controlling *Aspergillus flavus* growth with cultivated and wild garlic. In *Agronomy Abstracts* (p.122). Am. Soc. Agron., Madison, WI.

Keller, N.P., Cleveland, T.E., and Bhatnagar, D. 1992. A molecular approach towards understanding aflatoxin production. In D. Bhatnagar, E. B. Lillehoj, and D. K. Arora (Eds.), *Handbook of applied mycology* Vol. 5. (pp. 287-310). Marcel Dekker, New York.

Kilman, S. 1989. Spreading poison. *The Wall Street J.* 83 (37) Feb. 23, p. A6.

Klitch, M.A., Tiffany, L.H., and Knaphus, G. 1994. Ecology of the *Aspergilli* and litter. In J.W. Bennett and M.A. Klitch (Eds.), Aspergillus *biology and industrial applications* (pp. 329-353). Butterworth-Heineman, Boston, MA.

Landers, K.E., Davis, N.D., and Diener, U.L. 1967. Influence of atmospheric gases on aflatoxin production by *Aspergillus flavus* on peanuts. *Phytopathology* 57(10):1086-1090.

LaPrade, J.C. and Manwiller, A. 1976. Aflatoxin production and fungal growth on single cross corn hybrids inoculated with *Aspergillus flavus. Phytopathology* 66(5):675-677.

LaPrade, J.C. and Manwiller, A. 1977. Relation of insect damage, vector, and hybrid reaction to aflatoxin B_1 recovery from field corn. *Phytopathology* 67(4):544-547.

Lee, L.S., Bennett, J.W., Cucullu, A.F., and Ory, R.L. 1976. Biosynthesis of aflatoxin B_1: Conversion of versicolorin A to aflatoxin B_1 by *Aspergillus parasiticus. J. Agric. Food Chem.* 24(6):1167-1170.

Liang, S.-H., Skory, C.D., and Linz, J.E. 1996. Characterization of the function of the *ver-1A* and *ver-1B* genes, involved in aflatoxin biosynthesis in *Aspergillus parasiticus. Appl. Environ. Microbiol.* 62(12):4568-4575.

Lillehoj, E.B. 1983. Effect of environmental and cultural factors on aflatoxin contamination of developing corn kernels. In U. L. Diener, R. L. Asquith, and J. W. Dickens (Eds.), Aflatoxin and *Aspergillus flavus* in corn (pp. 27-34). *So. Coop Ser. Bull.* 279. Auburn University, Auburn, AL.

Lillehoj, E.B., Bockholt, A.J., Calvert, O.H., Findley, W.R., Guthrie, W.D., Horner, E.S., Josephson, L.M., King, S., Manwiller, A., Sauer, D.B., Thompson, D., and Widstrom, N.W. 1980. Aflatoxin in corn before harvest: Interaction of hybrid and locations. *Crop Sci.* 20(6):731-734.

Lillehoj, E.B. and Wall, J.H. 1987. Decontamination of aflatoxin-contaminated maize grain. In M.S. Zuber, E.B. Lillehoj, and B.L. Renfro (Eds.), *Aflatoxin in maize: A proceedings of the workshop* (pp. 260-279). CIMMYT, Mexico, D.F.

Lillehoj, E.B., Wall, J.H., Peterson, R.E., Shotwell, O.L., and Hesseltine, C.W. 1976. Aflatoxin contamination, fluorescence, and insect damage in corn infected with *Aspergillus flavus* before harvest. *Cereal Chem.* 53(4):505-512.

Lozovaya, V.V., Waranyuwat, A., and Widholm, J.M. 1998. β-1,3-glucanase and resistance to Aspergillus flavus infection in maize. *Crop Sci.* 38(5):1255-1260.

Lussenhop, J. and Wicklow, D.T. 1990. Nitidulid beetles (Nitidulidae Coleoptera) as vectors of *Aspergillus flavus* in preharvest maize. *Trans. Mycol. Soc. Japan* 31:63-74.

Mabrouk, S.S. and El-Shayeb, N.M.A. 1981. The effects of garlic on mycelial growth and aflatoxin formation by *Aspergillus flavus. Chem. Mikrobiol. Technol. Lebensm.* 7(2):37-41.

Magari, R. and Kang, M.S. 1997. Adaptability and stability in crop breeding. In M.S. Kang (Ed.), *Crop improvement for the 21st century* (pp. 113-125). Research Signpost, Trivandrum, India.

Mahanti, N., Bhatnagar, D., Cary, J., and Linz, J.E. 1996. Structure and function of *fas-1A*, a gene encoding a putative fatty acid synthetase directly involved in aflatoxin biosynthesis in *Aspergillus parasiticus*. *Appl. Environ. Microbiol.* 62(1):191-195.

Mahoney, N.E., Molyneux, R.J., and Mcgranahan, G.H. 1997. Effect of walnut constituents on *Aspergillus* growth and aflatoxin production. In J. Robens and J. Dorner (Eds.), *Aflatoxin elimination workshop: A decade of research progress 1988-1997* (p. 31). Food Safety and Health, USDA/ARS, Beltsville, MD.

Marsh, S.F. and Payne, G.A. 1984. Preharvest infection of corn silks and kernels by *Aspergillus flavus*. *Phytopathology* 74(11):1284-1289.

Mayo, N.S. 1891. Enzootic cerebritis of horses. *Kan. Agri. Expt. Stn. Bull.* 24:1-12.

McAlpin, C.E., Wicklow, D.T., and Platis, C.E. 1998. Genotypic diversity of *Aspergillus parasiticus* in an Illinois corn field. *Plant Dis.* 82(10):1132-1136.

McCormick, S.P., Bhatnagar, D., Goyens, W.R., and Lee, L.S. 1988. An inhibitor of aflatoxin synthesis in developing cottonseed. *Can. J. Bot.* 66(5):998-1002.

McKenzie, K.S., Kubena, L.F., Denvir, A.J., Rogers, T.D., Hitchens, G.D., Bailey, R.H., Harvey, R.B., Buckley, S.A., and Phillips, T.D. 1998. Aflatoxicosis in turkey poults is prevented by treatment of naturally contaminated corn with ozone generated by electrolysis. *Poultry Sci.* 77(8):1094-1102.

McMillian, W.W. 1983. Role of arthropods in field contamination. In U.L. Diener, R.L. Asquith, and J.W. Dickens (Eds.), *Aflatoxin and* Aspergillus flavus *in corn* (pp. 20-22). So. Coop. Ser. 279., Auburn University, AL.

McMillian, W.W. 1987. Relation of insects to aflatoxin contamination in maize growth in the southeastern USA. In M.S. Zuber, E.B. Lillehoj, and B.L. Renfro (Eds.), *Aflatoxin in maize: A Proceedings of the Workshop* (pp. 194-200). CIMMYT, Mexico, DF.

McMillian, W.W., Widstrom, N.W., and Wilson, D.M. 1982. Aflatoxin production on various popcorn genotypes. *Agron. J.* 74(1):156-157.

McMillian, W.W., Widstrom, N.W., and Wilson, D.M. 1993. Registration of GT-MASgk maize germplasm. *Crop Sci.* 33(4):882.

McMillian, W.W., Wilson, D.M., and Widstrom, N.W. 1985. Aflatoxin contamination of preharvest corn in Georgia: A six-year study of insect damage and visible *Aspergillus flavus*. *J. Environ. Quality* 14(2):200-202.

Moeder, W., Anegg, S., Thomas, G., Langebartels, C., and Sandermann Jr., H. 1999. Signal molecules in ozone activation of stress proteins in plants. In M.F. Smallwood, C.M. Calvert, and D.J. Bowles (Eds.), *Plant responses to environmental stress* (pp. 43-49). BIOS Scientific Publ., Oxford, U.K.

Moore, K.G., White, D.G., and Payne, G.A. 1997. Characterization of a chitinase from Tex6 inhibitory to *Aspergillus flavus*. In J. Robens and J. Dorner (Eds.), *Aflatoxin elimination workshop: A decade of research progress 1988-1997* (p. 39). Food Safety and Health, USDA/ARS, Beltsville, MD.

Moreno, O.J. and Kang, M.S. 1999. Aflatoxin in maize: The problem and genetic solutions. *Plant Breeding* 118(1):1-16.

Nagarajan, V. and Bhat, R.V. 1972. Factor responsible for varietal differences in aflatoxin production in maize. *J. Agric. Food Chem.* 20(6):911-914.

Neucere, J.N. 1992. Electrophoretic analysis of cationic proteins extracted from aflatoxin resistant/susceptible varieties of corn. *J. Agric. Food Chem.* 40(8):1422-1424.

Neucere, J.N., Cleveland, T.E., and Dischinger, G. 1991. Existence of chitinase activity in mature corn kernels (*Zea mays* L.). *J. Agric. Food Chem.* 39(7):1326-1328.

Neucere, J.N. and Godshall, M.A. 1991. Effects of base-soluble proteins and methanol soluble polysaccharides from corn on mycelial growth of *Aspergillus flavus*. *Mycopathologia* 113(2):103-108.

Neucere, J.N. and Zeringue, H.J. 1987. Inhibition of *Aspergillus flavus* growth by fractions of salt-extracted proteins from maize kernel. *J. Agric. Food Chem.* 35(5):806-808.

Nichols, T.E. 1983. Economic effects of aflatoxin in corn. In U.L. Diener, R.L. Asquith, and J.W. Dickens (Eds.), *Aflatoxin and* Aspergillus flavus *in corn.* So. Coop. Ser. Bull. 279 (pp. 67-71). Auburn University, Auburn, AL.

Norton, R.A. 1997. Interaction between sugars, sugar metabolites and triglycerides in the production of aflatoxin by *A. flavus* NRRL 3357. In J. Robens and J. Dorner (Eds.), *Aflatoxin elimination workshop: A decade of research progress 1988-1997* (pp. 21-22). Food Safety and Health, USDA/ARS, Beltsville, MD.

Ogbadu, G. and Bassir, O. 1979. Toxicological study of gamma-irradiated aflatoxins using the chicken embryo. *Toxicology and Applied Pharmacology* 51(2):379-382.

Ong, T.-M. 1975. Aflatoxin mutagenesis. *Mutation Res.* 32(1):35-53.

Papa, K.E. 1976. Linkage groups in *Aspergillus flavus*. *Mycologia* 68(1):159-165.

Papa, K.E. 1977. Genetics of aflatoxin production in *Aspergillus flavus* linkage between a gene for a high B_2 - B_1 ration and the histidine locus on linkage group VIII. *Mycologia* 69(6):1185-1190.

Papa, K.E. 1979. Genetics of *Aspergillus flavus* complementation and mapping of aflatoxin mutants. *Genet. Res.* 34(1):1-9.

Papa, K.E. 1980. Dominant aflatoxin mutants of *Aspergillus flavus*. *J. Gen. Microbiol.* 118(Pt. 1):279-282.

Payne, G.A. 1992. Aflatoxin in maize. *Crit. Rev. Plant Sci.* 10(5):423-440.

Payne, G.A. 1997. Characterization of inhibitors from corn seeds and the use of a new reporter construct to select corn genotypes resistant to aflatoxin accumulation. In J. Robens and J. Dorner (Eds.), *Aflatoxin elimination workshop: A decade of research progress 1988-1997* (pp. 66-67). Food Safety and Health, USDA/ARS, Beltsville, MD.

Payne, G.A. 1998. Process of contamination by aflatoxin producing fungi and their impact on crops. In K.K. Sinha and D. Bhatnagar (Eds.), *Mycotoxins in agriculture and food safety* (pp. 279-306). Marcel Dekker, Inc., New York.

Payne, G.A., Cassel, D.K., and Adkins, C.R. 1986. Reduction of aflatoxin contamination in corn by irrigation and tillage. *Phytopathology* 76(7):679-684.

Payne, G.A., Hagler, W.M., and Adkins, C.R. 1988. Aflatoxin accumulation in inoculated ears of field-grown maize. *Plant Dis.* 72(5):422-424.

Payne, G.A., Kamprath, E.J., and Adkins, C.R. 1989. Increased aflatoxin contamination in nitrogen-stressed corn. *Plant Dis.* 73(7):556-559.

Payne, G.A., Nystrom, G.J., Bhatnagar, D., Cleveland, T.E., and Woloshuk, C.P. 1993. Cloning of the *afl-2* gene involved in aflatoxin biosynthesis from *Aspergillus flavus. Appl. Environ. Microbiol.* 59(1):156-162.

Rambo, G.W., Tuite, J., and Caldwell, R.W. 1974. *Aspergillus flavus* and aflatoxin in preharvest corn from Indiana in 1971 and 1972. *Cereal Chem.* 51(5):595-604.

Rao, H.R.G. and Harein, P.K. 1972. Dichlorvos as an inhibitor of aflatoxin production on wheat, corn, rice, and peanuts. *J. Econ. Entomol.* 65(4):988-989.

Rehana, F., Basappa, S.C., and Murthy, V.S. 1979. Destruction of aflatoxin in rice by different cooking methods. *J. Food Sci. and Tech.* 16(3):111-112.

Robens, J. and Dorner, J. (Eds.) 1997. *Aflatoxin elimination workshop: A decade of research progress 1988-1997.* Food Safety and Health, USDA/ARS, Beltsville, MD.

Rodriguez-del-Bosque, L.A., Leos-Martinez, J., and Dowd, P.F. 1998. Effect of wounding and cultural practices on abundance of *Capophilus freemani* (Coleoptera Nitidulidae) and other microcoleopterans in maize in northeastern Mexico. *J. Econ. Entomol.* 91(4):796-801.

Rodriguez, J.G., Patterson, C.G., Potts, M.F., Poneleit, C.G., and Beine, R.L. 1983. In U.L. Diener, R.L. Asquith, and J.W. Dickens (Eds.), *Aflatoxin and Aspergillus flavus in corn* So. Coop. Ser. Bull. 279 (pp. 23-26). Auburn University, AL.

Russin, J.S., Guo, B.Z., Tubajika, K.M., Brown, R.L., Cleveland, T.E., and Widstrom, N.W. 1997. Comparison of kernel wax from corn genotypes resistant or susceptible to *Aspergillus flavus. Phytopathology* 87(5):529-533.

Schraudner, M., Ernst, D., Langebartels, C., and Sandermann, H. 1992. Biochemical plant responses to ozone. III. Activation of the defense-related proteins β-1,3-glucanase and chitinase in tobacco leaves. *Plant Physiol.* 99(4):1321-1328.

Schroeder, T., Zweifel, U., Sagelsdorff, P., Friederich, U., Luthy, J., and Schlatter, C. 1985. Ammoniation of aflatoxin-containing corn distribution, in vivo covalent deoxyribonucleic acid binding, and mutagenicity of ration products. *J. Agric. Food Chem.* 33(2):311-316.

Scott, G.E. and Zummo, N. 1988. Sources of resistance in maize to kernel infection by *Aspergillus flavus* in the field. *Crop Sci.* 28(3):504-507.

Scott, G.E. and Zummo, N. 1990a. Preharvest kernel infection by *Aspergillus flavus* for resistant and susceptible maize hybrids. *Crop Sci.* 30(2):381-383.

Scott, G.E. and Zummo, N. 1990b. Registration of Mp313E parental line of maize. *Crop Sci.* 30(6):1378.

Scott, G.E., and Zummo, N. 1992. Registration of Mp420 germplasm line of maize. *Crop Sci.* 32(6):1296.

Scott, G.E., Zummo, N., Lillehoj, E.B., Widstrom, N.W., Kang, M.S., West, D.R., Payne, G.A., Cleveland, T.E., Calvert, O.H., and Fortnum, B.A. 1991. Aflatoxin in corn hybrids field inoculated with *Aspergillus flavus*. *Agron. J.* 83(3):595-598.

Seedburo. 1995. Black light test. In *Seedburo Equipment Co. catalog* (p. 45). Seedburo, Chicago, IL.

Seenappa, M. and Nyagahungu, I.K. 1982. Retention of aflatoxin in *ugali* and bread made from contaminated flour. *J. Food and Tech.* 19(2):64-65.

Sharma, A., Padwal-Desai, S.R., Tewari, G.M., and Bandyopadhyay, C. 1981. Factors affecting antifungal activity of onion extractives against aflatoxin-producing fungi. *J. Food Sci.* 46(3):741-744.

Sharma, A., Tewari, G.M., Shrikhande, A.J., Padwal-Desai, S.R., and Bandyopadhyay, C. 1979. Inhibition of aflatoxin-producing fungi by onion extracts. *J. Food Sci.* 44(5):1545-1547.

Sharma, Y.K. and Davis, K.R. 1997. The effect of ozone on antioxidant responses in plants. *Free Rad. Biol. Med.* 23(3):480-488.

Shaver, D.L. 1983. Genetics and breeding of maize with extra leaves above the ear. In *Proc. Thirty-Eighth Annual Corn and Sorghum Research Conference* (pp. 161-180). American Seed Trade Assoc., Washington, DC.

Shotwell, O.L. and C.R. Hurburgh Jr., C.R. (Eds.) 1991. *Aflatoxin in corn: New perspectives.* Iowa State Univ. Press, Ames, IA.

Shroeder, H.W., Cole, R.J., Grigsby, R.D., and Hein, H. 1974. Inhibition of aflatoxin production and tentative identification of an aflatoxin intermediate "versicanol acetate" from treatment with dichlorvus. *Appl. Microbiol.* 27(2):394-399.

Skory, C.D., Chang, P.-K., Cary, J., and Linz, J.E. 1992. Isolation and characterization of a gene from *Aspergillus parasiticus* associated with the conversion of versicolorin A to sterigmatocystin in aflatoxin biosynthesis. *Appl. Environ. Microbiol.* 58(11):3527-3537.

Smallwood, M.F., Calvert, C.M., and Bowles, D.J. (Eds.) 1999. *Plant responses to environmental stress.* BIOS Scientific, Oxford, U.K.

Smart, M.G., Shotwell, O.L., and Caldwell, R.W. 1990. Pathogenesis in *Aspergillus* ear rot of maize: Aflatoxin B_1 levels in grain around round-inoculation sites. *Phytopathology* 80:1283-1286.

Taubenhaus, J.J. 1920. A study of black and yellow molds of ear corn. *Texas Agric. Exp. Stn. Bull.* 270:3-38.

Trail, F., Chang, P.-K., Cary, J., and Linz, J.E. 1994. Structural And functional analysis of the *nor-1* gene involved in the biosynthesis of aflatoxin by *Aspergillus parasiticus. Appl. Environ Microbiol.* 60(11):4078-4085.

Trail, F., Mahanti, N., Rarick, M., Mehigh, R., Liang, S.-H., Zhou, R., and Linz, J.E. 1995. Physical and transcriptional map of an aflatoxin gene cluster in

Aspergillus parasiticus and functional disruption of a gene involved early in the aflatoxin pathway. *Appl. Environ. Microbiol.* 61(7):2665-2673.

Tucker, D.H., Trevathan, L.E., King, S.B., and Scott, G.E. 1986. Effect of four inoculation techniques on infection and aflatoxin concentration of resistant and susceptible corn hybrids inoculated with *Aspergillus flavus*. *Phytopathology* 76(3):290-293.

Wallin, J.R. 1986. Production of aflatoxin in wounded and whole maize kernels by *Aspergillus flavus*. *Plant Dis.* 70(5):429-430.

Wallin, J.R., Widstrom, N.W., and Fortnum, B.A. 1991. Maize populations with resistance to field contamination by aflatoxin B_1. *J. Sci. Food. Agric.* 54(2):235-238.

White, D.G., Rocheford, T.R., Hamblin, A.M., and Forbes, A.M. 1997. Inheritance of molecular markers associated with and breeding for resistance to *Aspergillus* ear rot and aflatoxin production in corn using Tex6. In J. Robens and J. Dorner (Eds.), *Aflatoxin elimination workshop: A decade of research progress 1988-1997* (p. 61). Food Safety and Health, USDA/ARS, Beltsville, MD.

Wicklow, D.T., Hesseltine, C.W., Shotwell, O.L., and Adams, G.L. 1980. Interference competition and aflatoxin levels in corn. *Phytopathology* 70(8):761-764.

Wicklow, D.T., Horn, B.W., Burg, W.R., and Cole, R.J. 1984. Sclerotium dispersal of *Aspergillus flavus* and *Eupenicillium ochrosalmoneum* from maize during harvest. *Trans. Br. Mycol. Soc.* 83(Pt. 2):299-303.

Wicklow, D.T., Norton, R.A., and McAlpin, C.E. 1997. β-carotene inhibition of aflatoxin biosynthesis among *Aspergillus flavus* genotypes from Illinois corn. In J. Robens and J. Dorner (Eds.), *Aflatoxin elimination workshop: A decade of research progress 1988-1997* (p. 36). Food Safety and Health, USDA/ARS, Beltsville, MD.

Widstrom, N.W. 1979. The role of insects and other plant pests in aflatoxin contamination of corn, cotton, and peanuts—A review. *J. Environ. Qual.* 8(1):5-11.

Widstrom, N.W. 1996. The aflatoxin problem with corn grain. *Adv. Agronomy* 56:219-280.

Widstrom, N.W., McMillian, W.W., Beaver, R.W., and Wilson, D.M. 1990. Weather associated changes in aflatoxin contamination of preharvest maize. *J. Prod. Agric.* 3(2):196-199.

Widstrom, N.W., McMillian, W.W., Wilson, D.M., and Glover, D.V. 1984. Growth characteristics of *Aspergillus flavus* on agar infused with maize kernel homogenates and aflatoxin contamination of whole kernel samples. *Phytopathology* 74(8):887-890.

Widstrom, N.W., Sparks, A.N., Lillehoj, E.B., and Kwolek, W.F. 1975. Aflatoxin production by *Aspergillus flavus* and lepidopteran insect injury on corn in Georgia. *J. Econ. Entomol.* 68(6):855-856.

Widstrom, N.W., Wiseman, B.R., McMillian, W.W., Kwolek, W.F., Lillehoj, E.B., Jellum, M.D., and Massey, J.H. 1978. Evaluation of commercial and experimental three-way corn hybrids for aflatoxin B_1 production potential. *Agron. J.* 70(6):986-988.

Williams, W.P., Windham, G.L., and Davis, F.M. 1997. Aflatoxin accumulation in transgenic *Bt* corn hybrids after insect infestation. In J. Robens and J. Dorner (Eds.), *Aflatoxin elimination workshop: A decade of research progress 1988-1997* (p. 12). Food Safety and Health, USDA/ARS, Beltsville, MD.

Wilson, D.M., Gueldner, R.C., McKinney, J.K., Lievsay, R.H., Evans, B.D., and Hill, R.A. 1981. Effect of β-ionone on *Aspergillus flavus* and *Aspergillus parasiticus* growth, sporulation, morphology and aflatoxin production. *J. Am. Oil Chem. Soc.* 58(12):959A-961A.

Wilson, D.M., Huang, L.H., and Jay, E. 1975. Survival of *Aspergillus flavus* and *Fusarium moniliforme* in high-moisture corn stored under modified atmospheres. *Appl. Microbiol.* 30(4):592-595.

Wilson, D.M. and Jay, E. 1975. Influence of modified atmosphere storage on aflatoxin production in high moisture corn. *Appl. Microbiol.* 29(2):224-228.

Wilson, D.M., McMillian, W.W., and Widstrom, N.W. 1986. Use of *Aspergillus flavus* and *A. parasiticus* color mutants to study aflatoxin contamination of corn. In G.C. Llewellyn and C.E. O'Rear (Eds.), *Biodeterioration research 6* (pp. 284-288). The Cambrian New Ltd., Aberystwyth, UK.

Wilson, D.M. and Payne, G.A. 1994. Factors affecting *Aspergillus flavus* group infection and aflatoxin contamination of crops. In D. L. Eaton and J. D. Groopman (Eds.), *The toxicology of aflatoxins* (pp. 309-325). Academic Press, San Diego, CA.

Wilson, D.M., Walker, M.E., and Gascho, G.J. 1989. Some effects of mineral nutrition on aflatoxin contamination of corn and peanuts. In A. W. Englehard (Ed.), *Soilborne plant pathogens management of diseases with macro- and micro-elements* (pp. 137-151). APS Press, St. Paul, MN.

Windham, G.L. and Williams, W.P. 1998. *Aspergillus flavus* infection and aflatoxin accumulation in resistant and susceptible maize hybrids. *Plant Dis.* 82(3):281-284.

Wogan, G.N., Paglialunga, S., and Newberne, P.N. 1974. Carcinogenic effects of low dietary levels of aflatoxin B_1 in rats. *Food Cosmet. Toxicology* 12(5/6):681-685.

Woloshuk, C.P., Cavaletto, J.R., and Cleveland, T.E. 1997. Inducers of aflatoxin biosynthesis from colonized maize kernels are generated by an amylase activity from *Aspergillus flavus*. *Phytopathology* 87(2):164-169.

Wood, G.E. 1989. Aflatoxins in domestic and imported foods and feeds. *J. Assoc. Off. Anal. Chem.* 72(4):543-548.

Wright, M.S., Greene-McDowelle, D.M., Zeringue Jr., H.J., Bhatnagar, D., and Cleveland, T.E. 1997. Inhibitory effect of volatile aldehydes from aflatoxin-resistant varieties of corn on *Aspergillus parasiticus* growth and aflatoxin biosynthesis. In J. Robens and J. Dorner (Eds.), *Aflatoxin elimination workshop: A decade of research progress 1988-1997* (p. 38). Food Safety and Health, USDA/ARS, Beltsville, MD.

Wu, S., Kriz, A.L., and Widholm, J.M. 1994. Nucleotide sequence of a maize cDNA for a class II acidic β-1,3-glucanase. *Plant Physiol.* 106(4):1709-1710.

Yoshida, S., Kasuga, S., Hayashi, N., Ushiroguchi, T., Matsuura, H., and Nakagawa, S. 1987. Antifungal activity of ajoene derived from garlic. *Appl. Environ. Microbiol.* 53(3):615-617.

Yousef, A.E. and Marth, E.H. 1985. Degradation of aflatoxin M1 in milk in ultraviolet energy. *J. Food Protection* 48(8):697-698.

Yu, J., Cary, J.W., Bhatnagar, D., Cleveland, T.E., Keller, N.P., and Chu, F.S. 1993. Cloning and characterization of a cDNA from *Aspergillus parasiticus* encoding an O-methyltransferase involved in aflatoxin biosynthesis. *Appl. Environ. Microbiol.* 59(11):3564-3571.

Yu, J., Chang, P.-K., Cary, J.W., Bhatnagar, D., and Cleveland, T.E. 1997. *avnA*, a gene encoding a cytochrome P-450 monooxygenase, is involved in the conversion of averantin to averufin in aflatoxin biosynthesis in *Aspergillus parasiticus. Appl. Environ. Microbiol.* 63(4):1349-1356.

Zhang, Y., Kang, M.S., and Magari, R. 1997. Genetics of resistance to kernel infection by *Aspergillus flavus* in maize. *Plant Breeding* 116(2):146-152.

Zhang, Y., Simonson, J.G., Wang, G., Kang, M.S., and Morris, H.F. 1998. A reliable field-inoculation method for identifying aflatoxin-resistant maize. *Cereal Res. Commun.* 26(2):245-251.

Zuber, M.S. 1977. Influence of plant genetics on toxin production in corn. In J.V. Rodricks, C.W. Hesseltine, and M.A. Mehlman (Eds.), *Mycotoxin in human and animal health* (pp. 173-179). Pathotox Publishers, Park Forest South, IL.

Zuber, M.S., Calvert, O.H., Kwolek, W.F., Lillehoj, E.B., and Kang, M.S. 1978. Aflatoxin B$_1$ production in an eight-line diallel of *Zea mays* infected with *Aspergillus flavus. Phytopathology* 68(9):1346-1349.

Zuber, M.S., Calvert, O.H., Lillehoj, E.B., and Kwolek, W.F. 1976. Preharvest development of aflatoxin B$_1$ in corn in the United States. *Phytopathology* 66(9):1120-1121.

Zuber, M.S., Darrah, L.L., Lillehoj, E.B., Josephson, L.M., Manwiller, A., Scott, G.E., Gudauskas, R.T., Horner, E.S., Widstrom, N.W., Thompson, D.L., Bockholt, A.J., and Brewbaker, J.L. 1983. Comparison of open-pollinated maize varieties and hybrids for preharvest aflatoxin contamination in the Southern United States. *Plant Dis.* 67(2):185-187.

Zuber, M.S. and Lillehoj, E.B. 1979. Status of aflatoxin problem in corn. *J. Environ. Qual.* 8(1):1-5.

Zummo, N. and Scott, G.E. 1990. Relative aggressiveness of *Aspergillus flavus* and *A. parasiticus* on maize in Mississippi. *Plant Dis.* 74(12):978-981.

Zummo, N. and Scott, G.E. 1992. Interaction of *Fusarium moniliforme* and *Aspergillus flavus* on kernel infection and aflatoxin contamination in maize ears. *Plant Dis.* 76(8):771-773.

Zummo, N. and Scott, G.E. 1994. Pathogenicity of *Aspergillus flavus* group isolates in inoculated maize ears in Mississippi. In G.C. Llewellyn, W.V. Dashek, and O'Rear (Eds.), *Biodeterioration research 4* (pp. 217-224). Plenum Press, New York.

Chapter 5

Sorghum Improvement:
Past Achievements and Future Prospects

Prasanta K. Subudhi
Henry T. Nguyen
Michael L. Gilbert
Darrell T. Rosenow

INTRODUCTION

Sorghum [*Sorghum bicolor* (L.) Moench] is the fifth most important cereal crop providing food and fodder throughout the world (Doggett, 1988). It is a crop with extreme genetic diversity. Its adaptation to harsh environments, specifically heat and drought stress, accounts for its success throughout the semiarid regions of the world. It has numerous mechanisms that allow it to survive and be productive in these conditions. With the increasing pressure of population growth coupled with global warming and limited resources for food production, sorghum has the potential to be an even more important crop in the future because of its stress adaptation. Sorghum is a C4 monocot, very closely related to maize, being a member of the same tribe Andropogoneae with the same chromosome number ($2n = 20$). Despite its importance, the genetic characterization of this species is very limited. Mann and colleagues (1983) hypothesized that sorghum probably originated and was subjected to domestication more than 5,000 years ago in northeastern Africa. Recently, Wendorf and colleagues (1992) reported that carbonized seeds of sorghum, evacuated at Nabta Playa near the Egypt-Sudanese border, appear to be about 8,000 years old. Sorghum was moving in trade channels eastward over the Arabian Peninsula and across the Indian Ocean some 3,500 years ago. Over the centuries, this crop has been adapted to an array of conditions. Since the introduction of sorghum into the United States, initial selection by farmers and breeding efforts have transformed the species from a tall, tropically adapted, photoperiod-sensitive species into a short-stature, temperate cereal of great economic importance. Subsequently, the discovery of cytoplasmic genetic male sterility and development of hybrids contributed to the tremendous boost in sorghum productivity. Further genetic gain was achieved with the incorporation of new germplasm, selection for disease and insect resistance, as well as adoption of improved production practices. Considerable progress has been made in the development of breeding methodology for sorghum hybrid development and sorghum biotechnology, which will be reviewed in this chapter along with future prospects of this important cereal.

SORGHUM PRODUCTION, PRODUCTION AREAS, AND IMPORTANCE IN THE GLOBAL FOOD SYSTEMS

Sorghum production worldwide in the early 1960s was about 35 million tons but increased rapidly, reaching almost 70 million tons by 1978. Production then fluctuated substantially, reaching a peak of 77 million tons with 51 million hectares in 1985. Soon after that, sorghum production and acreage declined in the world. In the 1998 season, area sown and yield figures more or less followed production and declined as well, with about 44 million hectares sown and 61 million metric tons of grain with an average yield of about 1,378 kg/ha. World sorghum acreage, production, and productivity during the period from 1961 to 1998 are summarized in Table 5.1. With the exception of Africa, the area and production of all sorghum-growing coun-

TABLE 5.1. World Sorghum Acreage, Yield, and Production Between 1961 and 1998

Year	Area harvested (ha)	Yield (kg/ha)	Production (t)
1961	46,009,146	890	40,931,625
1965	47,393,461	989	46,867,471
1970	49,412,265	1,129	55,773,304
1975	46,871,861	1,321	61,904,718
1980	44,091,003	1,300	57,301,185
1985	50,854,564	1,524	77,515,354
1988	46,057,524	1,369	63,040,199
1990	41,552,820	1,369	56,899,394
1992	46,180,024	1,534	70,833,088
1994	44,417,154	1,359	60,378,390
1996	47,631,418	1,509	71,900,444
1998	44,296,542	1,378	61,044,434

Source: <http://www.fao.org>.

tries have been on decline. Area sown to sorghum in the United States began to decline in the last half of the 1980s, and in 1995 the area sown was approaching half of the area sown in 1986. This resulted in dramatic cutbacks in sorghum programs by the large seed companies. Part of this problem was due to factors such as allocation of acreage for maize, changes in the federal farm programs, surplus of feed grown worldwide, and low prices. The area sown to sorghum in the United States was 3.1 million hectares with production of 13.2 million tons in 1998. The statistics for the major sorghum-producing countries of the world during 1998 are shown in Table 5.2. The universal global trend has been toward increased productivity. Future increased productivity is possible mainly through improved agronomic practices and the cultivation of hybrids to exploit heterosis.

On a global basis, sorghum represents 4 percent of the total cereal production. Although this figure is small, there are many African countries in which sorghum production is of critical importance: Nigeria, Mali, Niger, Burkina Faso, Chad, Cameroon, Sudan, Ethiopia, Botswana, and Rwanda. In Africa as a whole, sorghum represents 18 percent of the total cereal production (Dendy, 1995). In regard to sorghum utilization in general, developing countries use it primarily as food, whereas developed countries use it as feed. Sorghum can be used industrially in place of other cereals and is nearly identical in food value. However, some of the colored pericarp types can cause off-color in food products, and the size and hardness of the grain require more processing and different procedures. Sorghum will likely

TABLE 5.2. Production, Area, and Productivity in Major Sorghum-Growing Countries in 1998

Country	Production (t)	Area (ha)	Yield (kg/ha)
Argentina	3,762,000	782,000	4,811
Australia	1,065,000	569,000	1,872
Burkina Faso	1,202,808	1,408,000	854
China	3,943,771	1,110,700	3,550
Ethiopia	1,083,230	981,710	1,103
India	7,989,700	10,683,900	748
Mali	559,260	571,722	978
Mexico	6,474,842	1,953,073	3,315
Nigeria	7,103,000	6,635,000	1,071
Sudan	4,781,000	6,405,000	746
United States	13,206,900	3,125,000	4,226
World	61,044,434	44,296,542	1,378

Source: <http://www.fao.org>.

never become a major industrial raw material in developed countries as long as abundant supply of other cereals is available. The use of grain as animal feed has been an important stimulus for the global use of sorghum (Dendy, 1995). Feed use was relatively less important until the mid-1960s when it expanded rapidly, particularly in North America. Feed utilization overtook food use for the first time in 1966. Over the past several decades, feed use has risen several folds. This use, up to 97 percent, has occurred in developed countries but also in some better-off developing countries, particularly in Latin America, where it accounts for about 80 percent of sorghum utilization.

EVOLUTIONARY GENETICS AND GERMPLASM DIVERSITY IN SORGHUM

Species Relationships

Sorghum is classified as a predominately self-pollinating species (Doggett, 1988), though a small amount of outcrossing (1 to 5 percent) has been significant in its evolution. Sorghum is a short-day plant and flowering is hastened by short daylength periods and high temperatures. The Spontanea complex of *Sorghum bicolor* (L.) Moench includes 17 different taxa or species (de Wet et al., 1970). Four of the species (*S. elliotii, S. hewisonii, S. niloticum, and S. sudanese*) represent stable hybrids between cultivated sorghums and members of Spontanea. The remaining species are subdivided among *aethiopicum, arundinaceum, verticilliflorum,* and *virgatum.* Cultivated sorghums probably originated from several different races of *verticilliflorum* (de Wet, 1976). The distribution and habitat of 16 species of Spontanea series is described in Table 5.3 (*S. pugionifolium* is allied with

TABLE 5.3. Distribution of Series Spontanea [*Sorghum bicolor* (L.) Moench]

Subspecies	Distribution and habitat
S. aethiopicum	Sudan to Mali; in dry, hot regions, along stream banks and in wadis
S. arundinaceum	Wet tropics along the Guinea coast, across the Congo to northern Angola; forest grass in low-lying areas, and along stream banks
S. brevicarinatum	Northeastern Kenya southward along the coast to the Usugara district of Tanzania; along stream banks and irrigation ditches
S. castaneum	Uganda; in low-lying, wet areas
S. elliotii	Uganda; weedy around villages
S. hewisonii	Sudan to Somaliland; weedy around villages and in cultivated fields
S. lanceolatum	Sudan westward to northern Nigeria and south to Uganda; swampy areas or along stream banks of otherwise dry areas
S. macrochaeta	Southern Sudan to the eastern Congo; along stream banks and lake shores

Subspecies	Distribution and habitat
S. niloticum	Southern Sudan to the eastern Congo and eastward to Kenya; along stream banks and weedy in cultivated fields
S. panicoides	Eastern Ethiopia; habitat not known
S. somaliense	Somaliland; water courses and damp areas
S. sudanense	Sudan to Egypt; along stream banks and weedy along irrigation channels
S. usambarense	Tanzania; humid habitats along rivers in Usambara district
S. verticilliflorum	Kenya to South Africa; as roadside weeds in damp areas, along stream banks and irrigated ditches, or as weeds in cultivated fields
S. virgatum	Sudan, northeastern Chad and Egypt; in dry regions along stream banks and irrigation ditches; desert grass
S. volelianum	Cameroons and Nigeria; tropical forest grass of river banks

Source: From de Wet et al. (1970).

S. somaliense). Domesticated sorghum has resulted through direct selection from principally one or two wild races in Africa (de Wet et al., 1970). Agronomically acceptable progeny can result from wild subspecies (*virgatum, arundinaceum, verticilliflorum, and aethiopicum*) × cultivated species crosses (Cox et al., 1984; Doggett and Majisu, 1968). Even though johnsongrass [*S. halepense* (L.) Pers.] and columbusgrass *(S. almum)* both have 40 chromosomes, these wild relatives cross at a low frequency with sorghum (20 chromosomes) (Gu et al., 1984).

Harlan and de Wet (1972) subdivided the cultivated sorghums into five morphologically distinct races: bicolor, guinea, caudatum, kafir, and durra. Intermediate races are designated, for example, as kafir-caudatum, durra-bicolor, etc. They speculated that the race durra and bicolor arose from the wild subspecies *aethiopicum*, that the kafirs arose from *verticilliflorum*, and that the guineas evolved from *arundinaceum*. Mann and colleagues (1983) provided a more updated version of modern sorghum-race evolution (see Table 5.4). Subraces or working groups (Murty et al., 1967) describe some of the variation within races and intermediate races and often refer to commonly used groups used by sorghum scientists as feterita, zerazera, kaura, kaoliang, milo, sorgo, sudangrass, etc. A refinement of the working groups as they fit with and complement the Harlan and de Wet race classification has been proposed by Dahlberg (2000).

Germplasm Collections and Diversity

The germplasm pool of the genus *Sorghum* is characterized by abundant diversity. The immense morphological diversity of the cultivated races of sorghum has resulted from variable climate and geographical exposure in

TABLE 5.4. Races of the Subspecies *Sorghum bicolor arundinaceum*

Race	Subspecies derivation	Characteristic	Diversification or progenitor for
Bicolor	*Aethipicum/ verticilliflorum* complex	Most primitive race, low yielding	Kaoliangs of China (possible contribution from *S. propinguum*)
Guinea	From bicolor through selection without introgression of wild races; *arundinaceum* is a possible indirect ancestor	Selected for adaptation to wet tropical habitats, oldest of specialized races, low yielding	Dominant sorghums of West Africa up to 5,000 mm rainfall
Caudatum	Selected directly from early race bicolor, associated with *verticilliflorum*	Common in areas receiving 250-1,300 mm rainfall, great adaptation to harsh conditions, high yielding	Dominant in Cameroon, Ethiopia, Sudan, Uganda, Chad, Nigeria, and western Kenya
Kafir	Derived from early bicolor, closely associated with the wild *verticilliflorum*, possible independent domestication	Relatively recent race, high yielding	Dominant in eastern and southeastern Africa
Durra	Selected from early bicolor via India, developed in Ethiopia and Sudan, associated with *aethiopicum*	Adapted to low rainfall areas north of equator between 10-15° N lat., compact panicle, white grain	Most important from Ethiopia to Mauritania and in India

Source: From Mann et al. (1983) and Doggett (1976).

which its wild ancestors evolved, coupled with selection pressure imposed by environment and by man for domestication. Many sources of exotic and unique germplasm have been discovered and utilized over the years for sorghum improvement. Traits such as yield, standability, greenbug resistance, midge resistance, disease resistance, and drought resistance have been found and incorporated into current germplasm and have resulted in tremendous improvements in the crop.

During the past few decades, tremendous effort has been made toward collection, preservation, and understanding of sorghum germplasm. The NPGS (National Plant Germplasm System) collections include 40,477 sorghum accessions (Eberhart et al., 1997). Of the 20,169 sorghum accessions in the base collection at NSSL (National Seed Storage Laboratory), USDA, Fort Collins, Colorado, 80 percent are currently in conventional storage at about –18° C while 20 percent are in cryostorage in vapor phase above liquid nitrogen at about –160°C. The International Crops Research Institute for the Semi-Arid Tropics (ICRISAT) located at Patancheru, near Hyderabad, India, has assembled a collection of 35,643 sorghum accessions. All the accessions are maintained and preserved in aluminum cans in the medium-term storage facility at about 4° C and 20 percent relative humidity. Freshly rejuvenated accessions with at least 90 percent viability and about 5 percent seed moisture content are being placed in moisture-proof aluminum foil packets that are vacuum sealed and stored in long-term storage at –20°C. Seventeen percent of the sorghum collections have been transferred to long-term storage.

Ninety percent of the sorghum collections have come from developing countries in the semiarid tropics. About 49 percent of the sorghum collections are from five countries: India, Ethiopia, Sudan, Cameroon, and Yemen. ICRISAT has maintained and continues to conserve 473 wild sorghum accessions from 23 taxa. Genetic stocks consisting of accessions identified as sources of resistance to major diseases, insect pests, striga, and stocks with genes for specific morphological, agronomic characteristics, and cytoplasmic male-sterile lines are also separately maintained (Prasada Rao and Mengesha, 1988).

To increase the genetic variability of any particular traits, breeders continue to search for new accessions that will provide the highest probability of identifying useful source materials with minimal screening. A list of candidate accessions often can be generated when appropriate information is in the database. An initial screening of a diverse but smaller subset may reduce time and cost while searching within the crop collection for the desired trait. The idea of developing such a subset was proposed by Frankel (1984) and further developed by Brown (1989, 1995). They suggested that a "core collection" should consist of a limited set of accessions derived from an existing germplasm collection, chosen to represent the genetic spectrum of the whole collection. The core should include as much genetic diversity as possible. The core subset is suggested to be about 10 percent of the crop collection but may vary from 5 percent for very large collections to 50 percent or more for very small collections, with about 3,000 suggested as a maximum number. Brown (1989) recommended using the stratified sampling method when establishing core collections. A sorghum core collection has been established at ICRISAT (Prasada Rao and Ramanatha Rao, 1995) by stratifying the total world collection geographically and taxonomically into subgroups. Accessions in each subgroup were then clustered into closely related groups based on characterization data, using principal component analysis. Representative accessions from each cluster were drawn in proportion to the total number of accessions present in that subgroup to form a sorghum core collection of 3,475 accessions (approximately 10 percent of the total world collection).

Utilization has been primarily limited to agronomically important traits and, in some cases, wild sources of germplasm. For example, use of zerazera sorghum has become widespread in the development of new, superior hybrids because of superior yield potential and grain quality (Duncan, Bramel-Cox, et al., 1991). Although utilization of the total collection has not been realized, several examples of successful utilization of germplasm are cited by Duncan, Bramel-Cox, et al. (1991). The classic example of germplasm utilization has been the USDA-Texas Agricultural Experiment Station (TAES) Sorghum Conversion program, which is described below in detail.

Sorghum Conversion Program

Due to the narrow genetic base and associated limitations imposed by the kafir × milo cross exploitation, sorghum breeders in the United States turned their attention to the world collection of sorghum. Hegari and feterita types introduced from the Sudan contributed much to germplasm development in the United States before 1965 (Webster, 1976), along with the kafir and milo (durra) types. The exotic sorghum germplasm possessed various useful attributes that could contribute substantially to yield, insect and disease resistance, drought resistance, improved grain quality, and other desirable traits in U.S. hybrids. However, further direct use of sorghum from exotic sources, which are adapted to tropical areas, was not possible due to growth characteristics, such as height and flowering. A majority (approximately 75 percent) of the accessions in the world collection was photoperiod sensitive and required short days for flowering and was, therefore, unadapted to the long days of the United States. Consequently, only a small portion of the total genetic diversity within the species was in usable form for sorghum improvement programs and for development of hybrids in the United States. In the mid-1960s, a program sponsored jointly by the Texas Agricultural Experiment Station and USDA-ARS was initiated to convert tropical-adapted alien sorghums to shorter and earlier types for use in temperate regions (Stephens et al., 1967). Through a backcross procedure, short, early genotypes possessing an average of 98 percent of the germplasm of the tropical parent could be recovered, manipulating the maturity and height genes (Quinby, 1971, 1974) by transferring them from a four-gene-dwarf Martin type (BTx406) to the exotic types. The conversion was accomplished by making initial crosses and backcrosses during short winter days in Puerto Rico and growing F_2 populations under long summer days in Texas where homozygous plants recessive for the desired height and maturity genes are selected and returned to Puerto Rico for another backcross cycle. The BTx406 is in normal (nonsterility-inducing) cytoplasm and by making the last backcross using the exotic line as the female, each converted line retained its own cytoplasm. Four backcrosses were thought to be necessary for complete conversion. Duncan, Bramel-Cox, et al., (1991) reviewed the program and its impact in detail. The conversion program was not, however, designed to identify combining ability or detailed breeding behavior other than to provide a preliminary assessment of overall genetic potential and certain other traits, such as disease or insect resistance, as identified during or after conversion. In addition, backcrossing was toward the germplasm source, to otherwise preserve its originality.

A total of 1,433 accessions have been submitted for conversion. Currently 653 converted lines have been released, 533 lines are listed in Duncan and Dahlberg (1993), 50 are listed in Rosenow, Dahlberg, et al. (1995), and 40 new lines were released by TAES and USDA during 1996 and 30 more in 1998. Twenty-seven new converted lines are currently ready for release. Among these released convertants (Duncan, Bramel-Cox, et al., 1991), 16 percent originated from Sudan, 4 percent from Uganda, 16 percent from Ethiopia, 24 percent from India, and 25 percent from Nigeria. This broader germplasm base is a significant source of resistance to diseases (downy mildew, head smut, anthracnose, and grain mold), pests (midge, chinch bug, and greenbug), abiotic stress (pre- and postflowering drought tolerance, salt/salinity tolerance, and acid-soil tolerance), and lodging resistance not previously available in the United States (Rosenow and Clark, 1987) (listed partially by Duncan, Bramel-Cox, et al., 1991). Several new cytoplasmic-genetic sterility systems have been identified in converted lines (Schertz and Ritchey, 1978; Schertz, 1983) and some A and B pairs have been released (Schertz, 1977; Schertz et al., 1981). Schuering and Miller (1978) have published an extensive list of fertility restorations (A1 cytoplasm) and sterility maintenance reactions for 1,750 accessions, including the converted lines and many partially converted lines. Several new food-type sorghums with breeding sources from converted lines have been released by TAES. They are white-seeded tan plant types that have good weathering resistance and grain (Miller et al., 1987). They should be very useful for poultry feed and to produce high-quality food products.

Materials developed in the conversion program to date have proven very useful. Private seed companies throughout the United States as well as sorghum breeders throughout the world, have made and are continuing to make extensive use of the partially converted and converted lines (Duncan, Bramel-Cox, et al., 1991). The extensive use of the zerazera group of converted sorghums has made major contributions to disease resistance, yield potential, and quality of the U.S. hybrid programs. The currently used tropically adapted photoperiod-insensitive sorghums in the southern United States and tropical areas of the world are primarily derivatives of zerazera-type sorghums. The first advanced parental lines with converted lines in their pedigree were released in 1974. After 1974, the germplasm made available through improved adaptation by conversion has slowly widened the germplasm pool. In addition to the slow reduction in relative importance of the early introductions, the number of introductions found in the pedigree of post-1974 parental lines increased from 11 in the early hybrid parental lines to 28 (Duncan, Bramel-Cox, et al., 1991). The increased use of exotic germplasm for sorghum improvement is reflected in the reduced reliance on the pre-1936 introductions in today's parental lines, the increasing number of introductions in present inbred line pedigrees, and 76 percent of recently released parental lines with exotic germplasm contributions. The predominance of two introductions, SC170, and SC110, in the publicly released present parental lines and their even more significant contribution to the commercial hybrid germplasm pool would indicate a continued need to diversify the sorghum germplasm pool. In the future, this situation may change dramatically, as exotic germplasm from both wild races and unadapted cultivated sorghums is better utilized to broaden the genetic base. Even though wild sorghums are not priority choices for yield genes, their potential for improving tolerance to stress environments could prove useful.

The caudatums and their derivatives have many useful traits and are relatively easy to use in breeding programs (Rosenow and Clark, 1987). Zerazera, a subgroup of caudatum, combines high-yield capability, wide adaptation, excellent disease resistance and grain quality, and superior sources of insect resistance. Other useful groups include caudatum, caudatum-kafir, caudatum-nigricans, caudatum-kaura, dobbs, caudatum-guineense, nigricans, caffrorum (kafirs), and durra. The Ethiopian durras are an excellent source for stay green, a postflowering drought-tolerance trait.

HISTORY OF SORGHUM IMPROVEMENT

The history of sorghum breeding is well documented. Duncan, Bramel-Cox, et al. (1991) summarized known introductions of sorghum into the United States, including broomcorn introduced by Benjamin Franklin in 1725, 'Chinese Amber' sweet sorgo in 1851, milo in 1879, 'Blackhull Kafir' in 1886, 'Feterita' in 1906, and 'Hegari' in 1908. The first mention of the value of a guinea kafir corn from West Africa occurred at the Philadelphia Agricultural Society in 1810. Early germplasm improvement activities of farmers and public agencies were confined to the introductions of new accessions from abroad, to reselection within existing strains for height, maturity, and morphological types, and to efforts to preserve the identity of cultivars already in existence. H.N. Vinal and A.B. Cron of the USDA began to hybridize sorghums in 1914 (Vinal and Cron, 1921; Vinal, 1926) and released 'Chiltex' and 'Premo,' which were the first cultivars from artificial hybridization to be grown by farmers. Later, several cultivars were released (summarized by Duncan, Bramel-Cox, et al. 1991) that were disease resistant and widely adapted with excellent harvesting characteristics, good standability, and medium maturity. As the value of sorghum in low-rainfall areas of the Great Plains was recognized, breeders in state agricultural experiment stations and in the Federal Agricultural Service began sorghum improvement in the 1920s. With the development of hybrids in the early 1950s, sorghum productivity jumped dramatically in the United States by the mid-1970s and, at present, hybrids occupy 100 percent of grain sorghum acreage in the United States and also large acreage in other sorghum-growing countries. Improved agricultural practices combined with improved germplasm that responded to these agronomic improvements also contributed to the increased yields. A single yellow endosperm sorghum cultivar, Kaura, introduced from Nigeria in the early 1950s, had a dramatic impact on sorghum improvement in the United States, as the yellow endosperm trait was incorporated into many parental lines. In the 1970s, yellow endosperm parental lines were the primary male parents in widely adapted, high-yielding bronze-grained (yellow by red) hybrids. In the United States, irrigated acres were normally used for other crops (maize, cotton) and the poorer and dryland acres were used for sorghum. This is still true today. Although sorghum yield leveled off during the 1970s due to its movement from irrigated to dryland areas, this does not truly reflect the progress made on sorghum improvement. Sorghum breeders have spent much effort working on drought tolerance and standability and actually made much progress in improving and stabilizing productivity of sorghum grown in dryland areas.

The availability of diverse germplasm resources has led to steady improvement in sorghum. Early work on utilization of sorghum germplasm was confined to pure-line selection within cultivated landrace populations in Africa and India that resulted in improved cultivars, some of which continue to be widely grown. Selection for dwarf cultivars was initiated in the United States to allow mechanized, combine harvest and then, followed by exploitation of cytoplasmic male sterility, permitted the production of combine-height commercial hybrids. Crossing and/or backcrossing between the adapted introductions and local germplasm have been used to derive improved self-pollinated varieties and parental lines (Prasada Rao et al., 1989). Many economically important traits have been identified and prioritized in different environments, and techniques for their systematic evaluation have been developed. A multidisciplinary approach has been followed to integrate traits for yield, response to biotic and abiotic stresses, and improvement in grain and forage quality. As a consequence, significant contributions to production have occurred in many aspects. These include change in grain-straw ratio (i.e., shorter, generally earlier plants are more responsive to management), commercial exploitation of heterosis, effective use of resistance and quality traits, interaction of disciplines to support integrated pest management, and crop management contributions including stand establishment, weed control, and response to fertilization and irrigation. Manipulation of many useful traits, such as increased seed number, large panicles, greater total plant weight, drought tolerance, disease resistance, greater plant height, longer maturity, greater leaf area indices, increased green leaf retention, and greater partitioning of dry matter, have contributed to increased yields (Miller and Kebede, 1984).

BREEDING TOOLS

Several breeding tools have been developed and applied for quick identification and incorporation of superior and unique germplasm into currently adapted types of sorghum for hybrid development. These methods, along with a vast array of available germplasm resources, accelerated the improvement of sorghum.

CMS (Cytoplasmic-Genetic Male Sterility) System

Sorghum is a self-pollinated species with perfect flowers. Because of these characteristics, the use of hybrid vigor was delayed until CMS was identified even though the advantages of hybrid vigor were recognized about as early as in maize (Conner and Karper, 1927). Since then, numerous workers have reported frequent occurrence of heterosis for grain yield over the better parent. Murty (1999) and Pedersen and colleagues (1998) have recently reviewed the progress and future outlook of sorghum hybrid development. The extent of heterosis was dependent upon the specific parental combinations and their genetic divergence. The genetic divergence was not correlated with geographic diversity (Joshi and Vashi, 1992). The yield increase in hybrids was consistent over a range of test conditions (Doggett, 1961). This superiority was mainly attributed to the increased number of grains per panicle. Hybrids also had a larger root system than their parents (Quinby, 1974). The first production of public hybrids began in 1955, and the TAES released seven hybrids that year involving Combine kafir-60 females × milo-kafir male pollinators. The introduction of an exotic sorghum (kaura from Nigeria) with yellow endosperm characteristics (carotenoid and xanthophyll pigments) and extensive breeding of yellow endosperm parental lines in the 1950s and 1960s provided additional improvement in combining ability for yield with the A-B heterotic parental group. The release of converted and partially converted exotics later helped develop many disease-, pest-, and drought-resistant hybrids.

Cytoplasmic-male sterility in sorghum was first reported by Stephens and Holland (1954) working at the Chillicothe Research Station in Texas, in crosses between Dwarf Yellow Sooner Milo and Texas Blackhull Kafir. The sterility was caused by an interaction of cytoplasm from milo with the genes from Kafir, and this cytoplasm is also known as A1 or milo cytoplasm. The milo-kafir system provided the basis for commercial hybrid development. The milo cytoplasm used in the first hybrids is still the main male sterility-inducing cytoplasm used today. With the heightened awareness of the benefits of using alternative sources of male sterility-inducing cytoplasms, scientists in India (Tripathi, 1979; Appathurai, 1964; Rao, 1962), Africa (Webster and Singh, 1964), and in the United States (Stephens and Holland, 1954; Ross, 1965; Ross and Hackerott, 1972; Schertz and Ritchey, 1978; Schertz and Pring, 1982) have identified many different sources of sterility-inducing cytoplasms. Cytoplasms from different sources, however, might not differ in the manner in which they induce male sterility. The main approach to determine sterility-inducing differences among the cytoplasms has been to cross each male-sterile by one male parent and to determine which F_1s differ in fertility restoration. From such studies, it has been determined that there are four distinct sterility-inducing cytoplasms (A1 to A4) and three others that are less distinct. A set of tester lines has been identified to distinguish A1, A2, A3, and A4 (Schertz et al., 1997). Others (Worstell et al., 1984; Schertz and Pring, 1982) have reported similar studies. Indian scientists have compared the cytoplasm isolated in the United States with those isolated in India, and the following relationships were revealed (Schertz et al., 1997): G1 (G2-s, msG1, G-1-G, G1A, G1a-A3) are analogous to A3 (Nilwa, IS1112). VZM-1 and VZM-2 are the same. M35-1 and M31-2 are identical. Hagpur A has milo cytoplasm. Restorers are difficult to find for the sterility-inducing cytoplasms identified in India. Some of the Indian CMS lines differ from milo in their response to different fertility restorer genes (Tripathi et al., 1980). Three of these six lines developed by Ross and Hackerott (1972) have been confirmed as differing from milo in fertility restoration reaction and mitochondria DNA sequences (Conde et al., 1982). In China, the lines that restored fertility of the A1 source were also restorers for the A2, but these did not restore fertility in A3 and A4 cytoplasms. Fertility restoration patterns of 60 hybrids, involving 10 male-sterile lines with diverse cytoplasms (A1, A2, A3, and G1) and six testers, were studied by Senthil and colleagues (1994). The order of hybrid sterility in the diverse cytoplasms increased from A1-A2-G1- and A3. Studies showed that G1 and A3 caused a very complete and unique sterility, which was not easy to restore. Schertz and colleagues (1989) also divided 22 male-sterile lines (Quinby, 1980; Rosenow et al., 1980) into seven cytoplasmic groups based on fertility of F_1 progeny derived from crosses with different fertility restorer genotypes.

The A1 system is still the most widely used for hybrid development. The other systems have various advantages and disadvantages both for breeding and seed production. The use of alternate cytoplasms opens up the possibilities of different heterotic groupings with potentially new alleles (Gilbert, 1994). In sorghum, most high-yielding restorers on A1 are maintainers in A2 and, thus, can be used as seed parents; however, there are very few complete restorers for the A2 system. A3 male steriles, though they shed no pollen, have plump yellow anthers, which cannot easily be distinguished from fertiles and have essentially no known restorers. A4 and 9E have other disadvantages. Maunder and Pickett (1959) recognized one nuclear gene and Erichsen and Ross (1963) identified the second nuclear gene that interacts with sterile cytoplasms to cause male sterility. Schertz and Stephens (1966) have designated the two genes as *Msc1* and *Msc2;* their recessive alleles cause male sterility.

Pring and Colleagues have proposed that restoration of the A3 cytoplasm is gametophytic. Consistent with a gametophytic mode is the observation of iodine stainability of pollen in sterile or partially restored lines. Segregation pattern for stainability within the population of an anther is consistent with Mendelian segregation and a possible three-gene model. It is clear that pollen abortion occurs very late in development.

The environment has an effect on the expression of sterility/fertility, more with some cytoplasms than with others. Plants with A3 cytoplasm grown in the greenhouse during the winter without supplemental light are more sterile than identical plants in the field. Also, A1 steriles are more sterile in short winter days in Puerto Rico than under long summer days in Texas. Murty (1993) proposed a system of hybrid production relying on the environment to make the female line fertile in selected plantings for seed production.

Hybrid Development

Selection of hybrid parents in sorghum is essentially a balance of selection for per se performance and combining ability (Andrews et al., 1997). Some of the requirements of seed parents per se are different from those of pollen parents. Values of lines per se are easier to select for than combining ability, which can be evaluated only through hybrid performance. Some traits of parental lines per se are well expressed in hybrids. Unfortunately, in the case of yield, the correlation between parents and their hybrids is unreliable. In the study by Gilbert (1989), when lines were randomly selected across three generations and two populations, the correlation with hybrid yields was generally very low and not significant. However, in another set of lines that were at least partially selected on the basis of combining ability, the correlation with hybrid yield was higher and highly significant.

Parental Line Criteria

Seed parents have a number of particular per se requirements besides stable and perfect male sterility (Andrews, 1987). The lines must be as high yielding as possible, low tillering with good head exsertion, good seed set, large seed size, and good seedling vigor. They also must possess a number of per se traits that have a good correlation with hybrid performance, such as height, maturity, and disease, pest, and lodging resistance. Pollen parents should completely restore male fertility, even under low temperature, but should have profuse, early, and prolonged pollen shed. Pollen parents are often high tillering, which ensures continued pollen supply and head number in hybrids. Pollen parents also have been shown to contribute to hybrid seed germination and vigor (Maunder et al., 1990). With the A1 CMS system in sorghum, high-temperature environments put more stress on the expression of male sterility and are thus useful in detecting A lines, which might break down and shed some pollen. Conversely, cooler temperature is useful in evaluating the capacity of a pollen parent to restore male sterility in hybrids.

Determination of Heterotic Affinities

The concept of heterotic groups is well developed in maize (Pollak et al., 1991). Historically in sorghum, members of the kafir and milo (durra) and caudatum races provided contrasting hybrid parents, the female parents being of the kafir group. When CMS was first discovered in sorghum, kafirs were found to carry nuclear genes that caused male sterility when put into milo cytoplasm. The milo and caudatum races carried nuclear genes for restoration of male fertility as well as contrasting genetic diversity, which produced good hybrid vigor. The situation now, however, has become complicated with the inclusion of other sources of diversity into both the B kafir pool and the R milo pool, although heterotic contrasts obviously exist (Gilbert, 1994).

The concept of heterotic patterns in respect to hybrid breeding is not so well developed in sorghum. The development of complementarity between gene pools, one of each parental type (male or R-line pool, and a seed parent or B-genome pool) is a logical notion for breeding for increased heterosis. The consequence of this is that when generating new lines, crosses are made within each pool or to unrelated germplasm. Testing for combining ability is, however, conducted using representatives of the opposite pools (as in reciprocal recurrent selection procedures), which, in the long term, preserves and enhances the combining abilities and fits well with the concept of parental pools. The discovery of alternate male sterile cytoplasms will effectively allow the construction of heterotic pools different from the R lines and B lines defined relative to A1 cytoplasm.

In many breeding programs, heterotic affinities are determined solely on the basis of whether the new accession is a restorer or maintainer in regard to the CMS system being used. However, a lot more information is actually needed for developing superior hybrids. A new accession should contribute new useful genes, either for yield or defensive traits, to its heterotic group and increased heterosis to hybrids with the counterpart group. For disease and pest resistance, resistance levels to various biotypes and segregation patterns among progeny from crosses with sources of resistance can provide useful information. For yield, however, only crosses with representatives of both heterotic groups can classify the usefulness of a new accession (Gilbert, 1994). Dudley (1984a,b,c,d; 1987a,b) outlined methods of evaluating materials to determine their breeding value. These methods assume that the key to success in any plant-breeding program is the choice of parental germplasm. Basically, the assumption is made that a breeder first chooses a hybrid to be improved and then chooses which inbred of the hybrid is to be improved. With this assumption, a donor line is chosen and breeding methodology is employed to add favorable alleles from the donor line to the parental inbred. A major problem is choosing from among the large number of diverse materials, a suitable donor line, and to which parent it should be crossed. Although the theory developed by Dudley was primarily for cross-pollinated crops, such as maize, Kramer (1987) applied these concepts to sorghum, making use of these new tools of genetic theory and functional heterotic group testers. From this work and in cooperation with Rosenow and K.F. Schertz of the Texas Agricultural Experiment Station, a number of potentially superior donor lines and heterotic group placement were derived. Gilbert (1989) tested the theory using both well-known materials (standard B lines such as Tx399, Tx378, etc.) and other lesser known materials derived from the drought-tolerance breeding project at Lubbock, Texas, and confirmed the heterotic relationship of known materials and identified favorable donors. Most breeding programs do not employ these complicated tests (except for resistance to well-known pest biotypes) and rely on the assumption that a new line with obvious per se differences is likely to contain genetic diversity, which may prove useful for yield.

Selection for Combining Ability

Testcrossing in sorghum is the accepted method for line evaluation and is practiced when the primary objective is hybrid development. However, it has been less extensively researched in sorghum than in maize. The general procedure is to cross the potential R lines to the male sterile version of the seed parent(s) (A line) and evaluate the resulting F_1 crosses for combining ability. Evaluating potential seed parent lines is more difficult, because they first must be male sterilized prior to extensive evaluation. Development of methods to determine the relative combining abilities of potential seed parents prior to male sterilization would greatly enhance the efficiency of sorghum-breeding programs. Some of these methodologies are explained below.

Ross and Kofoid (1978b) evaluated sorghum R lines in combination with two inbred seed parent lines and their F_1s. They noted a slight advantage for the F_1 versus the inbred lines as testers when general combining ability (GCA) was the selection criterion. In another study (Ross and Kofoid,

1978a), a broad-based population tester was used to evaluate six inbred seed parent lines; the use of testcrosses to screen seed parent lines in sorghum was advocated. Additionally, they noted that R lines could be extracted from the population tester if a specific seed parent × tester population cross warranted it. Hookstra and colleagues (1983) reached a similar conclusion when 19 A lines were topcrossed to six populations. Use of broad-based populations was advocated as testers to simultaneously evaluate the initial combining abilities of seed parent lines and populations. Mohamed (1980) and Chungu (1992), however, were unable to differentiate which of three types of testers (inbred lines, F_1s, or populations) were the most appropriate for evaluating combining abilities of inbred A or B lines in sorghum.

Schertz and Johnson (1984) proposed a method to identify the highest combining female lines prior to male sterilization. This required the production of related three-way crosses to predict the performance of one F_1 hybrid, e.g., single-cross hybrid (D × C) was predicted through the average performance of related three-way cross hybrids $\{(F \times D) \times E\}$, $\{(G \times D) \times E\}$, and $\{H \times D\} \times E$. Although the authors concluded that the method would be effective in identifying the highest combining seed parent lines, the requirement of three different three-way crosses to predict the performance of one single cross makes the procedure realistically somewhat impractical. Additionally, comparisons were confounded due to the genotype and maternal contribution of some of the inbreds used to produce the three-way cross hybrids (i.e., F, G, and H).

Lee and colleagues (1992) proposed a more efficient method for screening seed parent lines for combining ability prior to male sterilization. This method uses what are normally R lines in commercial A1 hybrids, sterilized in A3 cytoplasm. Potential seed parent lines (known to be maintainers but not yet sterilized in A1 CMS) are then used as a combination that is the nuclear reciprocal of the traditional A line × R line commercial hybrid in A1 cytoplasm. Paired comparisons between specific A1 (traditional) and reciprocal A3 (predictor) hybrids revealed few differences for grain yield, plant height, and days to midbloom. They indicated that the relative performance of hybrid progeny produced by an A1 seed parent line and an R1 male line can be well characterized by the relative performance of the reciprocal cross. In another study by Gilbert and colleagues (1991), the hybrids produced in the normal cytoplasm generally outyielded the hybrids produced in alternate cytoplasms. However, the important factor when testing for combining ability might not be a set of testers with equivalent yield. The important factor with regard to an alternate set of testers would seem to be their correlation of rankings when compared to the normal testers. This study also concluded that the alternate testers were a valid tool to be used in testing B-line material. Thus, the combining ability of potential seed parents can be evaluated before lengthy and time-consuming sterilization. This work has positive implication on the efficiency of sorghum breeding.

There is general acceptance that early generation testing assists in the selection of good combiners. Disagreement exists, however, whether it is worth expending testcrossing resources at a stage when a high proportion of the plants or families tested will be subsequently discarded on per se criteria. Many programs delay testing until the F_4, when more per se selection has been done, and then start with two widely adapted tester lines. Ideally, one would like to select for combining ability that would work with any new parent. This is unrealistic because a few tester lines do not represent any heterotic group well. On the other hand, it is easy to visualize improving an existing hybrid—the progeny selected from the crosses with one parent are best evaluated by using the other parent as the tester. However, this approach will build only on the specificity of one tester and does not broadly expand the potential of the breeding program.

Opinion is also divided about when early testing for combining ability in seed parent development should begin: before, during or after male sterile development. If A1 CMS is being used, these first testcrosses should be on a line known to carry excellent sterile cytoplasm, since that testcross will constitute the first step in the A-line development and provide the cytoplasm that will remain thereafter. In sorghum, if A1 lines have not been developed, it is necessary to sterilize R1 lines in A2 or A3 cytoplasm, to provide appropriate testers (from the "opposite" heterotic group) to directly evaluate the combining ability of the possible B lines. Thus, apart from combining ability, selection for many of the per se selection criteria for hybrid parents can be applied rapidly in the first two or three segregating generations. In an earlier study, Rosenow (1970) concluded that early generation testing was valid on R lines. Because of the inconsistency of inbred per se testing, perhaps a testcross program would be the superior avenue to pursue. Hybrid testing could begin as early in the development process as resources will allow and will generally correlate to results in later generations. In practice, most breeding programs adopt a compromise. Through experience or testing, a few of the best A lines and R lines with good general combining ability and wide adaptation are identified. One or two of these may be used for the initial screening of new early generation lines, then as numbers to be tested are reduced, more testers are used to both substantiate GCA and increase the discovery of added specific combining ability (SCA) expression. Early generation combining-ability tests are not intended to identify definitely the best combiners but to increase the probability of retaining them for detection in later tests. The first set of testcrosses may be screened only visually in one or two environments with possibly a 30 to 40 percent selection pressure. The choice of testers for the initial screening is, therefore, very important. They should be of contrasting parentage but known to combine well with a broad range of material.

Combining ability studies in sorghum have shown that the additive component of genetic variance is of primary importance for the majority of traits. GCA is primarily a function of additive gene action (which can be fixed through selection and inbreeding), whereas SCA depends on nonadditive gene action that can exist only in heterozygotes. Relative magnitude of GCA:SCA have been estimated to be as high as 31:1 for some traits (Beil and Atkins, 1967; Kambal and Webster, 1965; Collins and Pickett, 1972; Laosuwan and Atkins, 1977; Ross et al., 1983). This would indicate that broad-based testcrosses would be useful in selecting for combining ability. However, successful commercial hybrids would always be generated from a specific cross between two parental lines. Hybrid parents need to be genetically complemented for vigor and yield-associated traits but not for other often-recessive traits that would adversely affect height, maturity, grain qualities, or resistance.

Population Improvement

Population improvement involves the generation of broad-based gene pools and their improvement through recurrent selection. Favorable genes are concentrated through recurrent selection, resulting in increased mean of the population and superior performance of the best families (Hallauer, 1981). The tandem cycling of selection and recombination is particularly important for improvement of polygenic traits and for simultaneous improvement of several traits (Doggett, 1982). This method could increase the effective use of nonelite source materials, where the greater opportunities for recombination could break linkages between genes for the desired traits and unfavorable agronomic characteristics. An array of populations for long-term improvement of key agronomic traits or trait combinations and resistance to major insect pests are being developed and improved at the ICRISAT Asia Center, India (Rattunde et al., 1997). Both mass selection and progeny-based selection methods have been used to improve various traits of interest. Large responses of both grain (13.2 percent) and stover (16.4 percent) yields were observed in India while maturity was held constant. Mass selection was not effective for resistance to stem borer *(Chilo partellus)* (ICRISAT, 1988). Seven cycles of mass selection for white grain color and guinea glume and grain type, initially with mild selection intensity, achieved high frequencies of both traits in the guinea × caudatum grain-mold population.

In broad-based populations, genetic variation and potential long-term genetic gains are maximized. However, because these populations tend to have low means for critical agronomic traits, frequency of elite segregates is low.

Since these lines rarely possess the full component of required agronomic characteristics, these represent improved source materials useful in crossing with elite lines. Most elite populations are of narrow genetic base and thus reduce the opportunity for long-term genetic gains. These populations restrict the introduction of undesirable alleles for key agronomic traits during population development. For obtaining specific new traits from unadapted germplasm, introgression programs are used. In response to the need to access a wide range of germplasm, the sorghum-conversion program was initiated; this program provided a wide range of new diversity. ICRISAT has conducted extensive germplasm utilization including wild relatives, but much of its advanced material, which has been used by breeding programs worldwide, has significant zerazera content. Populations have contributed to R-line production, but pedigree selection is still the principal method for parent development.

The economic benefits of population improvement are ultimately realized when genetic material from these populations is used to develop lines and varieties for cultivation. Superior families developed and identified through recurrent selection provide a starting point for line development. Traditional pedigree-selection methods used during the inbreeding process produce pure lines for direct use as varieties or hybrid parental lines or, more frequently, as improved parental material for advancing pedigree-breeding activities.

Researchers are also pursuing introgression into elite populations, using population backcrossing to accommodate the contradictory goals of deriving elite lines and making long-term gains in the source population (Bramel-Cox and Cox, 1988; Menkir et al., 1994a,b). The conclusion was that the population approach could be useful for the utilization of both the wild relatives and most exotic (unadapted) cultivated sorghums. Effective introgression using a population backcross approach would require sampling enough plants during backcrossing to retain a maximum diversity of alleles from the exotic source. Also, when new source material is introgressed into a population already improved for that trait, screening at each level of backcrossing would insure retention of more genes for that trait when handling small population sizes.

Comstock and colleagues (1949) first suggested the use of reciprocal recurrent selection (RRS) to enhance heterosis between two populations. It maximizes the genetic divergence between the populations for loci with dominance effects, by basing selection on crosses generated with one parent from each population. These methods would be useful for crop species where hybrids are commercially viable and large interpopulation heterosis is expected or observed. Although this approach allows integration of long-term and short-term breeding objectives (Eyherabide and Hallauer, 1991), the use of RRS, especially full-sib RRS, is hampered by the sterility system used to enable random mating. All crosses (test and selection units) would be generated using a male-sterile line as the female for which no selfed seeds can be produced. Thus, from the selected full-sibs, only the male parents from each cross can be used as recombinational units, effectively reducing selection intensity and failing to capture genes from those female parents producing superior crosses.

An informal type of recurrent selection was practiced by R. E. Karper and Frank Gaines in Texas in the 1950s and 1960s in their yellow endosperm-breeding program. They intercrossed early generation progeny (even F_2 plants) extensively, always putting on a lot of selection pressure for grain yield and agronomic acceptability. Their breeding materials had wide adaptation as parental lines around the world, and were the dominant male parents of hybrids for many years, with some still currently in use. Breeding hybrid parents, especially in seed companies, has become increasingly dependent on crossing elite by elite lines, B × B lines, and R × R lines. This practice progressively narrows the genetic base of breeding populations and requires new traits, especially resistance, to be brought in by prebreeding, often backcrossing. The success of a backcrossing program greatly depends on the precision with which the desired trait can be identified and, thus, preserved in the backcrossing/introgression process. The tendency is to select for genes having major effects or tightly linked gene complexes. Thus, elite × raw germplasm crosses are not used in the expectation of immediate discovery of a new elite line; the most common short-term approach in breeding inbred parents has been elite × elite line crosses followed by pedigree selection. However, introgression seeking new genes must be a feature of any long-term balanced breeding program. In sorghum, the exotic (unadapted) germplasm content usually must be reduced to 12 percent or less by backcrossing to adapted parents before useful segregates occur (Maunder, 1992).

Seed Production System

One of the first things a breeder must determine about a new line to be used in the hybrid production programs is that line's ability to fully restore, or completely fail to restore, male fertility to a CMS line. Experience has shown that partial fertility restorers should be discarded immediately because they usually cannot be completely male sterilized and are not liable to be used as male parents (Schertz and Dalton, 1970). High-temperature environment is a good screening tool to ensure that only the best seed parents (A lines) are retained. The converse applies to selection for good restorer lines (R lines). Cool temperatures induce male sterility (and even female sterility under extreme conditions) in some hybrids made with poor restorer lines. Once the fertility-restoration reaction of a new germplasm source is determined, development of new parental lines is initiated. Since the merit of a line is now largely dependent on its performance in hybrid combinations, the means to select individuals or lines from within segregating populations for high combining ability become very important in parental line breeding. Line development usually follows traditional pedigree breeding procedure, but population approaches have also been successful (Doggett, 1988). One must also take into consideration the incidence of diseases while making the choice of CMS source. For example, it has been demonstrated that race 3 of *Sphacelotheca reiliana*, causing head smut in sorghum, is especially virulent for the A2 cytoplasmic male-sterility source (Rodriguez-Herrera et al., 1993).

The nuclear/cytoplasm genetic interaction is varied. In some instances, apparently, a single dominant nuclear gene restores fertility, and in others, as many as two or more major genes and several modifiers are involved in fertility restoration, and conversely, in instability of sterility. Because of this complexity and diversity, the development of females with stable sterility and males with dependable restoration of fertility is difficult. When choosing to use a cytoplasm other than milo, one should consider the advantage to be greater than the extra care needed to work with more than one cytoplasm. As more than a single CMS system is used in breeding programs, it is important to have a method to distinguish the cytoplasms, for example, good molecular markers to distinguish cytoplasms used in breeding and hybrid seed production. Several usable male sterility-inducing cytoplasms provide a degree of potential protection against associated hazards. More important, they provide the diversity needed to exploit more fully the germplasm diversity in hybrid development and production.

Future Outlook of Sorghum Hybrid Seed Production

In grain sorghum, only single-cross hybrids based on cytoplasmic-genetic male sterility (CMS) are used. Farmers in many countries generally prefer totally uniform sorghum hybrids. This preference does not pose severe problems because inbreeding depression is not important in this virtually inbreeding species, and inbred lines (especially seed parents) are relatively high yielding. Three-way hybrids (using F_1 seed parents to increase seed yields) have been tried, but the heterogeneity in hybrids is a negative factor to the farmers. Additionally, because biotic resistance may have evolved to operate well in a homozygous state in this species, it has been possible (with some exceptions, e.g., shoot fly resistance) to find and incorporate effective resistance to major pests and diseases into sorghum-inbred parental lines.

While in many of the sorghum-growing areas of the world significant progress has been achieved in sorghum breeding, there are some exceptions. Guinea sorghum has been difficult to improve, as have long-season, photoperiod-sensitive cultivars of tropical Africa. Although improvement in rainy-season sorghum in India was readily made, the development of sorghum cultivars for the post–rainy season has been slower. These more difficult situations present an interesting challenge.

Population improvement of sorghum is currently being conducted with diverse materials and objectives. In the future, population improvement will be a more important breeding tool, since it allows greater recombination and a larger number of favorable alleles than is possible with the same number of plants handled via pedigree methods. The documented gains in population means and superior families achieved through recurrent selection show the effectiveness of this general approach. The open-population approach with introgression via population backcrossing should improve the source populations within the context of an applied program in sorghum. This method would provide both source materials for immediate development of end products as well as longer-term genetic enhancement required for future gains.

Despite the theoretical basis for population improvement in sorghum and the documented improvements in population means, little evidence exists of useful parental lines extracted from populations. There were extensive efforts with population breeding in the 1970s by both private and public breeders, but many of those have been abandoned in favor of direct line × line crosses and progeny breeding. The closeness in relationships among cultivated sorghums and the weedy sorghums, along with the dominance of the weedy type traits, creates problems in random mating populations. Also, the ease with which populations can quickly become taller and late maturing is a problem. Populations as a germplasm source where certain specific traits can be accumulated, such as disease resistance, are very useful. However, for yield per se, populations have not proven very useful either for cultivar development in developing countries or in hybrids.

New genes are needed for long-term progress in breeding. For convenience, we will differentiate between variability that contributes purely to yield potential through better and more efficient growth and defensive traits that provide protection from pests, diseases, and abiotic stresses, such as drought, thereby allowing the plant to better realize its potential. Yield potential is important at all levels of production, perhaps relatively more so under stress conditions where limited resources must be utilized more efficiently. A large collection of sorghum exists but often with only some basic morphological descriptions. Lacking, with some exceptions for drought (Rosenow and Clark, 1995), is any systematic evaluation for important attributes, such as pest, disease, and stress resistance (Maunder, 1992) or for worth in breeding for yield potential.

Presently, all sorghum hybrids are based on milo cytoplasm. No detrimental effects have been observed so far with the use of this cytoplasm. The dangers of relying on only one cytoplasm were generally realized, and a good number of alternative sources of cytoplasm has been identified. However, the behavior of those cytoplasms is not well documented. Studies by Murty (1986, 1992) resulted in the development of diversified female parents. These studies have also led to the concept of two-line hybrids. The production of two-line hybrids in rice, making use of temperature and photoperiod-sensitive genetic male sterility (Yuan, 1990; Sun et al., 1993) and similar attempts in sorghum using season-sensitive cytoplasmic genetic male sterility (Murty, 1993, 1995), represent attempts in this direction. Murty (1993) developed male-sterile parents that were stable for male sterility during the winter season characterized by short photoperiod (<12 h) and low temperatures (<20°C). Breakdown of male sterility of these lines during the hot summer season with long photoperiod (>12 h) made them appropriate for use as uniline CMS lines. A two-line sweet sorghum has recently been reported (Murty, 1995).

One-line hybrids are possible through apomixis. Apospory type of apomixis has been described in sorghum (Hanna et al., 1970), which would make it useful if it could be perfected. The mechanism and frequency of apomixis were researched in detail in line R473 that resulted from the cross of IS 2942 × Aispuri in India (Rao and Narayana, 1968; Murty and Rao, 1972; Murty et al., 1984). The facultative apomixis in this line is complicated by cross sterility. The highly variable frequency of apomixis in R473 does not make it a promising line in its present form. Apomixis has been reported from tissue culture (Elkonin et al., 1995). Transferring obligate apomixis from *Cenchrus ciliaris* to sorghum, through protoplast fusion and regeneration, is proposed as a more feasible solution, but induction of suspension cultures and somatic hybridization have not yet been accomplished. Apomixis could have a major impact on hybrid production. It would allow breeders to use more rapidly and efficiently the germplasm available to produce hybrids.

BREEDING OBJECTIVES AND ACHIEVEMENTS

Significant improvement has been made for some traits, such as grain quality, plant color, maturity, resistance to greenbug, downy mildew, anthracnose, head smut, midge, and striga, whereas improvement of other traits, such as resistance to stem borer and head bugs, drought tolerance etc., has been slow and less successful. The conversion program sets an ideal example for many other crop species for widening the genetic base using the unadapted germplasms. Progress in sorghum breeding for improved adaptation and utilization has been tremendous, however, at the cost of increased susceptibility to some biotic and abiotic stresses. Emphasis on yield has narrowed the genetic base and reduced natural plant-defense mechanisms. Changes in phenotype and maturity as well as production in vast monocultures have exacerbated the problem. The progress made on different aspects of sorghum improvement is briefly summarized below.

Pests and Disease Resistance

Host-plant resistance (HPR), as a component of integrated pest management, has been employed for only a few of the many insect pests recorded as panicle pests of sorghum. Some important panicle pests are sorghum midge [*Stenodiplosis sorghicola* (Coquillett)], earhead bug (*Calocoris angustatus* Lethierry), African sorghum head bug [*Eurystylus oldi* (Poppius)], corn earworm [*Helicoverpa armigera* (Hubner) and *Helicoverpa zea* (Boddie)]. Major progress has been made in combining enhanced levels of HPR to sorghum midge with local adaptation in the breeding programs at ICRISAT Asia center (Sharma et al., 1994), Texas A & M University (Peterson et al., 1994), and in Australia (Henzell et al., 1994). Breeding for HPR to other insects has attracted less effort. There has been effort for developing cultivars with resistance to ear head bugs in India and to African sorghum head bugs in West Africa. Foliage pests of grain sorghum are greenbug *(Schizaphis graminum)*, yellow sugarcane aphid *(Sipha flava)*, chinch bug *(Blissus leucopterus)*, fall armyworm *(Spodoptera frugiperda)*, corn leaf aphid *(Rhopalosiphum maidis)*, and Bank's grass mites *(Oligonychus pratensis)*. In the United States, primary emphasis in developing sorghum resistant to insects has been on greenbug. Hybrids resistant to biotypes C, E, and I have been available to growers (Peterson et al., 1997). The other insect pests of sorghum in the United States are occasional or restricted to small geographical areas. Success has also been achieved in developing stem borer and shoot fly tolerant varieties in East Africa and India (Sharma, 1993). Quantitative sources of resistance to sugarcane aphid have been identified. At ICRISAT, more than 340 accessions of wild relatives of sorghum, belonging to sections Chaeto, Hetero, Stipo, Para, and sorghum, were evaluated for resistance to shoot fly (ICRISAT, 1988, 1989). Seven accessions showed very high levels of resistance to shoot fly, with one close to the immunity level. Techniques such as embryo rescue and protoplast fusion offer opportunities to transfer the resistance genes to varieties and hybrids.

Diseases of major importance are grain molds, charcoal rot, anthracnose, downy mildew, leaf blight, smuts, and ergot. Essential requirements for successful deployment and management of disease-resistance genes include: (a) access to collections of diverse host germplasm, (b) efficient disease-screening techniques, (c) effective resistance factors, (d) a knowledge of inheritance pattern, (e) a strategy of resistance deployment, and (f) an appropriate method for monitoring resistance. Priority diseases for different regions of the world have been identified and this must be continuously updated because diseases, pathogens, host cultivars, and cropping systems change over time and space. For example, before 1995, ergot had not been reported in the Americas, but by 1996 the disease had appeared in Brazil, Argentina, Bolivia, and Mexico (Reis et al., 1996). It has since spread throughout U.S. sorghum production areas, as well as to Australia (Bandyopadhyay et al., 1998). The genetic resources division of ICRISAT and Texas A & M University maintain lists of disease-resistant accessions and converted sorghum lines for specific and multiple disease resistance. Fortunately, sorghum landraces exhibit a wide range of diversity and the world collections of sorghum contain sources of resistance to most diseases. Significant progress has been made in breeding varieties and hybrids resistant to downy mildew, anthracnose, and leaf blight (Mukuru, 1992; Rosenow and Frederiksen, 1982; Thakur et al., 1997). However, progress has been slower in breeding for resistance to diseases, such as grain mold and stalk rot, in which gene effects are small.

Striga Resistance

Genetic resistance to striga, though limited, is available in sorghum germplasm, making host-plant resistance a feasible control measure. The following genotypes have been extensively evaluated with mixed results: IS9830, IS3167 (Framida), IS8577 (Dobbs), IS7777, SRN39, Tetron, P967083, and 555 (Ejeta et al., 1997). The conventional selection procedure of evaluating striga-infested plots has not been widely successful owing to the complexity of the biology of the host-parasite relationship and environmental factors. Future approaches to breeding for striga resistance will need to be based on a better understanding of the basic host-parasite biology and selection for host genotypes that lack an essential signal(s) for successful parasitism. A laboratory assay has been developed for screening genotypes for low-germination stimulant production. The components of striga resistance need to be identified by employing both conventional and nonconventional approaches, which can facilitate development of crop genotypes with durable resistance. Field resistance to striga has been shown to be quantitatively inherited. Research efforts at Purdue University in the United States and the University of Hohenheim, Germany, have been targeting the identification and eventual exploitation of QTL associated with striga resistance.

Grain Quality

There are three principal end uses for grain: feed, food, and brewing (Bramel-Cox et al., 1995). Since there is little incentive in either food or feed markets to demand higher nutritional quality, there is no motivation for applied breeders to put much emphasis on selection for nutritional quality.

Although genetic variability for feeding quality (nutritional value), including sources of high lysine in protein and high digestibility from waxy endosperm has been demonstrated in sorghum (Hamaker and Axtell, 1997), selection is not practiced for it except for avoiding tannin in the grain. Some important criteria are flour yield, starch properties, especially water absorption and retention, appearance of grain, taste, resistance to discoloration during ripening, endosperm hardness, tannin content, etc. In general, food-quality grain is assumed to have acceptable feed quality. Since food quality varies by region and use, quality criteria have been somewhat difficult to define and therefore to use. For industrial use and various food products, many quality parameters have been determined (Rooney et al., 1980, 1986, and 1997). Several of these criteria are related to rather simply inherited traits relative to grain, plant, and glume, making selection for them rather simple, but others are complicated and not well understood. In many cases, having white seeds, a tan plant color, an absence of testa, and good grain mold/ weathering resistance would enhance food quality and industrial utilization.

Sorghum for Forage and Fuel

Sorghums are unique species in their ability to be used as forage for livestock systems. A collaborative program between the Genetic Resources Unit of ICRISAT and the National Bureau of Plant Genetic Resources, Indian Council of Agricultural Research, has resulted in a comprehensive evaluation of forage sorghums in the world collection (Mathur et al., 1991, 1992). In the United States, the resources committed to forage sorghum improvement are declining, although recent interest in brown-midrib forages may reverse that trend. Forage sorghums differ widely in chemical composition and nutritive values, both of which are genetically controlled (Hoveland and Monson, 1980; Gourley and Lusk, 1978). Although most forage quality parameters appear to be quantitatively inherited (Bramel-Cox et al. 1995), several simply inherited characters like brown midrib, plant color, sweetness, juiciness, and even seed pericarp color, have significant impact on forage quality. Duncan (1996a) listed a number of traits, which should be incorporated into forage sorghum hybrids for quality enhancement. These include brown midrib, bloomlessness, glossiness, sweet stem, nonlignified cellulose, desirable carbohydrate level as indicated by a 3:1 ratio of acetate:propionate in rumenal fluid, low tannin content, low luteolin content and its derivatives, tan plant, low HCN-p, low arabinosyl-linked hemicellulose, high grain to stover ratio, dry stem, green leaf retention, ratooning of multiple cuttings, tropical adaptation, and acid-soil tolerance. Sorghum-forage breeders have a wealth of genes and genetic knowledge to work with that is not available to breeders working with many other forage species. However, incorporating all of these, plus high yield and agronomic acceptability, into hybrids would be an ambitious effort. Genetic engineering also offers tremendous potential in modifying quantitatively controlled traits for improving forage-quality component (Wheeler and Corbett, 1989).

Considerable effort has been spent developing sweet sorghum lines for ethanol production. However, the fuel industry has failed to develop around sweet sorghum in the United States. Recent research emphasis has begun to shift from sugar to biomass. Advances in fermentation technology accomplished with molecular biology (Vogel, 1996; Zhang et al., 1995) are making the above scenario economically feasible. If this industry develops, our challenge will be to develop sorghum forages that can provide biomass that is economically competitive at acceptable environmental, political, and cultural costs compared to other potential biomass species. Other developing industries, such as the fiberboard and paper industry, also could utilize sorghum stover as raw substrate.

Abiotic Stress Tolerance

Drought stress is a major constraint to sorghum production worldwide. Large genetic variations exist among sorghum lines for response to pre- and postflowering drought, and sources of resistance have been identified and utilized in breeding programs (Rosenow et al., 1997). Conventional breeding techniques with large field screening nurseries in semiarid environments have proven successful in screening and breeding for drought resistance. However, variability in rainfall and a large interaction between timing of stress and stage of plant growth often make screening difficult and slow. Walulu and colleagues (1994) determined that the stay-green trait from SC35 (IS12555), a durra from Ethiopia, was conditioned by a single gene, or two genes, primarily dominant in nature. This supports the research by Tenkouano and colleagues (1993) showing that nonsenescence and charcoal rot resistance were controlled by only a few genes. A recent study at ICRISAT by van Oosterom and colleagues (1996) on inheritance of staygreen found that slow senescence rate was dominant over fast rate and that inheritance of the onset of senescence under postflowering stress was additive.

Observations in the Texas A & M University sorghum-breeding program have indicated that resistance in some stay-green sources (SC35, SC33, SC56) is dominant in nature, whereas in others (SC599, Tx435, Tx2908, and B1—a BTx625 x B35 derivative), it is recessive; in still others (BQL41, QL36, and NSA440), it is partially dominant (Rosenow et al., 1997). A large number of sorghum parental A and R lines were classified for stay green and lodging and for their expression of dominance in the stay green trait in F_1 hybrids (Rosenow and Clark, 1995; Rosenow, Woodfin, et al., 1995). The best combination of resistance to pre- and postflowering stress and good grain yield in an F_1 hybrid often has come from a cross between a high stay-green female and a high preflowering drought-resistant high-yielding male parent. Preflowering stress resistance is primarily a dominant trait. Hybrid vigor itself appears to contribute a significant degree of tolerance to preflowering stress. Some introductions of photoperiod-insensitive sorghum, such as Ajabsido and Koro Kollo from Sudan, Segaolane from Botswana, and El Mota from Niger, possess outstanding preflowering drought resistance and can be used directly in breeding programs.

Although lodging of sorghum results in grain losses worldwide, it is very costly in regions where harvesting is done with machines. The causes of death and lodging are not well understood; the following three hypotheses, however, were proposed by Henzell and colleagues (1984): (1) plants die as a result of water deficit, i.e., a physiological breakdown due to dehydration; (2) pathogens cause death; or (3) death is caused by an interaction between physiological stress factors and pathogens. Several sources of resistance in sorghum, including drought-induced premature leaf and plant senescence (IS12555C, IS12568C, NSA 440, and SC599-11E [IS17459 derived]), charcoal rot (IS12555C, IS 12568C, KS19, and New Mexico 31), Fusarium stem rot (SC326-6 [IS3758 derived]), and B35 (IS12555 derived) are available. IS 12555C had been identified as resistant to three of the major causes of sorghum lodging (weak neck, stress-type stalk lodging, and after-freeze stalk breakage), suggesting that IS12555C or its appropriate derivatives would be useful in programs where improved standability is a primary objective. Johnson and colleagues (1997) reported that stay-green hybrids averaged more grain yield and had less lodging than the senescence-prone hybrids. More important, when entries in trials in which the senescence-prone hybrids had no lodging were averaged, the stay-green hybrids maintained yield advantage over the senescence-prone hybrids. These data indicate that competitive hybrids can be produced that have improved lodging resistance as a result of the stay-green trait. The stay-green trait in B35 has been shown to increase levels of resistance to lodging caused by charcoal rot, fusarium stalk rot, wind, after-freeze stalk breakage, and water deficiency during grain fill, while not significantly affecting grain yield potential. More information is needed on the inheritance of various drought and lodging resistance traits. Molecular markers have been identified for some pre- and postflowering drought resistance traits (Tuinstra et al., 1996, 1997; Crasta et al., 1999; Xu et al., 2000; Kebede et al., 2001). These markers are being utilized to facilitate transfer of the stay-green trait to improved agronomic types while maintaining competitive grain-yield potential, and at the same time eliminating the need for water-stressed breeding nurseries.

Photoperiod Insensitivity, Adaptation, and Yield

Photoperiod insensitivity allows breeding and development of cultivars to fit defined target environments. By removing confounding variation created by photoperiod sensitivity, yield and yield stability might be enhanced through critical selection for yield components. Maturity gene *Ma1* and the *Ma5-Ma6* gene interaction are responsible for the bulk of the sensitivity to photoperiod. As an adaptive trait, manipulation of maturity has provided highest yields in widely different regions of the world, i.e., early sorghums in drought-prone or short-duration seasons versus late-maturing sorghums in well-watered, longer-duration seasons (Miller et al., 1997).

Several workers have attempted to elucidate the heterotic response of sorghum for yield and related traits (reviewed by Miller and Kebede, 1984). The positive traits include increased plant height, heavier panicles, larger leaf area, increased vegetative weight, and greater length, volume, and growth rates of roots. However, the most important components of sorghum yield are number of kernels per panicle (Quinby, 1963; Stickler and Pauli, 1961), size of kernels (Eastin, 1970; Quinby, 1970), and number of panicles per unit area. The size and number of kernels have been shown to be strongly negatively correlated (Kirby and Atkins, 1968; Miller, 1976). Some breeding programs have successfully maximized that negative correlation by selecting female parents for hybrids with the highest number of kernels within some prejudged minimum range, then selecting male parents (R lines) with large kernels (without regard for numbers) (Miller et al., 1997). The importance of additive genetic variance, especially for grain yield, points to the potential for further yield improvement (Kambal and Webster, 1965; Niehaus and Pickett, 1966).

CURRENT PROGRESS IN BIOTECHNOLOGY FOR SORGHUM IMPROVEMENT

Plant biotechnology brings a powerful set of new molecular tools to the plant breeder's assistance. Marker-assisted breeding and gene transfer open up new horizons for crop improvement through safe and precise manipulation of useful traits. DNA markers allow geneticists and breeders to unambiguously map and follow numerous interacting genes simultaneously that determine complex traits. The mapped markers provide access to genes of interest for isolation/cloning even in the absence of any known product. Crops can now receive and express genes from any source. Despite the importance of the sorghum as a model for study of drought and heat tolerance among the grass relatives, biotechnology research in sorghum has been limited. Some progress has been made in genome mapping of this crop, and many genes related to growth, development, and important metabolic processes have been isolated and characterized (reviewed by Subudhi and Nguyen, 2000a). The DNA markers and genetic maps will be important tools for direct investigation of several facets of crop improvement and will provide vital links between plant breeding and basic plant biology (Lee, 1995). Recent reviews by Bennetzen (1997) and Subudhi and Nguyen (2000a) provide background on the roles and uses of modern molecular techniques for genetic improvement of sorghum. We discuss here briefly the prospects of sorghum improvement in the context of biotechnological advances.

Genetic Diversity Studies

Despite the existence of extensive germplasm collections and their partial characterization, very limited concrete information about which materials will be sources of novel and useful genes is available. For instance, the allelic diversity for genes not found in cultivated sorghum appears to be comparatively low in some wild relatives of sorghum (Oliveira et al., 1996). However, on the contrary, the study of Cui and colleagues (1995) revealed greater diversity in wild subspecies than in domestic accessions. To generate a comprehensive knowledge regarding the amount of genetic diversity in parental lines of the commercial sorghum hybrids to aid in increasing the effectiveness of future hybrid development programs, Ahnert and colleagues (1996) investigated the genetic relationships in a group of 58 restorer (R) and 47 sterility-maintainer (B) elite sorghum inbred lines using RFLPs and pedigree data. R lines revealed more diversity than B lines and are clustered into two main groups; one derived mainly from feterita and the other from zerazera, both from the caudatum race. Characterization of sorghum accessions at the DNA level can help identify the truly novel accessions by indicating fairly unambiguously overall relatedness and allelic novelty. Core collections can be identified that contain the majority of the diversity in a species, permitting a more detailed and cost-effective characterization of the potentially useful agronomic traits. Pedigree analysis with DNA markers will be used to identify the DNA segments that are the source of important traits currently used in improved varieties and landraces (Lee, 1995). Different male sterility-inducing cytoplasm types also could be fingerprinted and distinguished using sorghum mitochondrial DNA sequences (Xu et al., 1995; Conde et al., 1982; Pring et al., 1982; Chen et al., 1993; Baily-Serres et al., 1986a,b; Pring et al., 1995).

Genome Mapping and Molecular Breeding

Sorghum-genome mapping based on DNA markers began in the early 1990s; since then, several genetic maps of sorghum have been constructed

(Hulbert et al., 1990; Whitkus et al., 1992; Binelli et al., 1992; Melake-Berhan et al., 1993; Pereira et al., 1994; Chittenden et al., 1994; Ragab et al., 1994; Xu et al., 1994; Dufour et al., 1997). Recently, Subudhi and Nguyen (2000b) completely aligned all 10 linkage groups of all the major sorghum RFLP maps using a common recombinant inbred line (RIL) population and sorghum probes, along with many cereal anchor and maize probes, for greater and effective utilization of markers. The simplicity and speed of genetic analysis based on simple sequence repeats (SSRs) is well recognized. In sorghum, SSRs are not yet available in good number. Only 91 SSRs have been developed in sorghum to date for public use (Brown et al.,1996; Taramino et al., 1997; Kong et al., 1997).

A number of useful DNA markers for disease-resistance genes have been identified (for both single genes and multiple loci [QTL]) for head smut, anthracnose, rust, downy mildew, Acromonium wilt, virus disease (Tao et al., 1998b; Weerasuriya, 1995; Oh et al., 1993, 1994), and leaf blight resistance (Boora et al., 1999). Some mapped morphological traits are awns, mesocarp thickness (Tao et al, 1998a), juicy midrib, red coleoptile, red pericarp, etc. (Xu et al., 2000). In addition, several important agronomic traits have been mapped and these include traits associated with domestication (Paterson et al., 1998), flowering (Lin et al., 1995), tillering or rhizome characteristics (Paterson et al., 1995), grain quality, yield components and other traits (Rami et al., 1998), and drought-tolerance traits (Tuinstra et al.,1996, 1997: Crasta et al., 1999; Xu et al., 2000). These studies will not only facilitate transfer of these traits precisely and effectively using marker-assisted selection to agronomically acceptable cultivars but will also set the stage for map-based cloning of those genes. Many QTLs, depending on their relative effects and position, could be used as targets for marker-assisted selection and provide an opportunity for accelerating breeding programs. To isolate genes for agriculturally important traits, such as shattering, photosensitivity (Lin, 1998) and stay green (Xu et al., 2000), from sorghum by map-based gene cloning systems, BAC libraries from *Sorghum bicolor* and *Sorghum propinquum* have recently been constructed (Woo et al., 1994; Lin et al., 1999).

Comparative mapping information suggests that the conservation of the maize and sorghum genomes encompass sequence homology, colinearity, and function despite their divergence millions of years ago and subsequent evolution in different hemispheres with contrasting ecogeographical conditions. Thus, it provides a means to unify genomes of grass species and simplify molecular analysis of complex phenotypes (Bennetzen and Freeling, 1993). Map synteny can be used as a mechanism to determine whether two genetic processes in two different species are due to different alleles of the same genes. This, in turn, provides crop scientists with broader allelic variation than would be possible within any single species.

Sorghum has the potential to become a model system for understanding C4 plants. It has a relatively small genome among the grass relatives (Arumuganathan and Earle, 1991) and has the expected smaller distance between individual genes than its larger genome relative, maize (Avramova et al., 1996; Chen et al., 1997). Hence, map-based gene cloning and the development of a contiguous physical map are reasonable options in sorghum vis-à-vis most other grasses. Given its small genome and transformation competence, sorghum could become a model along with rice for understanding the molecular genetics of all grass species.

Tissue Culture and Genetic Engineering

Techniques in plant cell culture and plant genetic engineering have had very limited impact to date on sorghum-improvement programs. Sorghum is now amenable to in vitro manipulation. Immature embryos, young leaf bases, shoot apex, and immature inflorescence have been used in sorghum as explants for successful regeneration of plants (Bhaskaran and Smith, 1990). Despite repeated efforts, plant regeneration from single cells or protoplasts has not been realized in sorghum. In vitro anther culture for haploid plant production can provide a rapid method for cereal crop improvement; however, it has not, thus far, received much attention, as evidenced by the limited number of reports (Rose et al., 1986; Kumaravadivel and Rangasamy, 1994). The effectiveness and efficiency of using in vitro generated variability is often questioned on the grounds of instability and poor understanding. Thus, successful selection of desirable regenerants with improved traits depends on expression of a wide diversity for specific traits, trait stability across generations, adequate population size in the field to have a reasonable chance of visually selecting desirable variants, and a proper field environment that fosters trait expression (Duncan, 1996b; Duncan et al., 1995; Smith et al., 1993). Although somaclonal variation has had some success in providing usable germplasm (Duncan, 1996b; Duncan et al., 1992, 1995; Duncan, Isenhour, et al., 1991; Duncan, Waskom, et al., 1991; Isenhour et al., 1991; Wiseman et al., 1996; Miller et al., 1992), from the practical viewpoint, it is felt that there is already significant genetic variation that needs to be evaluated without generating more through cell culture.

Plant genetic engineering has the potential to bring totally novel traits into any crop species. Particle bombardment has been used to obtain transgenic sorghum plants using both immature embryos from explants of P898012 (Casas et al., 1993; Kononowicz et al., 1995) and inflorescence explants of SRN39 (Smith et al., 1997). Analyses of genetic data indicate that cotransformation can result in effective segregation of the selectable marker from the other transgenes. In spite of great potential for application of transformation technology in sorghum improvement, there are no programs to access the pool of genes that is available for genetic engineering due to the lack of a suitable transformation system. With the identification and cloning of agronomically important genes of sorghum in the near future, progress in developing a suitable and stable transformation system will be rewarding. However, care needs to be taken in the selection of genes and proper agronomic practices to prevent the spread of these genes into weedy species, such as johnsongrass, which readily cross with both grain and forage sorghums.

CHALLENGES AND FUTURE PROSPECTS

Sorghum is used in a wide variety of traditional foods in the semiarid tropics. However, its use as food has declined in urban areas, as wheat, rice, and maize products become more plentiful. The lack of a consistent supply of high grain-quality sorghum for processing severely limits its acceptance. Other constraints to the use of sorghum include its image as "a second class" crop, tannins, low-cost imported wheat, rice, maize, and government policies. Sorghum acreage has been on the decline. Both private and public programs and research efforts have been greatly reduced. Less work is being done in all areas of sorghum improvement today than in the past. Sorghum must find its unique place or it will continue to be only a substitute or companion crop to other feed grains. Several areas are being looked at to improve the future for sorghum. Increased effort should be made to improve the end-use quality, allowing value-added processing. This could improve farm income from identity-preserved types of sorghum. There is a great potential for sorghum to be utilized in a variety of ways. Unique attributes with regard to light-colored grain and tan plant types for food use are being studied. Progress has been made in blending sorghum and millet flour with flour of wheat and maize. The increased availability of good quality sorghum in Zambia has resulted in the increased use of sorghum by national brewers.

Looking to the future in areas where sorghum has been traditionally grown for food, the use of both grain and stover for animal feed should likely become a more prominent part of the crop improvement effort. Many genes for improving the forage quality of sorghum are available. Industrial uses, such as ethanol, are possible and are being utilized. As more traits with economic importance become available in the future, which might be important on a few acres but with a very high dollar return, sorghum could be an important crop for consideration. Also, as water continues to become less available and/or more marginal land is brought into cultivation, sorghum has the opportunity to become the crop of choice as its input cost is very low. However, because sorghum is not considered as important as other grains in the agricultural economy of any developed country, it is likely to continue to be ignored as targets of study and improvement. Therefore, sorghum needs attention and resources to compete with other major cereals. Sorghum continues to be a critically important subsistence food crop in Sub-Saharan Africa, where population growth rates are high. In these areas, rainfall is low and the environment is not well suited for other grain crops. Thus, sorghum will be a key in providing a stable agricultural system with adequate food in these areas. With enhanced characterizations and improved end-use utilization, sorghum could maintain or increase its contributions to world agriculture.

The challenges for the sorghum breeders in the future will be development of new hybrids with increased productivity and profitability. New technology and methods will have to be developed to accomplish better utilization of world germplasm collection and increased sustainability of sorghum production. Although yield gain has been dramatic in the past in developed countries, further improvement of yield will be more difficult to achieve in the future. The availability of vast genetic diversity preserved in world germplasm resources, coupled with improved breeding and biotechnological tools, should contribute to meeting these challenges. In the near future, sorghum will remain the primary crop on poor soils and under poor management conditions. Therefore, utilization and improvement of yield stability will depend on increased research efforts on tolerance to drought, temperature, and adverse soil conditions. Greater use of the local landraces in crosses with agronomically elite material will be needed to combine higher grain yield with tolerance to locally important biotic and abiotic stresses (Andrews and Bramel-Cox, 1993). Improved selection methods that enhance combining ability for grain and biomass yield, along with better information on heterotic patterns between germplasm accessions, are essential to increase sorghum production worldwide. Facultative apomixis of the aposporous type has now become available and may be useful in fixing heterosis in the near future.

Biotechnology is of increasing significance, is exciting, and holds much promise. It improves our understanding regarding the expression of agronomic traits; helps in tracking genes in the breeding process, transferring useful genes across infertility barriers, recognizing similarities and differences among germplasms; and, finally, hastens the product development process. However, biotechnology efforts and financial support for sorghum have generally been lagging behind other crops, such as maize and rice. Only a few sorghum programs are using these tools in breeding for precise and efficient incorporation of useful traits. Greater effort should be made toward identification of useful genes, mapping, and sequencing of sorghum genome and subsequent development of techniques to transfer reliably the useful genes. The priority in the future will be increased utilization of genes for the genetic enhancement of existing genotypes. The isolation and characterization of useful sorghum genes have facilitated efforts to genetically engineer this plant. Although success in sorghum transformation has been regarded as an academic exercise, no commercial products have been released to date. This is primarily because most of the genes with high dollar return (such as herbicide resistance) will not be incorporated into sorghum until the outcrossing issue with weedy-type sorghums is resolved. Sorghum readily outcrosses with several weedy species, such as johnsongrass. For this reason, a herbicide-tolerant sorghum would not likely receive regulatory approval and, therefore, is not currently being pursued in the private sector.

The potential of sorghum genetics research in improving other grass species is enormous (Bennetzen and Freeling, 1993; Moore et al., 1993; Devos and Gale, 1997; Moore, 1995). Sorghum, due to its simpler and smaller genome, could serve as a model to bridge the comparative analyses among the members of the grass family. It is expected that the sorghum genetic maps will be integrated with those of other species, such as maize, rice, barley, and wheat in the near future. This should permit exchange and mutual exploitation of information and materials, leading to success in map-based cloning of useful genes based on cross-species genome maps and, thereby, expediting advances in various aspects of basic biology and crop improvement. Greater use of genomics tools for germplasm management and gene manipulation through marker-assisted selection will no doubt accelerate the sorghum improvement programs. Finally, sorghum, as a drought-tolerant crop species with a relatively small genome size, will continue to be an excellent model for the investigation of genes involved in drought tolerance and plant adaptation to harsh climatic conditions.

Financial support to our sorghum-improvement program from the Texas Advanced Technology Research Program, Texas Higher Education Coordinating Board is greatly appreciated.

REFERENCES

Ahnert, D., Lee, M., Austin, D.F., Livini, C., Woodman, W.L., Openshaw, S.J., Smith, J.S.C., Porter, K., and Dalton, G. 1996. Genetic diversity among elite sorghum inbred lines assessed with DNA markers and pedigree information. *Crop Sci.* 36(5):1385-1392.

Andrews, D.J. 1987. Breeding pearl millet grain hybrids. In W.P. Feistritzer and A.F. Kelly (Eds.), *FAO/DANIDA, regional seminar on breeding and producing hybrid varieties, November 11-13, 1986* (pp. 83-109). Surabaya, Indonesia, FAO, Rome.

Andrews, D.J. and Bramel-Cox, P.J. 1993. Breeding cultivars for sustainable crop production in low input dryland agriculture in the tropics. In D.R. Buxton, R. Shibles,

R.A. Forsberg, B.L. Blad, K.H. Asay, G.M. Paulsen, and R.F. Wilson (Eds.), *International crop science I* (pp. 211-223). CSSA, Madison, WI.

Andrews, D.J., Ejeta, G., Gilbert, M., Goswami, P., Kumar, K.A., Maunder, A.B., Porter, K., Rai, K.N., Rajewski, J.F., Belum Reddy, V.S., Stegmeier, W., and Talukdar, B.S. 1997. Breeding hybrid parents. In *Proc. Intl. Conf. on Genetic Improvement of Sorghum and Pearl Millet, Sept. 23-27, 1996* (pp. 173-187). Lubbock, TX, INTSORMIL Publ. No. 97-5.

Appathurai, R. 1964. Diverse plasmons in male sterile sorghum. *Madras Agril. J.* 51(7): 276-278.

Arumuganathan, K. and Earle, E.D. 1991. Nuclear DNA content of some important species. *Plant Mol. Biol. Rep.* 9(3):208-218.

Avramova, Z., Tikhonov, A., SanMiguel, P., Jin, Y.K., Liu, C., Woo, S.S., Wing, R.A., and Bennetzen, J.L. 1996. Gene identification in a complex chromosomal continuum by local genomic cross-referencing. *The Plant J.* 10(6):1163-1168.

Baily-Serres, J., Dixon, L.K., Liddell, A.D., and Leaver, C.J. 1986a. Mitochondrial genome rearrangements leads to extension and relocation of the cytochrome c oxidase subunit I gene in sorghum. *Cell* 47(4):567-576.

Baily-Serres, J., Dixon, L.K., Liddell, A.D., and Leaver, C.J. 1986b. Nuclear-mito-chondrial interactions in cytoplasmic male-sterile *Sorghum Theor. Appl. Genet.* 73(2):252-260.

Bandyopadhyay, R., Frederickson, D.E., McLaren, N.W., Oduody, G.N., and Ryley, M.J. 1998. Ergot: A new disease threat to sorghum in the Americas and Australia. *Plant Dis.* 82(4):356-367.

Beil, G.M. and Atkins, R.E. 1967. Estimates of general and specific combining ability in F1 hybrids for grain yield and its components in grain sorghum, *Sorghum vulgare* Pers. *Crop Sci.* 7(3):225-228.

Bennetzen, J.L. 1997. The potential of biotechnology for the improvement of sorghum and pearl millet. In *Proc. Intl. Conf. on Genetic Improvement of Sorghum and Pearl Millet, Sept. 23-27, 1996* (p. 13-20). Lubbock, TX, INTSORMIL Publ. No. 97-5.

Bennetzen, J.L. and Freeling, M. 1993. Grasses as a single genetic system: Genome composition, colinearity, and compatibility. *Trends in Genet.* 9(8):259-261.

Bhaskaran, S. and Smith, R.H. 1990. Regeneration in cereal tissue culture: A review. *Crop Sci.* 30(6):1328-1336.

Binelli, G., Gianfranceschi, L., Pe, M.E., Taramino, G., Busso, C., Stenhouse, J., and Ottaviano, E. 1992. Similarity of maize and sorghum genomes as revealed by maize RFLP probes. *Theor. Appl. Genet.* 84(1-2): 10-16.

Boora, K.S., Frederiksen, R.A., and Magill, C.W. 1999. A molecular marker that segregates with sorghum leaf blight resistance in one cross is maternally inherited in another. *Mol. Gen. Genet.* 261(2):317-322.

Bramel-Cox, P.J., and Cox, T.S. 1988. Use of wild germplasm in sorghum improvement. In D. Wilkinson (Ed.), *Proceedings of the 43rd Annual Corn and Sorghum Industry Research Conference, Chicago, IL. 8-9 Dec. 1988* (pp.13-26). Am. Seed Trade Assoc., Washington, DC.

Bramel-Cox, P.J., Kumar, K.A., Hancock, J.D., and Andrews, D.J. 1995. Sorghum and millets for forage and feed. In D.A.V. Dendy (Ed.), *Sorghum and the millets: Chemistry and technology* (pp. 325-364). American Association of Cereal Chemists, Inc., St. Paul, MN.

Brown, A.H.D. 1989. The case for core collections. In A.H.D. Brown, O.H. Frankel, D.R. Marshall, and J.T. Williams (Eds.), *The use of plant genetic resources* (pp.136-156). Cambridge Univ. Press, Cambridge.

Brown, A.H.D. 1995. The core collections at the crossroads. In T. Hodgkin, A.H.D. Brown, Th.J.L. van Hintum, and E.A.V. Morales (Eds.), *Core collections of plant genetic resources* (pp. 3-19). IBPGRI, Sayce Publishing, and John Wiley and Sons, Chichester.

Brown, S.M., Hopkins, M.S., Mitchell, S.E., Senior, M.L., Wang, T.Y., Duncan, R.R., Gonjalez-Candelas, F., and Kresovich, S. 1996. Multiple methods for the identification of polymorphic simple sequence repeats (SSRs) in sorghum (*Sorghum bicolor* L. Moench). *Theor. Appl. Genet.* 93(1-2):190-198.

Casas, A.M., Konowicz, A.K., Zehr, U.B., Tomes, D.T., Axtell, J.D., Butler, L.G., Bressan, R.A., and Hasegawa, P.M. 1993. Transgenic sorghum plants via microprojectile bombardment. *Proc. Natl. Acad. Sci.* 90(23):11212-11216.

Chen, M., San Miguel, P., deOliveira, A.C., Woo, S., Zhang, H., Wing, R.A., and Bennetzen, J.L. 1997. Microcollinearity in the sh2-homologous regions of the maize, rice and sorghum genomes. *Proc. Natl. Acad. Sci.* 94(7):3431-3435.

Chen, Z., Muthukrishnan, S., Liang, G.H., Schertz, K.F., and Hart, G.E. 1993. A chloroplast DNA deletion located in RNA polymerase gene *rpoC2* in CMS lines of sorghum. *Mol. Gen. Genet.* 236(2-3):251-259.

Chittenden, L.M., Schertz, K.F., Lin, Y.-R., Wing, R.A., and Paterson, A.H. 1994. A detailed RFLP map of *Sorghum bicolor* × *S. propinquum*, suitable for high-density mapping, suggests ancestral duplication of sorghum chromosomes or chromosomal segments. *Theor. Appl. Genet.* 87(8):925-933.

Chungu, C. 1992. Tester choice in evaluating new parental lines in grain sorghum. Doctoral thesis, University of Nebraska, Lincoln.

Collins, F.C. and Pickett, R.C. 1972. Combining ability for grain yield, percent protein, and glycine/100g protein in a nine-parent diallel of *Sorghum bicolor* (L.) Moench. *Crop Sci.* 112(4):423-425.

Comstock, R.E., Robinson, H.F., and Harvey, P.H. 1949. A breeding procedure designed to make maximum use of both general and specific combining ability. *J. Am. Soc. Agron.* 41(8):360-367.

Conde, M.F., Pring, D.R., Schertz, K.F., and Ross, W.M. 1982. Correlation of mitochondrial DNA restriction endonuclease patterns with sterility expression in six male-sterile sorghum cytoplasms. *Crop Sci.* 22(3):536-539.

Conner, A.B. and Karper, R.E. 1927. *Hybrid vigor in sorghum.* Texas Experiment Station Bulletin No. 359. Texas A & M University, Texas.

Cox, T.S., House, L.R., and Frey, K.J. 1984. Potential of wild germplasm for increasing yield of grain sorghum. *Euphytica* 33(3):673-684.

Crasta, O.R., Xu, W., Rosenow, D.T., Mullet, J.E., and Nguyen, H.T. 1999. Mapping of post-flowering drought resistance traits in grain sorghum: Association of QTLs influencing premature senescence and maturity. *Mol. Gen. Genet.* 262(3):579-588.

Cui, Y.X., Xu, G.X., Magill, C.W., Schertz, K.F., and Hart, G.E. 1995. RFLP-based assay of *Sorghum bicolor* (L.) Moench genetic diversity. *Theor. Appl. Genet.* 90(6):787-796.

Dahlberg, J.A. 2000. Classification and characterization of sorghum. In C.W. Smith and R.A. Frederiksen (Eds.), *Sorghum: Origin, history, technology and production.* John Wiley and Sons, New York.

de Wet, J.M.J. 1976. Variability in *Sorghum bicolor.* In J.R. Harlan, J.M.J. de Wet, and A.E.L. Stemler (Eds.), *Origins of African plant domestication* (pp. 453-463). Mouton Publ., Hague, Netherlands

de Wet, J.M.J, Harlan, J.R., and Price, E.G. 1970. Origin of variability in the spontanea complex of *Sorghum bicolor. American Journal of Botany* 57(6):704-707.

Dendy, D.A.V. 1995. Sorghum and the millets: Production and importance. In D.A.V. Dendy (Ed.), *Sorghum and millet chemistry and technology* (pp.11-25). Amer. Assoc. of Cereal Chemists, Inc., St. Paul, MN.

Devos, K.M. and Gale, M.D. 1997. Comparative genetics in grasses. *Plant Mol. Biol.* 35(1-2):3-15.

Doggett, H. 1961. Yields of hybrid sorghums. *Expt. Agric.* 5(1):1.

Doggett, H. 1976. *Sorghum bicolor* (Gramineae-Andropogoneae). In N.W. Simmonds (Ed.), *Evolution of crop plants* (pp. 112-117). Longman Group, Limited. New York.

Doggett, H. 1982. A look back at the 70s. In L.R. House, L.K. Mughogho, and J.M. Peacock (Eds.), *Sorghum in the eighties: Proc. Intl. Symp. on sorghum, November 2-7, 1981.* ICRISAT, Patancheru, A.P., India.

Doggett, H. 1988. *Sorghum,* Second edition. John Wiley and Sons, Inc., New York.

Doggett, H. and Majisu, B.N. 1968. Disruptive selection in crop development. *Heredity* 23:1-23.

Dudley, J.W. 1984a. Identifying parents for use in a pedigree breeding program. In H.D. Loden and D. Wilkinson (Eds.), *Proceedings of the 39th Annual Corn and Sorghum Industry Research Conference, Chicago, II, 5-6 Dec* (pp. 176-188). Amer. Seed Trade Assoc., Washington, D.C.

Dudley, J.W. 1984b. A method of identifying lines for use in improving parents of a single cross. *Crop Sci.* 24:355-357.

Dudley, J.W. 1984c. A method of identifying population containing favorable alleles not present in elite germplasm. *Crop Sci.* 24:1053-1054.

Dudley, J.W. 1984d. Theory of identification and use of exotic germplasm in maize breeding program. *Maydica* 29(4):391-407.

Dudley, J.W. 1987a. Modification of methods for identifying inbred lines useful for improving parents of elite single crosses. *Crop Sci.* 27(5):944-947.

Dudley, J.W. 1987b. Modification of methods for identifying populations to be used for improving parents of elite single crosses. *Crop Sci.* 27(5):940-943.

Dufour, P., Deu, M., Grivet, L., D'Hont, A., Paulet, F., Bouet, A., Lanaud, C., Glaszmann, J.C., and Hamon, P. 1997. Construction of a composite sorghum genome map and comparison with sugarcane, a related complex polyploid. *Theor. Appl. Genet.* 94(3-4):409-418.

Duncan, R.R. 1996a. Breeding and improvement of forage sorghums for the tropics. *Adv. Agron.* 57:161-185.

Duncan, R.R. 1996b. Tissue culture-induced variation and crop improvement. *Adv. Agron.* 58:210-240.

Duncan, R.R., Bramel-Cox, P.J., and Miller, F.R. 1991. Contributions of introduced sorghum germplasm to hybrid development in the USA. In H.L. Shands and L.E. Wiesner (Eds.), *Use of plant introductions in cultivar development* (pp. 69-102). Part 1. CSSA Spec. Publ. 17. CSSA, Madison, WI.

Duncan, R.R. and Dahlberg, J.A. 1993. Cross-reference of PI/IS/SC numbers from the U.S. conversion program. *Sorghum Newsl.* 34: 72-80.

Duncan, R.R., Isenhour, D.J., Waskom, R.M., Miller, D.R., Nabors, M.W., Hanning, G.E., Petersen, K.M., and Wiseman, B.R. 1991. Registration of GATCCP100 and GATCCP101: Fall armyworm resistant hegari regenerants. *Crop Sci.* 31(1):242-244.

Duncan, R.R., Waskom, R.M., Miller, D.R., Hanning, G.E., Timm, D.A., and Nabors, M.W. 1992. Registration of GAC103 and GC104 acid-soil tolerant Tx430 regenerants. *Crop Sci.* 32(4):1076-1077.

Duncan, R.R., Waskom, R.M., Miller, D.R., Voigt, R.L., Hanning, G.E., Timm, D.A., and Nabors, M.W. 1991. Registration of GAC102 acid tolerant hegari regenerant. *Crop Sci.* 31(5):1396-1397.

Duncan, R.R., Waskom, R.M., and Nabors, M.W. 1995. *In vitro* screening and field evaluation of tissue-culture-generated sorghum [*Sorghum bicolor* (L.) Moench] for stress tolerance. *Euphytica* 85(1-3):373-380.

Eastin, J.D. 1970. Photosynthesis and translocation in relation to plant development. In N.G.P. Rao and L.R. House (Eds.), *Sorghum in seventies* (pp. 214-244). Oxford and IBH publishing Co., New Delhi.

Eberhart, S.A., Bramel-Cox, P.J., and Prasada Rao, K.E. 1997. Preserving genetic resources. In *Proceedings of the International Conference on Genetic Improvement of Sorghum and Pearl Millet, Sept. 23-27, 1996* (pp. 25-41). Lubbock, TX, INTSORMIL Publ. No. 97-5.

Ejeta, G., L., Butler, L.G., Hess, D.E., Obilana, T., and Reddy, B.V. 1997. Breeding for Striga resistance in sorghum. In *Proceedings of the International Conference on Genetic Improvement of Sorghum and Pearl Millet, Sept. 23-27, 1996* (pp. 504-516). Lubbock, TX, INTSORMIL Publ. No. 97-5.

Elkonin, L.A., Enaleeva, H.K., Tsvetova,, M.I., Belyalva, E.V., and Ishin, A.G. 1995. Partially fertile line with apospory obtained from tissue culture of male sterile plant of sorghum (*Sorghum bicolor* L. Moench). *Ann. Bot.* 76(4):359-364.

Erichsen, A.W. and Ross, J.G. 1963. Inheritance of colchicine induced male sterility in sorghum. *Crop Sci.* 3(4):335-338.

Eyherabide, G.H. and Hallauer, A.R. 1991. Reciprocal full-sib recurrent selection in maize: I. Direct and indirect responses. *Crop Sci.* 31(4):952-959.

Frankel, O.H. 1984. Genetic perspectives of germplasm conservation. In W.K. Arber, K.L. Limensee, P.J. Peacock, and P. Starlinger (Eds.), *Genetic manipulation: Impact of man and society* (pp. 161-170). Cambridge Univ. Press, Cambridge.

Gilbert, M.L. 1989. Utilization of alternate cytoplasmic sterility systems for increased efficiency in sorghum breeding. Doctoral dissertation, Texas Tech University, Lubbock, Texas.

Gilbert, M.L. 1994. Identification and search for heterotic patterns in sorghum. In *Proc. 49th annual corn and sorghum research conf., Dec. 7-8, 1994* (pp. 117-126). American Seed Trade Association, Washington, DC.

Gilbert, M.L., Kramer, N., Rosenow, D.T., and Nguyen, H.T. 1991. Strategies for sorghum inbred development in the 90's. In *Seventeenth Biennial Grain Sorghum Res. and Util. Conf. Feb 17-20, 1991* (pp. 11-14). Holiday Inn Civic Center, Lubbock, Texas.

Gourley, L.M. and Lusk, J.W. 1978. Genetic parameters related to sorghum silage quality. *J. Dairy Sci.* 61(12):1821-1827.

Gu, M.H., Ma, H.T., and Liang. G.H. 1984. Karyotype analysis of seven species in the genus *Sorghum. J. Hered.* 75(3):196-202.

Hallauer, A.R. 1981. Selection and breeding methods. In K.J. Frey (Ed.), *Plant breeding II* (pp. 3-55). The Iowa State University press, Ames, Iowa.

Hamaker, B.R. and Axtell, J.D. 1997. Nutritional quality of sorghum. In *Proceedings of the International Conference on Genetic Improvement of Sorghum and Pearl Millet, Sept. 23-27, 1996* (pp. 531-538). Lubbock, TX, INTSORMIL Publ. No. 97-5.

Hanna, W.W., Schertz, K.F., and Bashaw, E.C. 1970. Apospory in *Sorghum bicolor* (L.) Moench. *Science* 170(395):338-339.

Harlan, J.R. and de Wet, J.M.J. 1972. A simplified classification of cultivated sorghum. *Crop Sci.* 12(2):172-176.

Henzell, R.G., Dodman, R.L., Done, A.A., Brengman, R.L., and Mayers, E.P. 1984. Lodging, stalk rot, and root rot in sorghum in Australia. In *Sorghum root and stalk rots: A critical review. Proceedings of the consulative group discussion on research needs and strategies for control of sorghum root and stalk rot diseases, November 27–December 2, 1983, Bellagio, Italy* (pp. 209-218). ICRISAT, Patancheru, A.P. 502324, India.

Henzell, R.G., Franzman, B.A., and Brengman, R.L. 1994. Sorghum midge resistance research in Ausralia. *Int. Sorghum Millets Newsl.* 35:41-47.

Hookstra, G.H., Ross, W.M., and Mumm, R.F. 1983. Simultaneous evaluation of grain sorghum A-lines and random-mating populations with top cross. *Crop Sci.* 23(5):977-981.

Hoveland, C.S. and Monson, W.G. 1980. Genetic and environmental effects on forage quality. In C.S. Hoveland (Ed.), *Crop quality, storage, and utilization* (pp. 139-168). Amer. Soc. Agron. And Crop Sci. Amer., Madison, WI.

Hulbert, S.H., Richter, T.E., Axtell, J.D., and Bennetzen, J.L. 1990. Genetic mapping and characterization of sorghum and related crops by means of maize DNA probes. *Proc. Natl. Acad. Sci.* 87(11):4251-4255.

ICRISAT (International Crops Research Institute for the Semi-Arid Tropics) 1988. *Annual Report 1987.* ICRISAT, Patancheru, 502324, A.P. India.

ICRISAT (International Crops Research Institute for the Semi-Arid Tropics) 1989. *Annual Report 1988.* ICRISAT, Patancheru, 502324, A.P. India.

Isenhour, D.J., Duncan, R.R., Miller, D.R., Waskom, R.M., Hanning, G.E., Wiseman, B.R., and Nabors, M.W. 1991. Resistance to leaf feeding by the fall armyworm (Lepidoptera: Noctuideae) in tissue culture derived sorghums. *J. Econ. Entom.* 84(2):680-684.

Johnson J.W., Stegmeier, W.D., Andrews, D.J., Rosenow, D.T., Henzell, R.G., and Monk, R.L. 1997. Genetic resistance to lodging. In *Proc. Intl. Conf. on Genetic Improvement of Sorghum and Pearl Millet, Sept. 23-27, 1996* (pp. 481-489). Lubbock, TX, INTSORMIL Publ. No. 97-5.

Joshi, P. and Vashi, P.S. 1992. Mahalonobis generalized distance and genetic diversity in sorghum. *Indian J. Genet. Plant Breed.* 52(1):85-93.

Kambal, A.E. and Webster, O.J. 1965. Estimates of general and specific combining ability in grain sorghum, *Sorghum vulgare,* Pers. *Crop Sci.* 5(6):521-523.

Kebede, H., Subudhi, P.K., Rosenow, D.T., and Nguyen, H.T. 2001. Quantitative trait loci influencing drought tolerance in sorghum (*Sorghum bicolor* L. Moench). *Theor. Appl. Genet.* (In press).

Kirby, J.S. and Atkins, R.E. 1968. Heterotic response for vegetative and mature plant characters in grain sorghum, *Sorghum bicolor* (L.) Moench. *Crop Sci.* 8(3):335-339.

Kong, L., Dong, J., and Hart, G. 1997. Linkage mapping of simple sequence repeat loci in *Sorghum bicolor* L. Moench. Plant and Animal Genome V. Conference San Diego, CA. <http://www.intl-pag.org/pag/5/abstracts/p-5d-193.html>, January 12-16, 1997.

Kononowicz, A.K., Casas, A.M., Tomes, D.T., Bressan, R.A., and Hasegawa, P.M. 1995. New vistas are opened for sorghum improvement by genetic transformation. *African Crop Sci. J.* 3:171-180.

Kramer, N.W. 1987. Grain sorghum production and breeding-historical perspectives to future prospects. In *Proceedings of the 42nd Annual Corn and Sorghum Research Conference* (pp. 1-9).Washington DC.

Kumaravadivel, N. and Rangasamy, S.R. 1994. Plant regeneration from sorghum anther cultures and field evaluation of progeny. *Plant Cell Rep.* 13(5):286-290.

Laosuwan, P. and Atkins, R.E. 1977. Estimates of combining ability and heterosis in converted exotic sorghums. *Crop Sci.* 17(1):47-50.

Lee, M. 1995. DNA markers and plant breeding programs. *Adv. Agron.* 55:265-344.

Lee, R.D., Johnson, B.E., Eskridge, K.M., and Pedersen, J.F. 1992. Selection of superior female parents in sorghum utilizing A3 cytoplasm. *Crop Sci.* 32(4):918-921.

Lin, Y.R. 1998. Construction of a *Sorghum propinquum* BAC library, toward positional cloning of the sorghum shattering gene (*Sh1*) and the sorghum photoperiodic gene (*Ma1*). Doctoral dissertation. Texas A & M Univ., College Station, TX.

Lin, Y.R., Schertz, K.F., and Paterson, A.H. 1995. Comparative analysis of QTLs affecting plant height and maturity across the Poaceae, in reference to an interspecific sorghum population. *Genetics* 140(1):391-411.

Lin, Y.R., Zhu, L., Ren, S., Yang, J., Schertz, K.F., and Paterson, A.H. 1999. A *Sorghum propinquum* BAC library, suitable for cloning genes associated with loss-of-function mutations during crop domestication. *Mol. Breed.* 5(6):511-520.

Mann, J.A., Kimber, C.T., and Miller, F.R. 1983. The origin and early cultivation of sorghums in Africa. *Texas Agric. Exp. Stn. Bull. 1454.* Texas A & M Univ., College Station.

Mathur, P.N., Rao, K.E.P., Singh, I.P., Agrawal, R.C., Mengesha, M.H., and Rana, R.S. 1992. *Evaluation of forage sorghum germplasm, Part 2.* NBPGR-ICRISAT, New Delhi, India.

Mathur, P.N., Rao, K.E.P., Thomas, T.A., Mengesha, M.H., Sapra, R.L., and Rana, R.S. 1991. *Evaluation of forage sorghum germplasm, Part 1.* NBPGR-ICRISAT, New Delhi, India.

Maunder, A.B. 1992. Identification of useful germplasm for practical plant breeding programs. In H.T. Stalker and J.P. Murphy (Eds.), *Plant breeding in the 1990s* (pp. 147-169). CAB Intl. Wallingford, UK.

Maunder, A.B. and Pickett, R.C. 1959. The genetic inheritance of cytoplasmic genetic male sterility in sorghum. *Agron. J.* 51(1):47-49.

Maunder, A.B., Sedlak, D., Lambright, L., and Matekaitis, L. 1990. Influence of the pollinator on germination and seedling vigor in sorghum. *Sorghum Newsl.* 31:14.

Melake-Berhan, A., Hulbert, S.H., Butler, L.G., and Bennetzen, J.L. 1993. Structural and evolution of the genomes of *Sorghum bicolor* and *Zea mays*. *Theor. Appl. Genet.* 86(5):598-604.

Menkir, A., Bramel-Cox, P.J., and Witt, M.S. 1994a. Comparison of methods for introgressing exotic germplasm into adapted sorghum. *Theor. Appl. Genet.* 89(2-3):233-239.

Menkir, A., Bramel-Cox, P.J., and Witt, M.S. 1994b. Selection for agronomically acceptable inbred lines in adapted × exotic sorghum backcross populations. *Crop Sci.* 34(4):1084-1089.

Miller, D.R., Waskom, R.M., Duncan, R.R., Chapman, P.L., Brick, M.A., Hanning, G.E., Timm, D.A., and Nabors, M.W. 1992. Acid soil stress tolerance in tissue-culture derived sorghum lines. *Crop Sci.* 32(2):324-327.

Miller, F., Muller, N., Monk, R., Murty, D.S., and Obilana, A.B. 1997. Breeding photoperiod insensitive sorghum for adaptation and yield. In *Proceedings of the International Conference on Genetic Improvement of Sorghum and Pearl Millet, Sept. 23-27, 1996* (pp. 59-65). Lubbock, TX, INTSORMIL Publ. No. 97-5.

Miller, F.R. 1976. Twin-seeded sorghum hybrids-facts or fancy? In *Proceedings of the 31st Annual Corn and Sorghum Research Conference, Chicago, III, December 7-9* (pp. 24-37). Am. Seed Trade Assoc., Washington, D.C.

Miller, F.R. and Kebede, Y. 1984. Genetic contribution to yield gains in sorghum, 1950 to 1980. In W.R. Fehr (Ed.), *Genetic contribution to yield gains of five major crop plants* (pp. 1-14). CSSA Spec. Publ. No. 7, CSSA, ASA, Madison, WI.

Miller, F.R., Mann, J.M., and Rooney, L.W. 1987. Developing food type sorghums. In *Proc. 15th Biennial Grain sorghum Res. and Util. Conf. Feb. 15-17, 1987, Lubbock, TX* (pp. 57-61).

Mohamed, A.B. 1980. An evaluation of eight female lines of grain sorghum using three groups of testers. Master's thesis, Univ. of Nebraska, Lincoln.

Moore, G. 1995. Cereal genome evolution: Pastoral pursuits with 'Logo' genomes. *Current Opi. Genet. and Dev.* 5(6):717-724

Moore, G., Gale, M.D., Kurata, N., and Flavell, R.B. 1993. Molecular analysis of small grain cereal genomes: Current status and prospects. *Bio/Technology* 11(5):584-589.

Mukuru, S.Z. 1992. Breeding for grain mold resistance. In W.A.J. de Milliano, R.A. Frederiksen, and G.D. Bengston (Eds.), *Sorghum and millet diseases: A second world review* (pp. 273-285). ICRISAT, Patancheru, A.P. 502 324, India.

Murty, B.R., Arunachalam, V., Saxena, M.B.L., and Govil, J.N. 1967. Classification and catalogue of a world collection of sorghum. *Ind. J. Genet. Plant Breeding* 27:1-74.

Murty, U.R. 1986. Milo and non-milo cytoplasms in *Sorghum bicolor* (L.) Moench. II. Fertility restorers and sterility maintainers on milo cytoplasms. *Cereal Res. Comm.* 14(2):191-196.

Murty, U.R. 1992. A novel male sterility system in sorghum [*Sorghum bicolor* (L.) Moench]. *Curr. Sci.* 63(3):142-143.

Murty, U.R. 1993. Milo and non-milo cytoplasms in *Sorghum bicolor* (L.) Moench. IV. Commercial production and utilization of uniline CMS systems. *Crop Improvement* 20:190-196.

Murty, U.R. 1995. Breeding two-line hybrids in *Sorghum bicolor* (L.) Moench. *Cereal Res. Comm.* 23(4):397-402.

Murty, U.R. 1999. Hybrid seed production in sorghum. In A.S. Basra (Ed.), *Heterosis and hybrid seed production in agronomic crops* (pp. 119-148). The Food Products Press, Binghamton, New York.

Murty, U.R., Kirti, P.B., Bharathi, M., and Rao, N.G.P. 1984. The nature of apomixis and its utilization in the production of hybrids ("vybrids") in *Sorghum bicolor* (L.) Moench. *Z. Pflanzenzuech* 92(1):30-39.

Murty, U.R. and Rao, N.G.P. 1972. Apomixis in breeding grain sorghums. In N.G.P. Rao and L.R. House (Eds.), *Sorghum in the seventies* (pp. 517-523). Oxford and IBH Publ. Co., New Delhi.

Niehaus, M.H. and Pickett, R.C. 1966. Heterosis and combining ability in a diallel cross in *Sorghum vulgare* Pers. *Crop Sci.* 6(1):33-36.

Oh, B.J., Frederiksen, R.A., and Magill, C.W. 1994. Identification of molecular markers linked to head smut resistance gene (*Shs*) in sorghum by RFLP and RAPD analyses. *Phytopathol.* 84(8):830-833.

Oh, B.J., Gowda, P.S.B., Xu, G.W., Frederiksen, R.A., and Magill, C.W. 1993. Tagging Acremonium wilt, downy mildew and head smut resistance genes in sorghum using RFLP and RAPD markers. *Sorghum Newsl.* 34:34.

Oliveira, A.C., Richter, T., and Bennetzen, J.L. 1996. Regional and racial specificities in sorghum germplasm assessed with DNA markers. *Genome* 39(3):579-587.

Paterson, A.H., Schertz, K., Lin, Y.-R., and Li, Z. 1998. Case history in plant domestication: Sorghum, an example of cereal evolution. In A.H. Paterson (Ed.), *Molecular dissection of complex traits* (pp. 187-195). CRC Press, Boca Raton, FL.

Paterson, A.H., Schertz, K.F., Lin, Y.R., Liu, S.C., and Chang, Y.L. 1995. The weediness of wild plants: Molecular analysis of genes influencing dispersal and persistence of Johnsongrass, *Sorghum halepense* (L.) *Pers. Proc. Natl. Acad. Sci.* 92(13):6127-6131.

Pederson, J.F., Kaeppler, H.F., Andrews, D.J., and Lee, R.D. 1998. Sorghum. In S.S. Banga and S.K. Banga (Eds.), *Hybrid cultivar development* (pp. 344-356). Narosa Publishing House, New Delhi, India.

Pereira, M.G., Lee, M., Bramel-Cox, P.J., Woodman, W., Doebley, J., and Whitkus, R. 1994. Construction of an RFLP map in sorghum and comparative mapping in maize. *Genome* 37(2):236-243.

Peterson, G.C., Reddy, B.V.S., Youm, O., Teetes, G.L., and Lambright, L. 1997. Breeding for resistance to foliar- and stem feeding insects of sorghum and pearl millet. In *Proceedings of the International Conference on Genetic Improvement of Sorghum and Pearl Millet, Sept. 23-27, 1996* (pp. 281-302). Lubbock, TX, INTSORMIL Publ. No. 97-5.

Peterson, G.C., Teetes, G.L., and Pendleton, B.B. 1994. Resistance to the sorghum midge in the United States. *Int. Sorghum Millets Newsl.* 35:48-63.

Pollak, L.M., Torres-Cardona, S., and Sotomayor-Rios, A. 1991. Evaluation of heterotic patterns among Caribbean and tropical × temperate maize populations. *Crop Sci.* 31(6):1480-1483.

Prasada Rao, K.E. and Mengesha, M.H. 1988. Sorghum genetic resources-synthesis of available diversity and its utilization. In R.S. Paroda, R.K. Arora, and K.P.S. Chandel (Eds.), *Plant genetic resources: Indian perspectives* (pp. 159-179). National Bureau of Plant Genetic Resources, New Delhi, India.

Prasada Rao, K.E., Mengesha, M.E., and Gopal Reddy, V. 1989. International use of sorghum germplasm collection. In A.D.H. Brown, O.H. Frankel, D.R. Marshall, and J.T. Williams (Eds.), *The use of plant genetic resources* (pp. 49-67). Cambridge University Press, Cambridge, England.

Prasada Rao, K.E. and Ramanatha Rao, V. 1995. Use of characterization data in developing a core collection of sorghum. In T. Hodgkin, A.H.D. Brown, Th.J.L. van Hintum, and E.A.V. Morales (Eds.), *Core collections of plant genetic resources* (pp. 109-115). IBPGRI, Sayce Publishing, and John Wiley and Sons, Chichester.

Pring, D.R., Conde, M.F., and Schertz, K.F. 1982. Organelle genome diversity among male sterile sorghum cytoplasms. *Crop Sci.* 22(2):414-421.

Pring, D.R., Tang, H.V., and Schertz, K.F. 1995. Cytoplasmic male sterility and organelle DNAs of sorghum. In C.S. Levings III and I.K. Vasil (Eds.), *The molecular biology of plant mitochondria* (pp. 461-495). Kluwer Academic Publishers, Dordrecht.

Quinby, J.R. 1963. Manifestation of hybrid vigor in sorghum. *Crop Sci.* 3(4):288-291.

Quinby, J.R. 1970. Leaf and panicle size of sorghum parents and hybrids. *Crop Sci.* 10(3):251-254.

Quinby, J.R. 1971. *A triumph of research: Sorghum in Texas.* Texas A & M University Press, College Station, TX.

Quinby, J.R. 1974. *Sorghum improvement and genetics of growth.* Texas A & M University Press, College Station, TX.

Quinby, J.R. 1980. Interaction of genes and cytoplasms in male sterility in sorghum. In Loden, H.D. and Wilkinson, D. (Eds.), *Proceedings of the Annual Corn and Sorghum Research Conference* (pp. 175-184). American Seed Trade Association, Washington, DC.

Ragab, R.A., Dronavalli, S., Saghai Maroof, M.A., and Yu, Y.G. 1994. Construction of sorghum RFLP linkage map using sorghum and maize DNA probes. *Genome* 37(4):590-594.

Rami, J.F., Dufour, P., Trouche, G., Fliedel, G., Mestress, C., Davrieux, F., Blanchard, P., and Hamon, P. 1998. Quantitative trait loci for grain quality, productivity, morphological, and agronomical traits in sorghum (*Sorghum bicolor* L. Moench). *Theor. Appl. Genet.* 97(4):605-616.

Rao, N.G.P. 1962. Occurrence of cytoplasmic-genetic male sterility in some Indian sorghums. *India J. Genet. Plant Breed.* 22(3):257-259.

Rao, N.G.P. and Narayana, L.L. 1968. Apomixis in grain sorghum. *India J. Plant Breed.* 28(2):121-127.

Rattunde H.F.W., Weltzien R. E., Bramel-Cox, P.J., Kofoid, K., Hash, C.T., Schipprack, W., Stenhouse, J.W., and Presterl, T. 1997. Population improvement of pearl millet and sorghum: Current research, impact, and issues for implementation. In *Proceedings of the International Conference on Genetic Improvement of Sorghum and Pearl Millet, Sept. 23-27, 1996* (pp. 188-212). Lubbock, TX, INTSORMIL Publ. No. 97-5.

Reis, E.M., Mantle, P.G., and Hassan, H.A. 1996. First report in the Americas of sorghum ergot disease, caused by a pathogen diagnosed as *Claviceps africana.* *Plant Dis.* 80(4):463.

Rodriguez-Herrera R., Williams-Alanis, H., Aguirre-Rodriguez, J. 1993. Comparable performance of isogenic sorghums in A1 and A2 cytoplasms. III. Head smut. *Sorghum Newsl.* 34:22.

Rooney, L.W., Khan, M.N., and Earp, C.F. 1980. The technology of sorghum products. In E. Inglett and L. Munck (Eds.), *Cereals for food and beverages: Recent progress in cereal chemistry* (pp. 513-554). Academic Press, New York.

Rooney, L.W., Kirleis, A.W., and Murty, D.S. 1986. Traditional foods for sorghum: Their production, evaluation and nutritional value. In Y. Pomeranz (Ed.), *Advances in cereal science and technology* (pp. 317-353). Vol. VIII, American Assoc. of Cereal Chemistry, St. Paul, MN.

Rooney, L.W., Waniska, R.D., and Subramanian, R. 1997. Overcoming constraints to utilization of sorghum and millets. In *Proceedings of the International Conference on Genetic Improvement of Sorghum and Pearl Millet, Sept. 23-27, 1996* (pp. 549-557). Lubbock, TX, INTSORMIL Publ. No. 97-5.

Rose, J.B., Dunwell, J.M., and Sunderland, N. 1986. Anther culture of *Sorghum bicolor* (L.) Moench I: Effect of panicle pretreatment, anther incubation temperature and 2,4, D concentration. *Plant Cell Tissue Organ Cult.* 6(1):15-22.

Rosenow, D.T. 1970. An evaluation of early generation testing for combining ability of restorer lines in grain sorghum. Doctoral dissertation, Texas A & M University.

Rosenow, D.T. and Clark, L.E. 1987. Utilization of exotic germplasm in breeding for yield stability. In R.R. Duncan (Ed.), *Proceedings of the 15th Biennial Grain Sorghum Research and Utilization Conf., Lubbock, TX, February 15-17, 1987* (pp. 49-56). Natl. Grain Sorghum Producers Assoc., Abernathy, TX.

Rosenow, D.T. and Clark, L.E. 1995. Drought and lodging resistance for a quality sorghum crop. In *Proc. 50th annual corn and sorghum industry research conference, December 6-7, 1995, Chicago, IL* (pp. 82-97). Am. Seed Trade Assoc., Washington, DC.

Rosenow, D.T., Dahlberg, J.A., Peterson, G.C., Clark, L.E., Sotomayor-Rios, A., Miller, F.R., Hamburger, A.J., Madera-Torres, P., Quiles-Belen, A., and Woodfin, C.A. 1995. Release of 50 converted sorghum lines and 253 partially converted sorghum bulks. *Intl. Sorghum and Millets Newsl.* 36:19-31.

Rosenow, D.T., Ejeta, G., Clark, L.E., Gilbert, M.L., Henzell, R.G., Borrell, A.K., and Muchow, R.C. 1997. Breeding for pre- and post-flowering drought stress resistance in sorghum. In *Proc. Intl. Conf. on Genetic Improvement of Sorghum and Pearl Millet, Sept. 23-27, 1996* (pp. 400-411). Lubbock, TX, INTSORMIL Publ. No. 97-5.

Rosenow, D.T. and Frederiksen, R.A. 1982. Breeding for disease resistance in sorghum. In L.R. House, L.K. Mughogho, and J.M. Peacock (Eds.), *Sorghum in the eighties: Proceedings of the International Symposium on sorghum, November 2-7, 1981* (pp. 447-455). ICRISAT Center, India, Vol I.

Rosenow, D.T., Schertz, K.F., and Sotomayor, A. 1980. Germplasm release of three pairs (A and B) of sorghum lines with A2 cytoplasmic-genic sterility system. *Texas Agric Exp. Stn. Misc Publ* 1448:1-2.

Rosenow, D.T., Woodfin, C.A., Beder, M.S., Gebeyehu, G., McCosker, A.N., and Nguyen, H.T. 1995. Stay-green reaction of sorghum lines in F1 hybrids. In *Proceedings of the 19th Biennial Grain Sorghum Research And Utilization Conference, March 5-7, 1995, Lubbock, TX* (p. 49).

Ross, W.M. 1965. Cytoplasmic male sterility and fertility restoration of some major sorghum groups. In Hang, F. (Ed.), *Proceedings of the 16th Biennial Grain Sor-*

ghum Research and Utilization Conference, Lubbock Texas (pp. 57-62). Amarillo, Texas.

Ross, W.M., Gorz, H.J., Haskins, F.A., Hookstra, G.H., Rutto, J.K., and Ritter, R. 1983. Combining ability effects for forage residue traits in grain sorghum hybrids. *Crop Sci.* 23(1):97-101.

Ross, W.M. and Hackerott, H.L. 1972. Registration of seven iso-cytoplasmic sorghum germplasm lines. *Crop Sci.* 12(5):720.

Ross, W.M. and Kofoid, K.D. 1978a. A preliminary evaluation of test crosses in *Sorghum bicolor* (L.) Moench. *Maydica* 23(2):101-109.

Ross, W.M. and Kofoid, K.D. 1978b. Evaluation of grain sorghum R-lines with a single-cross vs. inbred line tester. *Crop Sci.* 18(4):670-672.

Schertz, K.F. 1977. Registration of A2T×2753 and B2T×2753 sorghum germplasm (Reg. No. GP 30 and 31). *Crop Sci.* 17(6):983.

Schertz, K.F. 1983. Potentials with new cytoplasmic male sterility systems in sorghum. *Proceedings of the Annual Corn and Sorghum Research Conference* 38:1-10.

Schertz, K.F. and Dalton, L.G. 1970. Sorghum. In Fehr W.R., Handley, H.H. (Eds.), *Hybridization of crop plants* (pp. 210-221). Am. Soc. Agron. and Crop Sci. Soc. America, Madison, WI.

Schertz, K.F. and Johnson, J.W. 1984. A method for selecting female parents of grain sorghum hybrids. *Crop Sci.* 24(3):492-494.

Schertz, K.F. and Pring, D.R. 1982. Cytoplasmic sterility systems in sorghum. In L.R. House, L.K. Mughogho, and J.M. Peacock (Eds.), *Sorghum in the eighties: Proceedings of the International Symposium on sorghum, November 2-7, 1981* (pp. 373-383). ICRISAT, Patancheru, A.P., India.

Schertz, K.F. and Ritchey, J.R. 1978. Cytoplasmic genic male sterility systems in sorghum. *Crop Sci.* 18(5):890-893.

Schertz, K.F., Rosenow, D.T., and Sotomayor, A. 1981. Registration of three pairs (A and B) of sorghum lines with A2 cytoplasmic-genic sterility systems. *Crop Sci.* 21(1):148.

Schertz, K.F., Sivaramakrishnan, S., Hanna, W.W., Mullet, J., Sun, Y., Murty, U.R., Pring, D.R., Rai, L.N., and Reddy, B.V.S. 1997. Alternate cytoplasms and apomixes of sorghum and pearl millet. In *Proceedings of the International Conference on Genetic Improvement of Sorghum and Pearl Millet, September 23-27, 1996* (pp. 213-223). Lubbock, Texas. INTSORMIL Publ. No. 97-5.

Schertz, K.F., Sotomayor-Rios, A., and Torres, C. 1989. Cytoplasmic male sterility: Opportunities in breeding and genetics. In Hang, F. (Ed.), *Proceedings of the 16th Biennial Grain Sorghum Research and Utilization Conference, Lubbock, Texas* (pp. 175-186).

Schertz, K.F. and Stephens, J.C. 1966. *Compilation of symbols, recommended revisions and summary of linkages for inherited characters of* Sorghum vulgare pers. Technical Monograph 3, Texas Agricultural Research Station, p. 42.

Schuering, J.F. and Miller, F.R. 1978. *Fertility restorers and sterility maintainers to the milo-kafir genetic cytoplasmic male sterility system in the sorghum world collection.* Texas Agric. Exp. Stn. MP-1367.

Senthil, N., Palanisamy, S., and Sreerangaswamy, S.R. 1994. Characterization of diverse cytosteriles of sorghum through fertility restoration studies. *Cereal Res. Comm.* 22(3):179-184.

Sharma, H.C. 1993. Host plant resistance to insects in sorghum and its role in integrated pest management. *Crop Prot.* 12(1):11-34.

Sharma, H.C., Doumbia, Y.O., Haidara, M., Scheuring, J.F., Ramaiah, K.V., and Beninati, M.F. 1994. Sources and mechanisms of resistance to sorghum head bug, *Eurystylus immaculatus*, Odh. in West Africa. *Insect Sci. Appl.* 15(1):39-48.

Smith, R.H., Bhaskaran, S., Hasegawa, P.M., Bressan, R.A., Rathore, K.S., and Duncan, R.R. 1997. *Sorghum improvement using plant tissue culture. In Proceedings of the International Conference on Genetic Improvement of Sorghum and Pearl Millet, September 23-27, 1996* (pp. 241-249). Lubbock, Texas. INTSORMIL Publ. No. 97-5.

Smith, R.H., Duncan, R.R., and Bhaskaran, S. 1993. In vitro selection and somaclonal variation for crop improvement. In D.W. Buxton (Ed.), *International crop science congress* (pp. 629-633). *Crop Sci. Soc. Amer.*, Madison, WI.

Stephens, J.C. and Holland, R.F. 1954. Cytoplasmic male sterility for hybrid sorghum seed production. *Agronom. J.* 46(1):20-23.

Stephens, J.C., Miller, F.R., and Rosenow, D.T. 1967. Conversion of alien sorghums to early combine genotypes. *Crop Sci.* 7(4):396.

Stickler, F.W. and Pauli, A.W. 1961. Influence of date of planting on yield and yield components in grain sorghum. *Agronom. J.* 53(1):20-22.

Subudhi, P.K., and Nguyen, H.T. 2000a. Biotechnology-New horizons. In W. Smith and R.A. Frederiksen (Eds.), *Sorghum, history, production and technology,* (pp. 349-397). John Wiley and Sons, New York.

Subudhi, P.K. and Nguyen, H.T. 2000b. Linkage group alignment of sorghum RFLP maps using a common RIL mapping population. *Genome* 43(2):240-249.

Sun, Z., Cheng, S., and Hu, S.I. 1993. Determination of critical temperatures and panicle development stage for fertility change of thermosensitive genic male sterile rice line "5460S." *Euphytica* 67(1-2):27-33.

Tao, Y.Z., Jordan, D.R., Henzell, R.G., and McIntyre, C.L. 1998a. Construction of a genetic map in sorghum RIL population using probes from different sources and its comparison with other sorghum maps. *Aust. J. Agric. Res.* 49(5):729-736.

Tao, Y.Z., Jordan, D.R., Henzell, R.G., and McIntyre, C.L. 1998b. Identification of genomic regions for rust resistance in sorghum. *Euphytica* 103(3):287-292.

Taramino, G., Tarchini, R., Ferrario, S., Lee, M., and Pe, M.E. 1997. Characterization and mapping of simple sequence repeats (SSRs) in *Sorghum bicolor. Theor. Appl. Genet.* 95(1-2):66-72.

Tenkouano, A., Miller, F.R., Frederiksen, R.A., and Rosenow, D.T. 1993. Genetics of non-senescence and charcoal rot resistance in sorghum. *Theor. Appl. Genet.* 85(5):644-648.

Thakur, R.P., Frederiksen, R.A., Murty, D.S., Reddy, B.V.S., Bandyopadhyay, R., Giorda, L.M., Odvody, G.N., and Claflin, L.E. 1997. Breeding for disease resistance in sorghum. In *Proceedings of the International Conference on Genetic Improvement of Sorghum and Pearl Millet, September 23-27, 1996* (pp. 303-315). Lubbock, TX, INTSORMIL Publ. No. 97-5.

Tripathi, D.P. 1979. Characterization of diverse cytoplasmic genetic male steriles in sorghum (*Sorghum bicolor* (L.) Moench). Doctoral thesis. Indian Agricultural Research Institute, New Delhi, India.

Tripathi, D.P., Mehta, S.L., Rana, B.S., and Rao, N.G.P. 1980. Characterization of diverse cytoplasmic genetic male steriles in sorghum (*Sorghum bicolor* (L.) Moench). *Sorghum Newsl.* 23:107-108.

Tuinstra, M.R., Grote, E.M., Goldsbrough, P.B., and Ejeta, G. 1996. Identification of quantitative trait loci associated with pre-flowering drought tolerance in sorghum. *Crop Sci.* 36(5):1337-1344.

Tuinstra, M.R., Grote, E.M., Goldsbrough, P.B., and Ejeta, G. 1997. Genetic analysis of post-flowering drought tolerance and components of grain development in *Sorghum bicolor* (L.) Moench. *Mol. Breed.* 3(6):439-448.

van Oosterom, E.J., Jayachandran, R., and Bidinger, F.R. 1996. Diallel analysis of the stay green trait and its components in sorghum. *Crop Sci.* 36(3):549-555.

Vinal, H.N. 1926. A method of crossing sorghums. *J. Hered.* 17(8):297-299.

Vinal, H.N. and Cron, A.B. 1921. Improvement of sorghums by hybridization. *J. Hered.* 12(10):435-443.

Vogel, K.P. 1996. Energy production from forages (American agriculture—back to the future). *J. Soil and Water Conserv.* 51(2):137-139.

Walulu, R.S., Rosenow, D.T., Webster, D.B., and Nguyen, H.T. 1994. Inheritance of the stay-green trait in sorghum. *Crop Sci.* 34(4):970-972.

Webster, O.J. 1976. Sorghum vulnerability and germplasm resources. *Crop Sci.* 16(4):553-556.

Webster, O.J. and Singh, S.P. 1964. Breeding behaviour and histological structure of a non-dehiscent anther character in *Sorghum vulgare* Pers. *Crop Sci.* 4(6):656-658.

Weerasuriya, Y.M. 1995. The construction of a molecular linkage map, mapping of quantitative trait loci, characterization of polyphenols, and screening of genotypes for Striga resistance in sorghum. Doctoral dissertation, Purdue University, West Lafayette, IN, USA.

Wendorf, F., Close, A.E., Schild, R., Wasylikowa, K., Housley, R.A., Harlan, J.R., and Krolik, H. 1992. Saharan exploitation of plants 8,000 bp. *Nature* 359(6397):721-724.

Wheeler, J.L. and Corbett, J.L. 1989. Criteria for breeding forages for improved feeding value: Results of a 146. Delphi survey. *Grass Forage Sci.* 44(1):77-83.

Whitkus, R., Doebly, J., and Lee, M. 1992. Comparative genome mapping of sorghum and maize. *Genetics* 132(4):1119-1130.

Wiseman, B.R., Isenhour, D.J., and Duncan, R.R. 1996. In vitro production of fall armyworm (*Spodoptera frugiperda*)-resistant maize and sorghum plants. In Y.P.S. Bajaj (Ed.), *Biotech. in agriculture and forestry, Vol. 36, Somaclonal variation in crop improvement II* (pp. 67-80). Springer-Verlag, Berlin.

Woo, S.S., Jiang, J., Gill, B.S., Paterson, A.H., and Wing, R.A. 1994. Construction and characterization of a bacterial artificial chromosome library for *Sorghum bicolor. Nucl. Acids Res.* 22(23):4922-4931.

Worstell, J.V., Kidd, H.J., and Schertz, K.F. 1984. Relationships among male sterility inducing cytoplasms of sorghum. *Crop Sci.* 24(1):186-189.

Xu, G.W., Cui, Y.X., Schertz, K.F., and Hart, G.E. 1995. Isolation of mitochondrial DNA sequences that distinguish male sterility inducing cytoplasms in *Sorghum bicolor* (L). Moench. *Theor. Appl. Genet.* 90(7-8):1180-1187.

Xu, G.W., Magill, C.W., Schertz, K.F., and Hart, G.E. 1994. A RFLP linkage map of *Sorghum bicolor* (L). Moench. *Theor. Appl. Genet.* 89(2-3):139-145.

Xu, W., Subudhi, P.K., Crasta, O.R., Rosenow, D.T., Mullet, J.E., and Nguyen, H.T. 2000. Molecular mapping of QTLs conferring stay-green in grain sorghum. *Genome* 43(3):461-469.

Yuan, L.P. 1990. Progress of two-line system of hybrid rice breeding. *Sci. Agric. Sin.* 23(1):1-6.

Zhang, M., Eddy, C., Deanda, K., Finkstein, M., and Picataggio, S. 1995. Metabolic engineering of a pentose metabolism pathway in ethanologenic *Zymomonas mobilis. Science* 267(5195):240-243.

Chapter 6

The Common Bean
and Its Genetic Improvement

Shree P. Singh

ORIGIN, DOMESTICATION, AND DIVERSITY

There are four natural groups of species related to *Phaseolus* that are native to the Americas. These are *Chiapasana, Minkelersia, Phaseolus,* and *Xanthotrichia* (Debouck, 1999). The section *Phaseolus* comprises over 30 different species (Debouck, 1991, 1999; Delgado Salinas, 1985; Maréchal et al., 1978; Westphal, 1974). Although four other species, namely lima (*P. lunatus* L.), scarlet runner (*P. coccineus* L.), tepary (*P. acutifolius* A. Gray), and year-long bean (*P. polyanthus* Greenman) were also domesticated (Gepts and Debouck, 1991), none is as important and popular worldwide as *P. vulgaris* L., the common bean. The common bean is grown on all continents except, of course, Antarctica, and occupies more than 90 percent of production areas sown to *Phaseolus* species in the world.

Diversity among *Phaseolus* species in relation to the common bean is organized into the primary, secondary, and tertiary gene pools (Debouck, 1999; Debouck and Smartt, 1995). The primary gene pool of each cultivated species comprises both the wild populations (i.e., the immediate ancestor of cultigens) and cultivars. The tertiary gene pool of common bean comprises *P. acutifolius* and *P. parvifolius.* While the two species can be crossed without any aid (Singh, Debouck, et al., 1998), crosses of common bean with these species require embryo rescue (Haghighi and Ascher, 1988; Mejía-Jímenez et al., 1994; Singh, Debouck, et al., 1998). The secondary gene pool of common bean includes three species, namely, *P. costaricensis, P. coccineus,* and *P. polyanthus.* The three species cross among themselves and each is crossed with the common bean without embryo rescue, particularly when the common bean is used as the female parent (Baggett, 1956; Camarena and Baudoin, 1987; Cheng et al., 1981; Park and Dhanvantari, 1987; Singh, Debouck, et al., 1997). However, hybrid progenies between crosses of common bean and any of the three species forming the secondary gene pool may be partially sterile, and it may be difficult to recover the desired stable common bean phenotypes (Wall, 1970). Unlike the common bean and species within its tertiary gene pool, members of the secondary gene pool are characterized by a significant amount of outcrossing (from 20 to 40 percent).

Since the 1950s, bean researchers have developed an improved understanding of the origin, domestication, and evolution during domestication in the common bean (Berglund-Brücher and Brücher, 1976; Brücher, 1988; Delgado Salinas et al., 1988; Evans, 1980; Gentry, 1969; Gepts and Debouck, 1991; Gepts et al., 1986; Kami et al., 1995; Kaplan, 1965, 1981; Khairallah et al., 1992; Koenig and Gepts, 1989; Koinange et al., 1996; Miranda Colín, 1967; Smartt, 1969,1988; Weiseth, 1954). Wild populations of common bean are distributed from northern Mexico (Chihuahua) to northeastern Argentina (San Luis) (Toro et al., 1990). The common bean is a noncentric crop with multiple domestication throughout the distribution range of its wild populations in Middle and Andean South America (Gepts et al., 1986). Hybrids between wild and cultivated beans are fully fertile and no known barriers exist for gene introgression and exchange (Koinange et al., 1996; Motto et al., 1978; Singh et al., 1995).

Evans (1973) was the first to recognize the two major groups of common bean germplasm: large-seeded Andean and small-seeded Middle American, and she further divided them into five races. Furthermore, unequivocal evidence for the existence of two gene pools was provided by (1) establishing the relationship between the seed size (small versus large), the Dl genes $(Dl-1$ versus $Dl-2)$ (Shii et al., 1980), and the F_1 hybrid incompatibility (Gepts and Bliss, 1985; Singh and Gutiérrez, 1984); (2) phaseolin seed proteins (Gepts et al., 1986); (3) allozymes (Singh, Nodari, et al., 1991); (4) morphological traits (Singh, Gutiérrez, et al., 1991); and (5) DNA markers (Becerra-Velásquez and Gepts, 1994; Khairallah et al., 1990). Singh (1989) described in considerable detail the patterns of variation among common bean cultigens. Moreover, based on morphological, biochemical, adaptive, and agronomic traits and geographical distribution in their primary centers of origin and domestication, Singh, Gepts, et al. (1991) further divided the Andean and Middle American gene pools into six races: Andean (all large-seeded) = Chile, Nueva Granada, and Peru; Middle American = Durango (medium-seeded semiclimber), Jalisco (medium-seeded climber), and Mesoamerica (all small-seeded).

GROWTH HABITS

Variation in growth habit appears to be continuous from determinate bush to indeterminate, extreme climbing types. Singh (1982), however, classified growth habits into four major classes using the type of terminal bud (vegetative versus reproductive), stem strength (weak versus strong), climbing ability (nonclimber versus strong climber), and fruiting patterns (mostly basal versus along entire stem length or only in the upper part). These are: type I = determinate upright or bush; type II = indeterminate upright bush; type III = indeterminate, prostrate, nonclimbing or viny semi-climbing; and type IV = indeterminate, strong climbers.

USAGE

The cultivars for green-pod harvests are also called French, garden, green, snap, or stringless beans, but here they will be referred to as snap beans. Fully developed green pods of these cultivars have reduced fibers or no fibers in the pod walls and sutures. In some Central and Eastern African and Latin American countries, the young tender leaves or flowers are also harvested as fresh vegetables. However, the largest production (>14 million hectares) and consumption are of dry beans, followed by a much lower level of production for snap bean cultivars. The natives in the highlands of the Andes (certain regions of Peru and Bolivia) roast dry beans. These special popping beans are known as ñuñas. In addition, green leaves, stems, and shelled pods are fed to cattle; and dry plant stubbles are used as feed for cattle, ploughed under to increase soil organic matter, or used as fuel for cooking.

The cultivars harvested for green-shelled beans are often large-seeded cream mottled, red mottled, pink mottled, or white mottled. The distinguishing characteristic of such cultivars is that the pods change color (turn red or purple, with or without stripes) when the fresh seed is ready to be harvested for consumption. Some small- and medium-seeded cultivars of Middle American origin also have attractive pod colors when maturing. The pods not harvested for green-shelled seeds are allowed to mature normally on the plant to be harvested as dry beans. Thus, these are dual-purpose cultivars.

Since the 1870s, and especially within the past 50 years, there has been a major effort to develop snap bean cultivars. Most of the work has been done in Europe and the United States. Dramatic changes and large variations in plant type, maturity, fruiting pattern, and the length, shape, color, fleshiness, and other pod characteristics of snap bean cultivars have resulted. Both determinate bush and indeterminate climbing snap bean cultivars exist. The latter permit multiple harvests (pickings) over a longer period of time and have a much higher yield per unit area of cropped land than their bush-type counterparts. However, they are labor-intensive and require substantial initial investments to install trellises or stakes required for climbing and full plant growth and development. The cultivation of climbing snap bean cultivars is popular in China, home gardens in Europe, winter sowings in

Florida (United States), and near larger cosmopolitan cities in Latin America and other developing nations.

Snap bean cultivars with flat or cylindrical pods, yellow (waxy types) or green colors, and long or short pods, are all used for fresh, frozen, and canning purposes. As far as cylindrical types are concerned, there is a growing demand for smaller, thinner, and darker green pods, especially in France. For details regarding snap beans, readers should refer to Myers and Baggett (1999). The remainder of the discussion in this chapter will refer to dry beans.

Consumer preferences for seed size, color, shape, and brilliance of dry beans vary a great deal (Singh, 1992, 1999; Voysest and Dessert, 1991). But, commonly, they are grouped in small- (< 25g/100-seeds), medium- (25 to 40g/100-seeds), and large-seeded (>40g/100-seeds) types. It is not uncommon to find differences in taste among regions within a country. For example, in northeastern Brazil, light-colored cream and cream-striped beans are preferred, whereas in the southern region, black beans predominate. Similarly, medium-seeded, light-colored beans of different types (e.g., 'bayos', pintos, 'Flor de Mayo', etc.) are preferred in the central highlands of Mexico, whereas in the southern coastal part of the country, small black beans are favored. In Canada and the United States, pinto, red kidney, and small white (navy or pea) beans are preferred. In Africa and Asia, large-seeded beans of various colors, though not black, are popular.

In Latin America, the highest per capita consumption of dry beans is in Brazil and Mexico with more than 13 kg per year. In central and eastern African countries (e.g., Rwanda and Burundi), per capita consumption is over 40 kg per year.

DRY BEAN PRODUCTION

As noted earlier, dry beans are presently grown annually on more than 14 million ha worldwide (Singh, 1999). From its origin and domestication regions in Andean South and Middle America, dry bean production and consumption have expanded into other parts of the Americas (from about 35°S to >50°N latitude and from sea level to >3,000 m altitude), Africa, Asia, Europe, and other parts of the world (Gepts and Bliss, 1988). The Americas (8.3 million ha) are the largest common bean *(P. vulgaris)* producing region. Moreover, Brazil is the largest producer (5.2 million ha) and consumer in the world today.

In the Americas, common bean production in Brazil (2.8 million t) is followed by the United States (1.3 million t), Mexico (1 million t), Argentina, Canada, Colombia, Nicaragua, Honduras, Guatemala, El Salvador, Peru, Haiti, Ecuador, Chile, Venezuela, and the Dominican Republic. Among European and Asian countries (2.1 million ha), China (1.3 million ha), Iran, Japan, and Turkey are the major producers of the common bean. In Africa (3.3 million ha), Burundi, Ethiopia, Malawi, Republic of South Africa, Rwanda, Tanzania, Uganda, and Zimbabwe, among others, form the list of important common bean-producing nations. In Europe (and Asia), Albania, Belarus, Bulgaria, Croatia, Greece, Italy, Moldova Republic, Poland, Romania, Spain, Ukraine, and Yugoslavia are major producers (with a total of 0.6 million ha) of common beans (Singh, 1999).

The common bean is a short-day crop (White and Laing, 1989), and its growth and development are favored by mildly cool environments. Thus, in environments with 16 to 18°C mean growing temperatures with about 12 h daylength, and free from abiotic and biotic stresses, most cultivars complete their growing cycle from germination to seed maturity in 100 to 120 days. However, photoperiod-insensitive cultivars that are successfully grown at higher latitudes (>14 h daylength) in Canada, the United States, Europe, Japan, and other parts of the world have either evolved or been developed by breeding. Most cultivars grown in the highlands of Mexico, Central America, and the Andes are often highly sensitive to long photoperiod and high temperatures and will not complete their growing cycle under long-photoperiod conditions.

Below 2,000 m elevation in tropical and subtropical Latin America and Africa, dry beans are grown twice during the year, often coinciding with the prevalent bimodal rainfall patterns of these regions. In temperate regions at higher latitudes, dry beans are usually grown either as a spring or summer crop. In many regions (e.g., the Middle East, Iran, Europe, Canada, the United States, Chile, and hilly regions of India, Nepal, and Pakistan), dry beans are cultivated in the spring or summer together with other cool-season food legumes, such as chickpeas (*Cicer arietinum* L.), peas (*Pisum sativum* L.), lentils (*Lens culinares* L.), and broad beans (*Vicia faba* L.). Often the latter cool-season food legumes are sown several weeks earlier; bean plantings are done later when soil and air temperatures are relatively higher. Thus, while rainfed cultivation (e.g., in Argentina, central highlands of Mexico, southwestern Canada, and northeast and midwestern United States) may be practiced, especially in areas with more than 400 mm annual rainfall, the dry bean may require supplemental irrigation for secured harvests and higher yields. It may also be grown entirely as an irrigated crop (e.g., in Chile; California, Idaho, and Washington, United States; western Asia; and April to July plantings in central Brazil). In regions with warm or hot summers, beans are grown in the autumn (e.g., northeastern Argentina), spring (Indo-Gangetic plains of India), or winter (in many countries in Africa, Brazil, and the Caribbean).

Whereas in warm tropical regions, the crop is harvested within approximately 75 days, in the highlands (above 2,000 m elevation) of the Andes

(Colombia, Ecuador, Bolivia, and Peru), the climbing bean crop may often take more than 250 days to mature for harvest. In the humid highlands of Guatemala and Mexico, and in Asturias, Spain, climbing cultivars usually take approximately 150 days to mature. At higher latitudes in temperate climates, dry beans of growth habits I, II, and III are harvested within 100 to 120 days from planting.

Cultivars of growth habits I, II, and III are grown in monoculture as well as under different relay, strip, and intercropping systems throughout the world (Singh, 1992; Woolley et al., 1991). The type IV climbing cultivars always require support. Thus, these are grown either in association with maize (*Zea mays* L.), cassava (*Manihot esculenta* Crantz), and other crops, or they are grown on trellises or stakes. Type IV climbing cultivars are popular in regions such as Asturias, Spain, Antioquia, Colombia, or Florida, where highly priced snap beans (McClasan in Florida) or dry bean cultivars (cultivar Cargamanto in Colombia or Faba Granja in Spain) are grown for higher yields and multiple harvests. Although intercropping is often more profitable (Francis and Sanders, 1978) and may be favored for sustainable farming, yield reductions occur with intercropping for cultivars of all growth habits (Clark and Francis, 1985).

Although dry beans are grown in a wide range of soil types, light loamy soils with a pH between 5.5 to 7.0 and rich in organic matter are more suitable for good crop production. A 100- to 120-day crop with a seed yield of 2,500 kg·ha^{-1} will usually remove 60 to 80 kg of soil nitrogen and 40 kg of phosphorus. In acidic soils that are deficient in nitrogen and phosphorus and contain toxic levels of aluminum and/or manganese, it is essential to use appropriate corrective measures. These measures include the adequate use of lime and fertilizers rich in nitrogen and phosphorus, as well as other major and minor elements (Howeler, 1980; Thung, 1990; Thung and de Oliveira, 1998; Thung and Rao, 1999). Similarly, in somewhat alkaline soils, deficiency of microelements (zinc, iron, and boron) is common and use of gypsum, sulfur, and fertilizers rich in these elements is necessary.

As farmers move toward good conservation practices, they are increasingly using minimum or no tillage and organic and green manure, crop rotation, and integrated pest-management practices. These practices enhance and conserve soil fertility and moisture and control soil erosion, weeds, diseases and insects. Moreover, since in most traditional bean-growing regions nodulation and nitrogen fixation are common, use of inoculants with the most effective and competitive *Rhizobium* strains may be promoted while minimizing use of nitrogenous fertilizers at the time of sowing or restricting nitrogen to foliar use only. For high yields (>2,000 kg·ha^{-1}), chemical fertilizers are used in most bean production areas.

PRODUCTION PROBLEMS

The common bean suffers from both abiotic and biotic production constraints (Graham, 1978; Graham and Ranalli, 1997; Schwartz and Pastor-Corrales, 1989; Singh, 1992; Thung and de Oliveira, 1998; Thung and Rao, 1999; Wall, 1973; White et al., 1988; Wortmann et al., 1998; Zaumeyer and Meiners, 1975; Zaumeyer and Thomas, 1957; Zimmermann et al., 1988). Among the abiotic constraints, low soil fertility, in general, and especially nitrogen and phosphorus deficiency are most common. Aluminum and manganese toxicities are frequently problems too (Howeler, 1980; Araya and Beck, 1995; Thung, 1990; Thung and de Oliveira, 1998; Thung and Rao, 1999; Wortmann et al., 1998). Similarly, some form of water stress or drought is a widespread phenomenon throughout most bean production regions (White et al., 1988). In Latin America, drought is common and can be frequent in northeastern Brazil, coastal Peru, and in the central and northern highlands of Mexico. Complete crop failures under dryland conditions are not uncommon in these areas. In regions where the crop is planted toward the end of the rainy season (e.g., September to December in Central America), moderate water stress frequently occurs.

High temperatures (>30° C day and/or 20° C night) during anthesis and seed set in tropical lowlands (below 650 m elevation) and during summer at higher latitudes (e.g., California, Colorado, Idaho, Nebraska, Washington, and Wyoming in the United States), especially when relative humidity is low, can severely limit bean production. Recurring low temperatures (below 10° C) as well as frost during the beginning and end of the growing season in the highlands (above 2,000 m elevation) of Latin America and at higher latitudes (e.g., in the United States and Canada) can also reduce bean yields.

Among bacterial diseases, common bacterial blight [caused by *Xanthomonas campestris* pv. *phaseoli* (Smith) Dye] is a widespread problem from tropical to temperate bean-growing environments (Saettler, 1989; Yoshii et al., 1976). In relatively cooler and wetter areas, halo blight [caused by *Pseudomonas syringae* pv. *phaseolicola* (Burkh.)] and bacterial brown spot (caused by *Pseudomonas syringae* pv. *syringae* van Hall) may cause severe yield losses (Saettler and Potter, 1970).

Among fungal diseases, angular leaf spot [caused by *Phaeoisariopsis griseola* (Sacc.) Ferr.], anthracnose [caused by *Collitotrichum lindemuthianum* (Sacc. and Magnus) Lams-Scrib.], and rust [caused by *Uromyces appendiculatus* (Pers.) Ung.] are considered the most widely distributed foliar fungal diseases that cause severe yield losses of common bean in Latin America, Africa, and other parts of the world. Various root rots (Abawi, 1989; Keenan et al., 1974) in most bean-growing environments, web blight [caused by *Thanatephorus cucumeris* (Frank) Donk.] in the warm humid tropics, and white mold [caused by *Sclerotinia sclerotiorum* (Lib.) de Bary] (Kerr et al., 1978) or ascochyta blight [caused by *Phoma exigua* var. *diversispora* (Bub.) Boerma] in cool wet regions, occasionally become severe on the common bean.

Among viral diseases, bean common mosaic (BCM, caused by a potyvirus) in most bean-production regions and bean golden mosaic (BGM, caused by a geminivirus) in tropical and subtropical Central America, coastal Mexico, the Caribbean, Brazil, and Argentina (Bird et al., 1973; Costa and Cupertino, 1976; Gámez, 1971; Morales and Niessen, 1988) cause severe yield losses in the common bean. Sugar beet curly top in the northwestern United States and bean yellow mosaic in some European countries and the Middle East, North Africa, and Asia can also cause severe yield losses in susceptible cultivars. For more information about diseases, readers should refer to Schwartz and Pastor-Corrales (1989), Zaumeyer and Meiners (1975), and Zaumeyer and Thomas (1957).

Among insects, leafhoppers *Empoasca kraemeri* Ross and Moore (in the tropics and subtropics) and *E. fabae* Harris (in the temperate and cooler environments) are the most widely distributed pests in bean fields, especially in relatively drier areas. Bean pod weevil (*Apion godmani* Wagner and *A. aurichalceum*) causes severe damage to bean pods and seeds in the highlands of Mexico, in Guatemala, El Salvador, Honduras, and Nicaragua (Garza et al., 1996; Guevara-Calderón, 1961). In the highlands of Mexico and in the United States, Mexican bean beetles (*Epilachna varivestis* Mulsant) also cause severe leaf damage, especially in late maturing cultivars. Bean fly (*Ophiomyia phaseoli* Tryon) is by far the most damaging insect to beans in Africa (Karel and Autrique, 1989; Karel and Matee, 1986; Wortmann et al., 1998). The bean weevil *Zabrotes subfasciatus* Boheman (in warm tropical and subtropical environments) and *Acanthoscelides obtectus* (Say) (in cool and temperate environments) cause severe problems when dry beans are not properly stored. Additional information on bean insects can be found elsewhere (Altieri et al., 1978; Cardona, 1989; Karel and Autrique, 1989; Schoonhoven and Cardona, 1980).

Many broadleaf and grassy weeds invade bean fields. The composition of the weed population and the most dominant weeds in bean fields vary from region to region and depend upon several factors. These factors include the growing environments (dry versus wet; warm versus cool), agronomic management of not only the standing bean crop but also other crops grown on the farm, and the history of the fields being used for bean cultivation. Other important factors are the cropping systems, tillage systems (e.g., minimum or no tillage versus conventional methods), growth habit and competitive ability of bean cultivars, planting density (i.e., row spacing and spacing between plants within rows), moisture availability, and the control measures. For estimates of yield losses caused by major biotic and abiotic stresses, readers should refer to Singh (1999).

TRAITS DEFICIENT IN COMMON BEAN

Considering common bean production problems, it may be safe to say that while useful genes for plant architectural and phenological traits and adequate levels of resistance to some factors, such as angular leaf spot, anthracnose, BCM, and BGM, are found in cultigens (Bannerot, 1965; Beebe et al., 1981; Menezes and Dianese, 1988; Morales and Niessen, 1988; Pastor-Corrales et al., 1995, 1998; Schuster et al., 1983; Schwartz et al., 1982; Singh and Muñoz, 1999; Stavely, 1984; White and Laing, 1989), the levels of resistance to other factors are not adequate. Diseases and pests for which present resistance is not adequate include the following: common bacterial blight, halo blight, bacterial brown spot, ascochyta blight, web blight, white mold, bean fly, leafhoppers, and bruchids.

Resistance to bruchids was found in wild *P. vulgaris* (Cardona and Kornegay, 1989; Schoonhoven et al., 1983). Variation was also reported for a few photosynthesis traits in wild common bean (Lynch et al., 1992). *Phaseolus polyanthus* is particularly known for its resistance to ascochyta blight (Schmit and Baudoin, 1992) as well as to white mold (Hunter et al., 1982). *Phaseolus coccineus* has long been known as a source of resistance to anthracnose (Hubbeling, 1957), root rots (Wilkinson, 1983), white mold (Abawi et al., 1978), bean yellow mosaic (Baggett, 1956), and BGM (Beebe and Pastor-Corrales, 1991, CIAT, 1986). *Phaseolus costaricensis* might be interesting for its resistance to BGM (Singh, Roca, et al., 1997). Good sources of tolerance to leafhoppers exist in *P. acutifolius* (CIAT, 1995, 1996). Also, high levels of resistance to common bacterial blight (Coyne et al., 1963; Schuster et al., 1983; Singh and Muñoz, 1999) and bruchids (C. Cardona, unpublished) are found in some accessions of *P. acutifolius*.

HISTORY OF IMPROVEMENT

In Latin America, the first organized breeding program was at the Instituto Agronomico de Campinas (IAC), São Paulo, Brazil. Common bean breeding was probably initiated at IAC in the 1920s (Vieira, 1967, 1988; Voysest, 1983). Pure-line selections in local landraces also were made in Mexico in the 1940 to 1970 period, leading to the release of black bean cultivars, such as 'Actopan,' 'Antigua,' 'Jamapa,' and others (Voysest, 1983). In El Salvador, the black bean cultivars of the 'Porrillo' series were selected in local and introduced germplasm in the 1960s and 1970s. Similarly, a series of selections in landraces was done in Costa Rica. Some of these selections were later introduced and released in Brazil (Voysest, 1983). 'ICA Pijao,' released in Colombia in the 1960s, appears to have been the first popular cultivar developed through hybridization and selection.

In the United States, California, Idaho, Michigan, Nebraska, and New York have the longest history of bean improvement. Genetics and breeding of beans in Nebraska was initiated by R. A. Emerson before 1890 (D. P. Coyne, personal communication, at a conference, August 4, 2000). Michigan was the first state to employ a full-time bean breeder in 1906. In addition to California, Idaho, Michigan, Nebraska, and New York, cultivars have also been developed and released by public or private breeders in Colorado, North Dakota, and Washington. In Canada, evaluation and selection of common bean cultivars was initiated in the 1880s and 1890s (Park and Buzzell, 1995). This work was followed by the development of cultivars using hybridization and selection in the 1920s. For example, breeding programs in Alberta, Ontario, and Saskatchewan have used hybridization and selection to develop and release small-seeded cultivars. Manitoba has initiated a breeding program recently. For history of bean improvement in other regions of the world, readers should refer to Singh (1992).

SELECTION METHODS

Until the first quarter of the twentieth century in Canada and the United States, and the 1950s and 1960s in Latin America and Africa, selection between and within locally available and introduced landraces of the common bean was the predominant method of improvement. This procedure resulted in the release of many important cultivars. These cultivars had a significant impact on bean production for many years. For example, the cultivar Robust was released in Michigan in 1915 (Robertson and Frazier, 1978); 'Actopan,' 'Antigua,' and 'Jamapa' in Mexico; 'Porrillos' in El Salvador; and 'Rio Tibagi' and 'Carioca' in Brazil (Voysest, 1983). Later, the recurrent backcross (Alberini et al., 1983; Miranda et al., 1979; Pompeu, 1980, 1982), pedigree (Kelly, Hosfield, Varner, Uebersax, Brothers, et al., 1994; Kelly, Hosfield, Varner, Uebersax, Haley, et al., 1994), and mass-pedigree (Beebe et al., 1993; Grafton et al., 1993; Singh et al., 1989, 1993) methods and their modifications were used. More recently, congruity backcrossing (Haghighi and Ascher, 1988; Mejía-Jiménez et al., 1994; Urrea and Singh, 1995), single-seed descent (SSD) (Kelly et al., 1989; Urrea and Singh, 1994), recurrent (Beaver and Kelly, 1994; Duarte, 1966; Kelly and Adams, 1987; Singh et al., 1999; Sullivan and Bliss, 1983), and gamete (Singh, 1994, 1998; Singh, Cardona, et al., 1998) selection methods have been used for common bean improvement.

Early generation selection for seed yield in common bean was suggested by Singh and colleagues (1990) and Singh and Urrea (1995). Urrea and Singh (1994) found that the F_2-derived family method of selection was superior to the SSD and bulk methods commonly used for advancing early generations of hybrid populations. From early generation yield tests (F_2 to F_4), Singh and Terán (1998) identified high- and low-yielding populations that eventually produced high- and low-yielding advanced generation (F_7) lines.

Evaluation and selection methods used in early generations of F_1- or F_2-derived families depend upon the objectives of the program, environments, and resources available. For example, in temperate environments where only one field-crop season per year is feasible, breeders often advance material one or two generations in greenhouses or off-season nurseries out of the production region (Singh et al., 1999). Hence, meaningful evaluation for agronomic traits is not feasible during each generation and families could be advanced by taking single-seed or single-pod bulks from all plants within each family. The F_1- or F_2-derived F_4 or F_5 families are then grown in the field for evaluation and selection of promising families and for development of advanced generation lines of commercial value. In tropical and subtropical environments of Latin America (e.g., from coastal Mexico to Brazil) where two or more field nurseries are grown during the year, evaluation and selection for biotic and abiotic stresses and other agronomic traits in each generation help eliminate undesirable recombinants. Promising populations and families thus identified are used to develop lines possessing multiple desirable traits (Singh, Cardona, et al., 1998). Use of biochemical and molecular markers to select families and populations that are harvested in bulk in early segregating generations (F_2 to F_5) may not be feasible now because of the prohibitive costs of screening a large number of plants in each generation. For biotic and abiotic stresses that cannot be screened simultaneously, different locations and nurseries may be required for different generations to select promising populations and families within populations (Singh, Pastor-Corrales, et al., 1991; Singh, Gutiérrez, et al., 1992; Singh, Cardona, et al., 1998).

For traits controlled by recessive genes, such as susceptibility to leaf chlorosis induced by the geminivirus causing BGM (Urrea et al., 1996; Velez et al., 1998) and for seed characteristics, intensive selection in early segregating generations should be avoided. The frequency of desirable recombinants is very low and there is a danger of losing potentially useful recombinants that might arise in later generations. For such traits, it is preferable to initiate evaluation and selection in the F_4 generation. Moreover, before planting the F_5 seeds of selected promising families, all noncommercial seeds harvested from the F_4 plants are discarded. The F_5 is space planted under heavy and uniform pressure from diseases, such as angular leaf spot, anthracnose, rust, and common bacterial blight. Only plants possessing desirable levels of resistance, plant, maturity, and seed characteristics should be selected and harvested individually. These are then progeny tested (i.e., plant-to-row sowing) in the F_6. Those found uniform (or true breeding) for all desirable traits are bulk-harvested for seed increase in the F_7 and subsequent evaluations.

Use of a separate, complementary nursery for each of the major biotic and abiotic stresses, and adaptation, seed yield, and seed quality helps eliminate all entries that are inferior to the available elite lines and cultivars (Singh, 1992). Only superior lines of potential commercial value are advanced for further intensive evaluations in diverse environments and growers' fields and for seed increase. Thus, truly superior lines are selected as new cultivars for eventual release and use by growers, processors, traders, consumers, and the agricultural industry.

Most, if not all, commonly used crop-breeding methods have been employed with the common bean. In spite of this work, objective data comparing the efficiency of different selection methods, with some exceptions (Beaver and Kelly, 1994; Gutiérrez and Singh, 1992; Singh and Terán, 1998; Urrea and Singh, 1994, 1995), are scarce. Given the immense diversity in bean types (market classes), growing environments, production problems, breeding objectives, and resources available to common bean researchers, no breeding method will probably be equally suitable for all circumstances. Therefore, breeders must use a separate selection method or combine two or more methods to suit their needs.

No attempts are made here to describe or review various breeding methods currently in use with the common bean, because their details can be found in standard plant-breeding books and research articles (e.g., Fouilloux and Bannerot, 1988). Based on my experience in common bean improvement, the recurrent and congruity inbred-backcrossing, recurrent selection, and gamete selection methods or their modifications could be suggested for the introgression of useful genes from alien germplasm, for parental development and gene pyramiding, and for cultivar development, respectively. Because gamete selection (Singh, 1994, 1998) facilitates the integration of marker-assisted and conventional selection methods for simultaneous improvement of multiple qualitative and quantitative agronomic traits, and it is comparatively less known, a brief account of this technique will be given later in the chapter.

GENETIC PROGRESS ACHIEVED

Major achievements realized through breeding in Canada and the United States include earliness, adaptation to higher latitude, high yield, upright plant type, combination of *bc-3* and *I* genes for resistance to BCM, and rust and anthracnose resistance (Adams, 1982; Coyne et al., 1994; Grafton et al., 1993, 1997, 1999; Kelly, Hosfield, Varner, Uebersax, Brothers, et al., 1994; Kelly, Hosfield, Varner, Uebersax, Haley, et al., 1994; Kelly, Hosfield, Varner, Uebersax, Afandor, et al., 1995; Myers et al., 1991; Park et al., 1999). In the tropics and subtropics of Latin America, substantial progress has been achieved with efforts to incorporate resistance to several diseases and certain traits. These diseases and traits include BCM, BGM, common bacterial blight, leafhopper, bruchid, and apion, and upright plant type in red and black beans for Central America (Beebe and Pastor-Corrales, 1991; Beebe et al., 1993; Kornegay and Cardona, 1990, 1991). Most of these traits, with the exception of apion resistance (because it was not required), also have been bred into cream, cream-striped, and beige types for Brazil. In addition, high levels of tolerance to water stress and resistance to angular leaf spot and anthracnose were bred into Brazilian bean types (Alberini et al., 1983; Miranda et al., 1979; Pompeu, 1980, 1982; Singh, 1995b; Singh, Gutiérrez, et al., 1991; Singh, Urrea, et al., 1992; Singh, Debouck, et al., 1998; Thung et al., 1993).

Although desirable genes for individual traits have been incorporated into improved breeding lines and cultivars, resistance to more than two or three production-limiting factors has seldom been combined with high-yield and desirable plant, seed, and adaptation characteristics into one cultivar. This result is largely due to breeding efforts that have often focused on a single trait. Furthermore, a team approach and integrated genetic improvements were not practiced. These factors are the reasons that most released cultivars have not become popular and the impact of common bean breeding has not been adequately realized. Nonetheless, in the past few years, at institutions such as CIAT, where a team of researchers from different disciplines has worked together, it has been possible to breed simultaneously for multiple traits in some bean types, such as carioca and mulatinho (Singh, Urrea, et al., 1992; Singh, Cardona, et at., 1998; Thung et al., 1993).

DIFFICULTIES AND CHALLENGES

Lack of adoption of improved common bean cultivars is not uncommon in most countries. This inaction occurs because new cultivars may not consistently outyield the existing cultivars, may be late maturing, may not possess the desired seed characteristics, or may carry resistance to only one or two factors at most, whereas many production-limiting factors may occur in the same region. Thus, the full potential of improved cultivars is often not realized in growers' fields.

A large number of germplasm accessions is available for the primary, secondary, and tertiary gene pools of common bean (Debouck, 1999). Despite this, the genetic base of cultivars within each market class is extremely narrow (Adams, 1977; McClean et al., 1993; Voysest et al., 1994). This is because only a small fragment (<5 percent) of the available genetic diversity has been used globally (Singh, 1992) despite nearly a century of organized common bean improvement (Robertson and Frazier, 1978), and some bean breeders continue to breed cultivars for only one or two traits at a time. In addition, emphasis is placed on individual breeding projects instead of integrated genetic improvement.

Significant advances have been made in germplasm collection, characterization, and screening methods, including identification of useful genes and different types of markers for their direct and indirect selection. Alternative recombination and selection methods have also been developed. On the other hand, there has been a growing concern for conservation and efficient use of natural resources including agrobiodiversity, sustainable agriculture, and production and consumption of healthy and so-called organic food. Moreover, to sustain families on the farm and keep pace with ever-growing demographic developments, yield per unit of cropped land must be maximized, production costs reduced, and farming made attractive to increasingly demanding younger generations.

Thus, it is essential to develop broadly adapted, high-yielding, high-quality common bean cultivars that are less dependent on water, fertilizers, pesticides, and labor in the shortest time possible. This will be possible only by doing the following: maximizing genetic gains from selection for specific traits, and simultaneously improving the maximum number of agronomic traits by accumulating all the favorable alleles from the crop's cultivated races, gene pools, and wild populations forming its primary, secondary, and tertiary gene pools.

INTEGRATED GENETIC IMPROVEMENT

Breeding Objectives

Incorporating high yield (>2,500 kg/ha) and high seed quality in early maturing (<100 days from planting to harvest maturity), upright cultivars suitable for direct mechanical harvest has been a long-term breeding objective of most programs around the world. In temperate North America, other desirable traits include resistance to BCM, rust, common bacterial blight, halo blight, bacterial brown spot, anthracnose, white mold, and root rot for breeding programs seeking widely adapted bean cultivars. Water stress, low soil fertility, and cold and high temperature tolerance may also be required for at least some of the northern and northwestern United States and southern and southwestern Canada where beans are grown without irrigation.

Diseases of strategic importance for which resistance is required for Central America, Cuba, Mexico, and Venezuela include BCM, BGM, common bacterial blight, anthracnose, angular leaf spot, and rust. For Brazil and Argentina, white mold would be added to the above list. In addition, root rot and/or nematode resistance may be required for intensive large-scale mono-culture farming systems in Brazil and for coastal Peru. For insect problems, resistance to leafhoppers and bruchids for the entire region and bean pod weevil in Central American and Mexican cultivars is being sought. Some degree of tolerance to water stress and low soil fertility, especially to acidic soils deficient in phosphorus, nitrogen, and minor elements and sometimes possessing toxic levels of aluminum and manganese is also needed for Brazil, Costa Rica, Cuba, and Mexico, among others. The availability of early-maturing cultivars (< 70 days to maturity) with all of the above men-tioned resistances would facilitate increased adoption of improved cultivars in most countries.

To bring research and development benefits to subsistence farmers and the poorest of the poor who are growing climbing bean landraces in intercropping systems in Latin America and elsewhere, it is imperative to undertake their improvement. Most of the objectives for resistance breeding for biotic and abiotic stresses may be similar to those for less aggressive growth habit types I, II, and III cultivars for monoculture cropping systems mentioned earlier. In addition, due consideration would need to be given to intergenotypic and interspecific competition among major crops prevalent in the region for intercropping (Clark and Francis, 1985; Francis, 1981; Francis et al., 1978; Hamblin and Zimmermann, 1986; Kawano and Thung, 1982; Woolley et al., 1991).

The Three-Tiered Breeding Approach

For the integrated genetic improvement of common bean cultivars, a three-tiered breeding approach suggested by Kelly and colleagues (1997, 1998, 1999) will be described. This involves: (1) useful gene introgression from alien germplasm, (2) pyramiding genes for specific traits and parental development, and (3) cultivar development.

Gene Introgression from Alien Germplasm

Large differences in genetic distance occur between different *Phaseolus* species and *P. vulgaris* (Debouck, 1999; Debouck and Smartt, 1995) and be-tween gene pools and races within the common bean cultigens (Gepts and Bliss, 1985; Singh, 1989; Singh, Gepts, et al., 1991). Because of this and be-cause different breeding methods and strategies are required, introgression of useful genes from each major distantly related cultivated race, gene pool, wild population, and alien species from the secondary and tertiary gene pools must be accomplished separately.

Examples of gene introgression achieved between races within a gene pool include (1) upright plant architectural traits introgressed from race

Mesoamerica to race Durango (Grafton et al., 1997, 1999; Kelly and Adams, 1987; Kelly et al., 1990; Kelly, Hosfield, Varner, Uebersax, Miklas, et al., 1992; Kelly, Hosfield, Varner, Uebersax, Haley, et al., 1992; Park et al., 1999), (2) transfer of *Apion godmani* resistance from race Jalisco to race Mesoamerica (Beebe et al., 1993), and (3) introgression of recessive *bgm-1* gene resistance to BGM from race Durango to race Mesoamerica (Morales and Singh, 1993; Urrea et al., 1996; Velez et al., 1998).

Useful genes have only rarely been transferred between the Andean and Middle American gene pools. Nonetheless, there are a few recent examples of introgression of useful genes between these two gene pools, including resistance to BCM, BGM, rust, anthracnose, and leafhoppers. For example, both dominant *I* and recessive *bc-3* gene resistances for BCM have been transferred from small-seeded race Mesoamerica cultivars to large-seeded Nueva Granada types (F. J. Morales, 1996, personal conversation as co-workers at CIAT, Cali, Colombia). Similarly, resistance to leafhoppers has been introgressed from race Mesoamerica to Nueva Granada (Kornegay and Cardona, 1990). High levels of resistance to leaf chlorosis caused by BGM and controlled by the *bgm-1* gene were initially found in race Durango. This resistance has been transferred into large-seeded red and red-mottled Andean beans of race Nueva Granada for the Caribbean countries (Beaver et al., 1999) and into snap beans (McMillan et al., 1998).

Resistance to bruchids *(Z. subfasciatus)* was not found in thousands of accessions of cultivated common bean that were screened (Schoonhoven and Cardona, 1982). Only a few wild bean populations from the highlands of Mexico were found to be resistant (Schoonhoven et al., 1983). This extremely high level of resistance to bruchids has been successfully transferred from the Mexican wild bean populations to a range of cultivars (Cardona and Kornegay, 1989; Cardona et al., 1989, 1990).

Researchers have successfully used backcrossing to the recurrent common bean parent for introgression of resistance to common bacterial blight from *P. coccineus,* a member of the secondary gene pool (Freytag et al., 1982; Miklas et al., 1994; Park and Dhanvantari, 1987). High levels of resistance to common bacterial blight from the initial common × tepary bean (tertiary gene pool) cross (Coyne et al., 1963), as well as from the subsequent interspecies crosses (McElroy, 1985; Scott and Michaels, 1992; Singh and Muñoz, 1999) have been transferred to the common bean. However, introgression of resistance to leafhoppers, bruchids, heat, and water stress still needs to be achieved. Use of congruity backcrossing, production of large interspecies hybrid progenies from plant-to-plant paired pollinations at each step of crossing, and development of a large number of inbred lines before subjecting them to appropriate screenings may be helpful. Similarly, molecular marker-assisted recurrent and congruity inbred-backcrossings would expedite and permit more focused introgression of useful genes from alien-cultivated and wild germplasm forming the common bean's primary, secondary, and tertiary gene pools.

Pyramiding Genes for Specific Traits and Parental Development

For traits for which there is convincing and unequivocal evidence for the existence of two or more complementary useful genes, pyramiding those genes into a common genotype would help maximize the character expression or gains from selection. Pyramiding would also broaden the genetic base of cultivars and increase the durability of resistance, especially when it is caused by variable pathogens (e.g., angular leaf spot, anthracnose, BCM, and rust diseases). In such circumstances, it is essential to pyramid all available useful genes from within and across cultivated races and gene pools and wild populations of common bean, and from its secondary and tertiary gene pools. Often gene pyramiding into a common bean genotype follows after successful gene introgression from an alien germplasm has been accomplished.

Pyramiding multiple genes for specific traits has been achieved in these situations: (1) from within a given cultivated race of common bean (e.g., leafhopper resistance, Kornegay et al., 1989), (2) from among races within a primary gene pool (e.g., BCM, Kelly, Afandor, et al., 1995; seed yield and water stress tolerance, Singh, 1995a, 1995b; Singh et al., 1989, 1993, 1999), and (3) from across the two cultivated primary gene pools within *P. vulgaris* (e.g., rust resistance, Stavely and Grafton, 1989; Wood and Keenan, 1982), and (4) from across *Phaseolus* species (e.g., common bacterial blight, Singh and Muñoz, 1999).

Cultivar Development

A separate integrated genetic improvement program for cultivars of each major market class and production region is essential. High-yielding cultivars with the maximum expression of each trait and a combination of the maximum number of desirable traits must be sought in each successive breeding cycle. The salient features common to the integrated genetic improvement of cultivars of all market classes, using the gamete selection method (Singh, 1994, 1998; Singh, Cardona, et al., 1998) will be briefly discussed here.

Parental Selection and Programming Crosses. All commercial cultivars, elite lines, and reliable donor parents of necessary genes (including those obtained from introgression from alien germplasm and pyramiding useful complementary genes) are selected for hybridization. While selecting parents for specific cross combinations, due consideration is given to seed, plant, and adaptation characteristics of the parents, in addition to resistance to biotic and abiotic stresses and other useful traits that they might possess.

For each market class, all parents or the maximum number of possible parents must be similar to the specific cultivar groups under improvement in growth habit, maturity, seed color, and size, and all must be well adapted. Thus, each cross is made among only high-yielding, well-adapted, elite recipient and donor parents. Use of distantly related, poorly adapted, or noncommercial parents—especially parents belonging to another gene pool—generally are not used for immediate development of superior cultivars of specific market classes.

The other factor that must be considered is that the simultaneous improvement of the maximum number of traits is sought in each selection cycle. Thus, when the necessary genes for each of the major traits of interest are found in separate parents, biparental crosses and backcrosses are not adequate. A few multiple-parent crosses are preferred over a large number of single crosses and backcrosses. This is to reduce the time required for cultivar development and to combine all desirable alleles in the first step. Although comparatively more time is spent during hybridization to generate multiple-parent crosses, the process allows production of recombinants with favorable alleles for multiple traits, something that is not possible through single crosses and backcrosses without repeated cycles of selection for specific traits, one at a time. For example, if simultaneous selection for resistance to anthracnose, BCM, bean rust, and bruchids is sought, a four-way cross involving all four donor parents is made first. The double-cross F_1 hybrid thus developed then serves as the pollinator parent for the cultivar or elite lines to be improved. Often it is advisable to assure between 10 to 25 percent genetic contribution from each donor-parent of useful genes in a final multiple-parent cross. Moreover, a large number of plant-to-plant pollinations are made at each step of multiple-parent cross development to assure adequate sampling of gametes and genetic contribution of each parent involved in the final crosses.

Selection in Multiple-Parent F_1. When multiple-parent F_1 crosses are made, gamete selection as proposed by Singh (1994, 1998), using dominant and codominant morphological, biochemical, and DNA-based markers in heterogametic and heterogeneous crosses during hybridization and development of multiple-parent F_1 crosses helps assure accumulation of all necessary alleles early on. This process reduces the population size in the subsequent segregating generations (i.e., F_2 onwards) and maximizes the probability of developing new, high-yielding superior cultivars possessing the maximum number of desirable traits (Singh, Cardona, et al., 1998). The selected F_1 plants from the final multiple-parent crosses are each harvested separately to develop F_1-derived F_2 families.

Selection Among F_1-Derived Families. Gamete selection in the F_1, when combined with early generation (F_2 to F_4) evaluation and selection, helps identify promising populations and families within populations. These are then used to develop superior lines for subsequent evaluations and new cultivar selection. Both qualitative and quantitative traits between and within F_1-derived families are selected, using complementary nurseries for each major biotic and abiotic stress. Seed yield is determined in diverse environments representative of the bean production regions. The effectiveness of this selection strategy among F_1-derived families for resistance to angular leaf spot, anthracnose, BGM, common bacterial blight (CBB), and leafhoppers, in F_2 to F_4 of multiple-parent populations was demonstrated by Singh, Cardona, et al. (1998). All 127 F_1-derived families from one population (BZ 9780) were discarded by the F_4 generation. Out of 460 F_1-derived families from another population (GX 9792), only 17 families that segregated for all five resistances were retained through the F_4 generation. Although evaluation and selection for seed yield was not practiced in that study, Singh and Terán (1998) demonstrated the feasibility and effectiveness of yield testing in early generations of the F_1-derived families between and within populations of common bean. Thus, by combining gamete selection in multiple-parent F_1 crosses and judicious evaluation and selection in F_1-derived families in early generations, one should expect increased efficiency and maximize gains from selection.

Development and Evaluation of Lines. To develop advanced breeding lines possessing multiple desirable traits and to conform seed, plant, and adaptation characteristics of common bean, only selected promising families are used. The F_5 seed harvested from the F_4 plants of promising families, after discarding all noncommercial seeds, is space planted under appropriate disease pressure (e.g., anthracnose, BCM, and/or rust) for evaluation and selection of the maximum number of resistant plants with commercially acceptable seeds of good culinary quality.

All selected plants in the F_5 are harvested individually for progeny tests (i.e., plant-to-row sowing) in the F_6 and seed increase in F_7. These experimental lines are evaluated through adaptation and complementary nurseries for each major biotic and abiotic stress. Evaluation includes seed quality characteristics and yield trials across contrasting sites, including in the growers' fields, to identify new cultivars (Singh, 1992).

CONCLUSIONS AND PROSPECTS

There are over 29,000 cultivated and more than 1,300 wild accessions of *P. vulgaris* housed in the germplasm bank at CIAT, Cali, Colombia, and elsewhere. At CIAT, the number of accessions belonging to its secondary, tertiary, and quaternary gene pools, respectively, is 1,049, 325, and 248 (Debouck, 1999). Despite this diversity, the genetic base of commercial cultivars of specific market classes is narrow (Adams, 1977; McClean et al.,

1993; Voysest et al., 1994). Moreover, the average global yield of common bean remains low (<900 kg·ha⁻¹) and its production suffers from a wide range of both biotic and abiotic constraints, some causing up to 100 percent yield losses.

In the Latin American tropics and subtropics, a large proportion (>22,000 accessions) of cultivated accessions of common bean have been evaluated systematically for their reaction to important diseases, insects, and response to long photoperiod, water stress, and low soil fertility. Similar evaluations of wild accessions of common bean and all other related species have lagged behind. The situation may be more accentuated in the temperate environments of North America, Europe, Africa, and Asia. Nonetheless, it is widely known that common bean cultigens are deficient in many agronomic traits, including resistance to storage insects, leafhoppers, ascochyta blight, common bacterial blight, and white mold. Useful genes imparting resistance to these factors have been identified in wild beans (primary gene pool) and related species forming the secondary and tertiary gene pools.

Until recently, common-bean breeders and geneticists have tended to confine hybridization and selection within a specific market class. Moreover, despite nearly a century of breeding history (Robertson and Frazier, 1978), less than 5 percent of the available germplasm has been used in hybridization programs (Singh, 1992). To maximize and sustain bean production, it is essential to develop high-yielding, high-quality cultivars that are less dependent on water, fertilizer, pesticides, and manual labor. This need warrants a comprehensive integrated genetic improvement program in which all favorable alleles from cultivated and wild populations of common bean's primary, secondary, and tertiary gene pools are accumulated in superior cultivars of each major market class. To facilitate this task, it will be essential to carry out three major breeding activities. These activities are: (1) introgression of useful genes from distantly related races and gene pools of cultivated common bean and from its wild populations and related species (i.e., interspecific hybridization), (2) pyramiding of favorable alleles for specific traits from different species (i.e., parental development), and (3) cultivar development for specific market classes of beans. Germplasm-recombination and selection methods will vary depending upon the genetic distance between parents, breeding objectives, and available resources. Nonetheless, recurrent and congruity inbred-backcrossing, recurrent selection, and gamete selection methods are suggested, respectively, for each of the above three breeding activities.

Availability of an efficient and repeatable transformation system for *P. vulgaris* and integrated linkage maps, use of the knowledge of genetics of domestication and evolution, and development and use of marker-assisted selection should expedite and facilitate need-based, integrated genetic improvement of the common bean.

The diversity in common bean types, their multipurpose usage in a variety of food preparations, their wide adaptation, ability to be grown in different cropping systems, and ability to fix biological nitrogen make them a valuable food crop worldwide. These properties, in addition to the highly nutritive protein content of dry beans, when combined with cereal grains and their ability to help prevent cancer and lower cholesterol, should help increase the demand for them on a global basis. In developed countries, such as the United States, the per capita bean consumption has been increasing steadily. As awareness of the above properties of the common bean increases, consumption is likely to continue to rise. Similarly, diversification and publicity for processed bean products (e.g., chips for snacks, instant beans for cream or soups), availability of precooked beans in different forms, and internationalization of food habits could increase global bean production and consumption. Other important factors are the use of beans in baby foods, the reduction of flatulence, antinutritional factors, and cooking time.

REFERENCES

Abawi, G.S. 1989. Root rots. In H.F. Schwartz and M.A. Pastor-Corrales (Eds.), *Bean production problems in the tropics,* Second edition (pp. 105-157). CIAT, Cali, Colombia.

Abawi, G.S., Provvidenti, R., Crosier, D.C., and Hunter, J.E. 1978. Inheritance of resistance to white mold disease in *Phaseolus coccineus. J. Hered.* 69(3):200-202.

Adams, M.W. 1977. An estimation of homogeneity in crop plants with specific reference to genetic vulnerability in the dry bean *Phaseolus vulgaris* L. *Euphytica* 26(3):665-679.

Adams, M.W. 1982. Plant architecture and yield breeding in *Phaseolus vulgaris* L. *Iowa State J. Res.* 56(3):225-254.

Alberini, J.L., Kranz, M.W., Oliari, L., and Bianchini, A. 1983. 'IAPAR 5—Rio Piquiri' e 'IAPAR 7—Rio Vermelho,' novas variedades de feijoeiro para o estado do Paraná. *Pesq. Agropec. Bras.* 18(4):393-397.

Altieri, M.A., Francis, C.A., van Schoonhoven, A., and Doll, J.D. 1978. A review of insect prevalence in maize (*Zea mays* L.) and bean (*Phaseolus vulgaris* L.) polyculture systems. *Field Crops Res.* 1(1):33-49.

Araya, R. and Beck, D. (Eds.), 1995. *Memoria del Taller Internacional sobre Bajo Fosforo en el Cultivo de Frijol.* Universidad de Costa Rica, San Jose, Costa Rica.

Baggett, J.R. 1956. The inheritance of resistance to strains of bean yellow mosaic virus in the interspecific cross *Phaseolus vulgaris* × *P. coccineus. Plant Dis. Rep.* 40(7):702-707.

Bannerot, H. 1965. Résultats de l'infection d'une collection de haricots par six races physiologiques d'anthracnose. *Ann. Amélior. Plant.* 15(2):201-222.

Beaver, J.S. and Kelly, J.D. 1994. Comparison of selection methods for dry bean populations derived from crosses between gene pools. *Crop Sci.* 34(1):34-37.

Beaver, J.S., Zapata, M., and Miklas, P.N. 1999. Registration of PR9443-4 dry bean germplasm resistant to bean golden mosaic, common bacterial blight, and rust. *Crop Science* 39(4):1262.

Becerra-Velásquez, V.L. and Gepts, P. 1994. RFLP diversity of common bean (*Phaseolus vulgaris*) in its centres of origin. *Genome* 37(2):256-263.

Beebe, S.E., Bliss, F.A., and Schwartz, H.F. 1981. Root rot resistance in common bean germplasm of Latin American origin. *Plant Dis.* 65(6):485-489.

Beebe, S., Cardona, C., Diaz, O., Rodríguez, F., Mancía, E., and Ajquejay, S. 1993. Development of common bean (*Phaseolus vulgaris* L.) lines resistant to the pod weevil, *Apion godmani* Wagner, in Central America. *Euphytica* 69(1-2):83-88.

Beebe, S.E. and Pastor-Corrales, M.A. 1991. Breeding for disease resistance. In A. van Schoonhoven and O. Voysest (Eds.), *Common beans: Research for crop improvement* (pp. 561-617). C.A.B. Int., Wallingford, U.K. and CIAT, Cali, Colombia.

Berglund-Brücher, O. and Brücher, H. 1976. The South American wild bean (*Phaseolus aborigineus* Burk) as ancestor of the common bean. *Econ. Bot.* 30(3):257-272.

Bird, J., Sánchez, J., and Vakili, N.G. 1973. Golden-yellow mosaic virus of beans (*Phaseolus vulgaris*) in Puerto Rico. *Phytopathology* 63(12):1435. (Abstr).

Brücher, H. 1988. The wild ancestor of *Phaseolus vulgaris* in South America. In P. Gepts (Ed.), *Genetic resources of* Phaseolus *beans* (pp. 185-214). Kluwer, Dordrecht, Netherlands.

Camarena, F. and Baudoin, J.P. 1987. Obtention des premiers hybrides interspécifiques entre *Phaseolus vulgaris* et *Phaseolus polyanthus* avec le cytoplasme de cette dernière forme. *Bull. Rech. Agron. Gembloux* 22(1):43-55.

Cardona, C. 1989. Insects and other invertebrate bean pests in Latin America. In H.F. Schwartz and M.A. Pastor-Corrales (Eds.), *Bean production problems in the tropics*, Second edition (pp. 505-570). CIAT, Cali, Colombia.

Cardona, C. and Kornegay, J. 1989. Use of wild *Phaseolus vulgaris* to improve beans for resistance to bruchids. In S. Beebe (Ed.), *Current topics in breeding of common beans* (pp. 90-98). CIAT, Cali, Colombia.

Cardona, C., Kornegay, J., Posso, C.E., Morales, F., and Ramírez, H. 1990. Comparative value of four arcelin variants in the development of dry bean lines resistant to the Mexican bean weevil. *Entomol. Expt. Appl.* 56(2):197-206.

Cardona, C., Posso, C.E., Kornegay, J., Valor, J., and Serrano, M. 1989. Antibiosis effects of wild dry bean accessions on the Mexican bean weevil and the bean weevil (Coleoptera: Bruchidae). *J. Econ. Entomol.* 82(1):310-315.

Cheng, S.S., Bassett, M.J., and Quesenberry, K.H. 1981. Cytogenetic analysis of interspecific hybrids between common bean and scarlet runner bean. *Crop Sci.* 21(1):75-79.

CIAT. 1986. *Bean program annual report.* Centro Internacional de Agricultura Tropical, Cali, Colombia.

CIAT. 1995. *Bean program annual report.* CIAT, Cali, Colombia.

CIAT. 1996. *Bean program annual report.* CIAT, Cali, Colombia.

Clark, E.A. and Francis, C.A. 1985. Bean-maize intercrops: A comparison of bush and climbing bean growth habits. *Field Crops Res.* 10(2):151-166.

Costa, C.L. and Cupertino, F.P. 1976. Avaliacaõ das perdas na producaõ do feijoeiro causadas pelo virus do mosaico dourado. *Fitopat. Bras.* 1(1):18-25.

Coyne, D.P., Nuland, D.S., Lindgren, D.T., and Steadman. 1994. 'Chase' pinto dry bean. *HortScience* 29(1):44-45.

Coyne, D.P., Schuster, M.L., and Al-Yasiri, S. 1963. Reaction studies of bean species and varieties to common blight and bacterial wilt. *Plant Dis. Rep.* 47(5):534-537.

Debouck, D.G. 1991. Systematics and morphology. In A. van Schoonhoven and O. Voysest (Eds.), *Common beans: Research for crop improvement* (pp. 55-118). C.A.B. Int., Wallingford, U.K. and CIAT, Cali, Colombia.

Debouck, D.G. 1999. Diversity in *Phaseolus* species in relation to the common bean. In S.P. Singh (Ed.), *Common bean improvement in the twenty-first century* (pp. 25-52). Kluwer, Dordrecht, Netherlands.

Debouck, D.G. and Smartt, J. 1995. Beans, *Phaseolus* spp. (Leguminosae-Papilionoideae). In J. Smartt and N.W. Simmonds (Eds.), *Evolution of crop plants,* Second edition (pp. 287-294). Longman, London, U.K.

Delgado Salinas, A. 1985. Systematics of the genus *Phaseolus* (Leguminosae) in North and Central America. Doctoral dissertation, University of Texas, Austin, Texas.

Delgado Salinas, A., Bonet, A., and Gepts, P. 1988. The wild relative of *Phaseolus vulgaris* in Middle America. In P. Gepts (Ed.), *Genetic resources of* Phaseolus *beans* (pp. 163-184). Kluwer, Dordrecht, Netherlands.

Duarte, R.A. 1966. Responses in yield and yield components from recurrent selection practiced in a bean hybrid population in three locations in North and South America. *Diss. Abstr.* 27:1339B-1340B.

Evans, A.M. 1973. Commentary upon: Plant architecture and physiological efficiency in the field bean. In D. Wall (Ed.), *Potential of field beans and other food legumes in Latin America* (pp. 279-286). CIAT, Cali, Colombia.

Evans, A.M. 1980. Structure, variation, evolution, and classification in *Phaseolus.* In R.J. Summerfield and A.H. Bunting (Eds.), *Advances in legume science* (pp. 337-347). Royal Botanic Gardens, Kew, England.

Fouilloux, G. and Bannerot, H. 1988. Selection methods in the common bean (*Phaseolus vulgaris*). In P. Gepts (Ed.), *Genetic resources of* Phaseolus *beans* (pp. 503-542). Kluwer, Dordrecht, Netherlands.

Francis, C.A. 1981. Development of plant genotypes for multiple cropping systems. In K.J. Frey (Ed.), *Plant breeding II* (pp. 179-215). Iowa State Univ. Press, Ames, Iowa.

Francis, C.A., Prager, M., and Laing, D.R. 1978. Genotype × environment interactions in climbing bean cultivars in monoculture and associated with maize. *Crop Sci.* 18(2):242-246.

Francis, C.A. and Sanders, J.H. 1978. Economic analysis of bean and maize systems: Monoculture versus associated cropping. *Field Crops Res.* 1(4):319-335.

Freytag, G.F., Bassett, M.J., and Zapata, M. 1982. Registration of XR-235-1-1 bean germplasm. *Crop Sci.* 22(6):1268-1269.

Gámez, R. 1971. Los virus del frijol en Centroamérica. I. Transmisión por moscas blancas (*Bemisia tabaci* Genn.) y plantas hospedantes del virus del mosaico dorado. *Turrialba* 21(1):22-27.

Garza, R., Cardona, C., and Singh, S.P. 1996. Inheritance of resistance to the bean-pod weevil (*Apion godmani* Wagner) in common beans from Mexico. *Theor. Appl. Genet.* 92(3/4):357-362.

Gentry, H.S. 1969. Origin of the common bean, *Phaseolus vulgaris*. *Econ. Bot.* 23(1):55-69.

Gepts, P. and Bliss, F.A. 1985. F₁ hybrid weakness in the common bean: Differential geographic origin suggests two gene pools in cultivated bean germplasm. *J. Hered.* 76(6):447-450.

Gepts, P. and Bliss, F.A. 1988. Dissemination pathways of common bean (*Phaseolus vulgaris*, Fabaceae) deduced from phaseolin electrophoretic variability. II. Europe and Africa. *Econ. Bot.* 42(1):86-104.

Gepts, P. and Debouck, D. 1991. Origin, domestication, and evolution of the common bean (*Phaseolus vulgaris* L.). In A. van Schoonhoven and O. Voysest (Eds.), *Common beans: Research for crop improvement* (pp. 7-53). C.A.B. Int., Wallingford, U.K., and CIAT, Cali, Colombia.

Gepts, P., Osborn, T.C., Rashka, K., and Bliss, F.A. 1986. Phaseolin protein variability in wild forms and landraces of the common bean (*Phaseolus vulgaris*): Evidence for multiple centers of domestication. *Econ. Bot.* 40(4):451-468.

Grafton, K.F., Chang, K.C., Venette, J.R., and Vander Wal, A.J. 1993. Registration of 'Norstar' navy bean. *Crop Sci.* 33(6):1405-1406.

Grafton, K.F., Venette, J.R., and Chang, K.C. 1997. Registration of 'Maverick' pinto bean. *Crop Sci.* 37(5):1672.

Grafton, K.F., Venette, J.R., and Chang, K.C. 1999. Registration of 'Frontier' pinto bean. *Crop Sci.* 39(3):876-877.

Graham, P.H. 1978. Problems and potentials of field beans (*Phaseolus vulgaris* L.) in Latin America. *Field Crops Res.* 1(4):295-317.

Graham, P.H. and Ranalli, P. 1997. Common bean (*Phaseolus vulgaris* L.). *Field Crops Res.* 53(1/2):131-146.

Guevara-Calderón, J. 1961. El combate del picudo del ejote mediante la combinaicón de variedades resistentes e insecticidas. *Agric. Tec. Mex.* 1(12):17-19.

Gutiérrez, J.A. and Singh, S.P. 1992. Effect of alternative methods of advancing early generation populations of common bean in the tropics. *Turrialba* 42(4):482-486.

Haghighi, K.R. and Ascher, P.D. 1988. Fertile intermediate hybrids between *Phaseolus vulgaris* and *P. acutifolius* from congruity backcrossing. *Sex. Plant Reprod.* 1(1):51-58.

Hamblin, J. and Zimmermann, M.J.O. 1986. Breeding common bean for yield in mixtures. *Plant Breed. Rev.* 4:245-272.

Howeler, R.H. 1980. Nutritional disorders. In H.F. Schwartz and G.E. Gálvez (Eds.), *Bean production problems: Disease, insect, soil and climatic constraints of Phaseolus vulgaris* (pp. 341-362). CIAT, Cali, Colombia.

Hubbeling, N. 1957. New aspects of breeding for disease resistance in beans (*Phaseolus vulgaris* L.). *Euphytica* 6(1):111-141.

Hunter, J.E., Dickson, M.H., Boettger, M.A., and Cigna, J.A. 1982. Evaluation of plant introductions of *Phaseolus* spp. for resistance to white mold. *Plant Dis.* 66(4):320-322.

Kami, J., Becerra Velásquez, V., Debouck, D.G., and Gepts, P. 1995. Identification of presumed ancestral DNA sequences of phaseolin in *Phaseolus vulgaris. Proc. Natl. Acad. Sci. USA.* 92(4):1101-1104.

Kaplan, L. 1965. Archeology and domestication in American *Phaseolus* (beans). *Econ. Bot.* 19(4):358-368.

Kaplan, L. 1981. What is the origin of the common bean? *Econ. Bot.* 35(2):240-254.

Karel, A.K. and Autrique, A. 1989. Insects and other pests in Africa. In H.F. Schwartz and M.A. Pastor-Corrales (Eds.), *Bean production problems in the tropics,* Second edition (pp. 455-504). CIAT, Cali, Colombia.

Karel, A.K. and Matee, J.J. 1986. Yield losses in common beans following damage by bean fly *Ophiomyia phaseoli* Tryon Diptera: Agronyzidae. *Annu. Rep. Bean Improv. Coop.* 29(1):115-116.

Kawano, K. and Thung, M. 1982. Intergenotypic competition and competition with associated crops in cassava. *Crop Sci.* 22(1):59-63.

Keenan, J.G., Moore, H.D., Oshima, N., and Jenkins, L.E. 1974. Effect of bean root rot on dry and pinto bean production in southwestern Colorado. *Plant Dis. Rept.* 58(8):890-892.

Kelly, J.D. and Adams, M.W. 1987. Phenotypic recurrent selection in ideotype breeding of pinto beans. *Euphytica* 36(1):69-80.

Kelly, J.D., Adams, M.W., Saettler, A.W., Hosfield, G.S., Varner, G.V., Beaver, J.S., Uebersax, M.A., and Taylor, J. 1989. Registration of 'Mayflower' navy bean. *Crop Sci.* 29(6):1571-1572.

Kelly, J.D., Adams, M.W., Saettler, A.W., Hosfield, G.S., Varner, G.V., Ubersax, M.A., and Taylor, J. 1990. Registration of 'Sierra' pinto bean. *Crop Sci.* 30(3):745-746.

Kelly, J.D., Afanador, L., and Haley, S.D. 1995. Pyramiding genes for resistance to bean common mosaic virus. *Euphytica* 82(2):207-212.

Kelly, J.D., Hosfield, G.L., Varner, G.V., Uebersax, M.A., Afanador, L.K., and Taylor, J. 1995. Registration of 'Newport' navy bean. *Crop Sci.* 35(6):1710-1711.

Kelly, J.D., Hosfield, G.L., Varner, G.V., Uebersax, M.A., Brothers, M.E., and Taylor, J. 1994. Registration of 'Huron' navy bean. *Crop Sci.* 34(5):1408.

Kelly, J.D., Hosfield, G.L., Varner, G.V., Uebersax, M.A., Haley, S.D., and Taylor, J. 1994. Registration of 'Raven' black bean. *Crop Sci.* 34(5):1406-1407.

Kelly, J.D., Hosfield, G.L., Varner, G.V., Uebersax, M.A., Miklas, P.N., and Taylor, J. 1992. Registration of 'Alpine' great northern bean. *Crop Sci.* 32(6):1509-1510.

Kelly, J.D., Hosfield, G.L., Varner, G.V., Uebersax, M.A., Wassimi, N., and Taylor, J. 1992. Registration of 'Aztec' pinto bean. *Crop Sci.* 32(6):1509.

Kelly, J.D., Kolkman, J.M., and Schneider, K.A. 1997. Breeding for high yield in common bean. In S.P. Singh and O. Voysest (Eds.), *Taller de mejoramiento de frijol para el siglo XXI: Bases para una estrategia para América Latina* (pp. 187-204). CIAT, Cali, Colombia.

Kelly, J.D., Kolkman, J.M., and Schneider, K.A. 1998. Breeding for yield in dry bean (*Phaseolus vulgaris* L.). *Euphytica* 102(3):343-356.

Kelly, J.D., Schneider, K.A., and Kolkman, J.M. 1999. Breeding to improve yield. In S.P. Singh (Ed.), *Common bean improvement in the twenty-first century* (pp. 185-222). Kluwer, Dordrecht, Netherlands.

Kerr, E.D., Steadman, J.R., and Nelson, L.A. 1978. Estimation of white mold disease reduction of yield and yield components of dry edible beans. *Crop Sci.* 18(2):275-279.

Khairallah, M.M., Adams, M.W., and Sears, B.B. 1990. Mitochondrial DNA polymorphisms of Malawian bean lines: Further evidence for two major gene pools. *Theor. Appl. Genet.* 80(6):753-761.

Khairallah, M.M., Sears, B.B., and Adams, M.W. 1992. Mitochondrial restriction fragment length polymorphisms in wild *Phaseolus vulgaris* L.: Insights on the domestication of the common bean. *Theor. Appl. Genet.* 84(7/8):915-922.

Koenig, R. and Gepts, P. 1989. Allozyme diversity in wild *Phaseolus vulgaris*: Further evidence for two major centers of genetic diversity. *Theor. Appl. Genet.* 78(6):809-817.

Koinange, E.M.K., Singh, S.P., and Gepts, P. 1996. Genetic control of the domestication syndrome in common bean. *Crop Sci.* 36(4):1037-1045.

Kornegay, J. and Cardona, C. 1990. Development of an appropriate breeding scheme for tolerance to *Empoasca kraemeri* in common bean. *Euphytica* 47(2):223-231.

Kornegay, J. and Cardona, C. 1991. Breeding for insect resistance in beans. In A. van Schoonhoven and O. Voysest (Eds.), *Common beans: Research for crop improvement* (pp. 619-648). C.A.B. Int., Wallingford, U.K. and CIAT, Cali, Colombia.

Kornegay, J.L., Cardona, C., Esch, J.V., and Alvarado, M. 1989. Identification of common bean lines with ovipositional resistance to *Empoasca kraemeri* (Homoptera: Cicadellidae). *J. Econ. Entomol.* 82(2):649-654.

Lynch, J., González, A., Tohme, J.M., and García, J. 1992. Variation in characters related to leaf photosynthesis in wild bean populations. *Crop Sci.* 32(3):633-640.

Maréchal, R., Mascherpa, J.-M., and Stainier, F. 1978. Etude taxonomique d'un groupe complexe d'especes des generes *Phaseolus* et *Vigna* (Papilionaceae) sur la base de donées morphologiques et polliniques, traitées par l'analyse informatique. *Boissiera* 28(1):1-273.

McClean, P.E., Myers, J.R., and Hammond, J.J. 1993. Coefficient of parentage and cluster analysis of North American dry bean cultivars. *Crop Sci.* 33(1):190-197.

McElroy, J.B. 1985. Breeding for dry beans, *P. vulgaris* L., for common bacterial blight resistance derived from *Phaseolus acutifolius* A. Gray. Doctoral dissertation, Cornell University, Ithaca, NY.

McMillan Jr., R.T., Davis, M.J., McLaughlin, H.J., and Stavely, J.R. 1998. PCR evaluation of fourteen bean golden mosaic virus (BGMV) resistant snap bean germplasm lines for the presence of the virus. *Annu. Rep. Bean Improv. Coop.* 41:31-32.

Mejía-Jiménez, A., Muñoz, C., Jacobsen, H.J., Roca, W.M., and Singh, S.P. 1994. Interspecific hybridization between common and tepary beans: Increased hybrid embryo growth, fertility, and efficiency of hybridization through recurrent and congruity backcrossing. *Theor. Appl. Genet.* 88(3/4):324-331.

Menezes, J.R. and Dianese, J.C. 1988. Race characterization of Brazilian isolates of *Colletotrichum lindamuthianum* and detection of resistance to anthracnose of *Phaseolus vulgaris*. *Phytopathology* 78(6):650-655.

Miklas, P.N., Beaver, J.S., Grafton, K.F., and Freytag, G.F. 1994. Registration of TARS VCI-4B multiple disease resistant dry bean germplasm. *Crop Sci.* 34(5):1415.

Miranda Colín, S. 1967. Origen de *Phaseolus vulgaris* L. (frijol común). *Agrociencia* 1(2):99-109.

Miranda, P., Mafra, R.C., Correia, E.B., and De-Queiroz, M.A. 1979. IPA-79-19 uma nova variedade de feijao mulatinho (*Phaseolus vulgaris* L.) para Pernambuco. *Pes. Agrop. Pernamb.* 3(1):105-111.

Morales, F.J. and Niessen, A.I. 1988. Comparative responses of selected *Phaseolus vulgaris* germplasm inoculated artificially and naturally with bean golden mosaic virus. *Plant Dis.* 72(12):1020-1023.

Morales, F.J. and Singh, S.P. 1993. Breeding for resistance to bean golden mosaic virus in an interracial population of *Phaseolus vulgaris* L. *Euphytica* 67(1):59-63.

Motto, M., Sorresi, G.P., and Salamini, F. 1978. Seed size inheritance in a cross between wild and cultivated common beans (*Phaseolus vulgaris* L.). *Genetica* 49(1):31-36.

Myers, J.R. and Baggett, J.R. 1999. Improvement of snap bean. In S.P. Singh (Ed.), *Common bean improvement in the twenty-first century* (pp. 289-329). Kluwer, Dordrecht, Netherlands.

Myers, J.R., Hayes, R.E., and Kolar, J.J. 1991. Registration of 'UI 906' black bean. *Crop Sci.* 31(6):1710.

Park, S.J. and Buzzell, R.I. 1995. Common bean. In A.E. Slinkard and D.R. Knott (Eds.), *Harvest of gold: The history of field crop breeding in Canada* (pp. 1-16). University of Saskatchewan, Saskatoon, Canada.

Park, S.J. and Dhanvantari, B.N. 1987. Transfer of common blight (*Xanthomonas campestris* pv. *phaseoli*) resistance from *Phaseolus coccineus* Lam. to *P. vulgaris* L. through interspecific hybridization. *Can. J. Plant Sci.* 67(3):685-695.

Park, S.J., Kiehn, F., and Rupert, T. 1999. AC Ole common bean. *Can. J. Plant Sci.* 79(1):107-108.

Pastor-Corrales, M.A., Jara, C., and Singh, S.P. 1998. Pathogenic variation in, sources of, and breeding for resistance to *Phaeoisariopsis griseola* causing angular leaf spot in common bean. *Euphytica* 103(2):161-171.

Pastor-Corrales, M.A., Otoya, M.M., Molina, A., and Singh, S.P. 1995. Resistance to *Colletotrichum lindemuthianum* isolates from Middle America and Andean South America in different common bean races. *Plant Dis.* 79(1):63-67.

Pompeu, A.S. 1980. Yields of French bean lines (*Phaseolus vulgaris* L.) of the rosinha and roxinho groups resistant to *Colletotrichum lindemuthianum*. *Bragantia* 39(1):89-97.

Pompeu, A.S. 1982. Catu, Aete-3, Aroana-80, Moruna-80, Carioca-80 e Ayso: Novos cultivares de feijoeiro. *Bragantia* 41(2):213-218.

Robertson, L.S. and Frazier, R.D. (Eds.), 1978. Dry bean production: Principles and practices. *Michigan State University Ext. Bull. E-1251.* East Lansing, MI.

Saettler, A.W. 1989. Common bacterial blight. In H.F. Schwartz and M.A. Pastor-Corrales (Eds.), *Bean production problems in the tropics,* Second edition (pp. 261-283). CIAT, Cali, Colombia.

Saettler, A.W. and Potter, H.S. 1970. Chemical control of halo bacterial blight in field beans. P.1-8, *Michigan Agric. Exp. Sta. Res. Rept. 98.* East Lansing, Michigan.

Schmit, V. and Baudoin, J.P. 1992. Screening for resistance to *Ascochyta* blight in populations of *Phaseolus coccineus* L. and *P. polyanthus* Greenman. *Field Crops Res.* 30(2):155-165.

Schuster, M.L., Coyne, D.P., Behre, T., and Leyna, H. 1983. Sources of *Phaseolus* species resistance and leaf and pod differential reactions to common blight. *HortScience* 18(6):901-903.

Schwartz, H.F. and Pastor-Corrales, M.A. (Eds.) 1989. *Bean production problems in the tropics,* Second edition. CIAT, Cali, Colombia.

Schwartz, H.F., Pastor-Corrales, M.A., and Singh, S.P. 1982. New sources of resistance to anthracnose and angular leaf spot of beans (*Phaseolus vulgaris* L.). *Euphytica* 31(3):741-754.

Scott, M.E. and Michaels, T.E. 1992. *Xanthomonas* resistance of *Phaseolus* interspecific cross selections confirmed by field performance. *HortScience* 27(4):348-350.

Shii, C.T., Mok, M.C., Temple, S.R., and Mok, D.W.S. 1980. Expression of developmental abnormalities in hybrids of *Phaseolus vulgaris* L.: Interaction between temperature and allelic dosage. *J. Hered.* 71(4):218-222.

Singh, S.P. 1982. A key for identification of different growth habits of frijol *Phaseolus vulgaris* L. *Annu. Rep. Bean Improv. Coop.* 25:92-95.

Singh, S.P. 1989. Patterns of variation in cultivated common bean (*Phaseolus vulgaris*, Fabaceae). *Econ. Bot.* 43(1):39-57.

Singh, S.P. 1992. Common bean improvement in the tropics. *Plant Breed. Rev.* 10:199-269.

Singh, S.P. 1994. Gamete selection for simultaneous improvement of multiple traits in common bean. *Crop Sci.* 34(2):352-355.

Singh, S.P. 1995a. Selection for seed yield in Middle American versus Andean × Middle American interracial common-bean populations. *Plant Breed.* 114(3):269-271.

Singh, S.P. 1995b. Selection for water-stress tolerance in interracial populations of common bean. *Crop Sci.* 35(1):118-124.

Singh, S.P. 1998. Uso de marcadores y selección de gametos para el mejoramiento simultáneo de caracteres múltiples de frijol (*Phaseolus vulgaris* L.) para Mesoamérica y el Caribe. *Agron. Mesoam.* 9(1):1-9.

Singh, S.P. 1999. Production and utilization. In S.P. Singh (Ed.), *Common bean improvement in the twenty-first century* (pp. 1-14). Kluwer, Dordrecht, Netherlands.

Singh, S.P., Cajiao, C., Gutiérrez, J.A., García, J., Pastor-Corrales, M.A., and Morales, F.J. 1989. Selection for seed yield in inter-gene pool crosses of common bean. *Crop Sci.* 29(5):1126-1131.

Singh, S.P., Cardona, C., Morales, F.J., Pastor-Corrales, M.A., and Voysest, O. 1998. Gamete selection for upright carioca bean with resistance to five diseases and a leafhopper. *Crop Sci.* 38(3):666-672.

Singh, S.P., Debouck, D.G., and Roca, W.M. 1997. Successful interspecific hybridization between *Phaseolus vulgaris* L. and *P. costaricensis* Freytag and Debouck. *Annu. Rep. Bean Improv. Coop.* 40:40-41.

Singh, S.P., Debouck, D.G., and Roca, W.M. 1998. Interspecific hybridization between *Phaseolus vulgaris* L. and *P. parvifolius* Freytag. *Annu. Rpt. Bean Improv. Coop.* 41:7-8.

Singh, S.P., Gepts, P., and Debouck, D.G. 1991. Races of common bean (*Phaseolus vulgaris*, Fabaceae). *Econ. Bot.* 45(3):379-396.

Singh, S.P. and Gutiérrez, J.A. 1984. Geographical distribution of *DL1* and *DL2* genes causing hybrid dwarfism in *Phaseolus vulgaris* L., their association with seed size, and their significance to breeding. *Euphytica* 33(2):337-345.

Singh, S.P., Gutiérrez, J.A., Molina, A., Urrea, C., and Gepts, P. 1991. Genetic diversity in cultivated common bean. II. Marker-based analysis of morphological and agronomic traits. *Crop Sci.* 31(1):23-29.

Singh, S.P., Gutiérrez, J.A., Urrea, C.A., Molina, A., and Cajiao, C. 1992. Location-specific and across-location selections for seed yield in populations of common bean, *Phaseolus vulgaris* L. *Plant Breed.* 109(4):320-328.

Singh, S.P., Lépiz, R., Gutiérrez, J.A., Urrea, C., Molina, A., and Terán, H. 1990. Yield testing of early generation populations of common bean. *Crop Sci.* 30(4):874-878.

Singh, S.P., Molina, A., and Gepts, P. 1995. Potential of wild common bean for seed yield improvement of cultivars in the tropics. *Can. J. Plant Sci.* 75(4):807-813.

Singh, S.P., Molina, A., Urrea, C.A., and Gutiérrez, J.A. 1993. Use of interracial hybridization in breeding the race Durango common bean. *Can. J. Plant Sci.* 73(3):785-793.

Singh, S.P. and Muñoz, C.G. 1999. Resistance to common bacterial blight among *Phaseolus* species and common bean improvement. *Crop Sci.* 39(1):80-89.

Singh, S.P., Nodari, R., and Gepts, P. 1991. Genetic diversity in cultivated common bean. I. Allozymes. *Crop Sci.* 31(1):19-23.

Singh, S.P., Pastor-Corrales, M.A., Molina, A., Urrea, C., and Cajiao, C. 1991. Independent, alternate, and simultaneous selection for resistance to anthracnose and angular leaf spot and effects on seed yield in common bean (*Phaseolus vulgaris* L.). *Plant Breed.* 106(4):312-318.

Singh, S.P., Roca, W.M., and Debouck, D.G. 1997. Ampliación de la base genética de los cultivares de frijol: Hibridación interespecífica en especies de *Phaseolus.* In S.P. Singh and O. Voysest (Eds.), *Taller de mejoramiento de frijol para el siglo XXI: Bases para una estrategia para América Latina* (pp. 9-19). CIAT, Cali, Colombia.

Singh, S.P. and Terán, H. 1998. Population bulk versus F$_1$-derived family methods of yield testing in early generations of multiple-parent interracial and inter-gene pool crosses of common bean. *Can J. Plant Sci.* 78(3):417-421.

Singh, S.P., Terán, H., Muñoz, C.G., and Takegami, J.C. 1999. Two cycles of recurrent selection for seed yield in common bean. *Crop Sci.* 39(2):391-397.

Singh, S.P. and Urrea, C. 1995. Inter- and intraracial hybridization and selection for seed yield in early generations of common bean, *Phaseolus vulgaris* L. *Euphytica* 81(1):131-137.

Singh, S.P., Urrea, C.A., Molina, A., and Gutiérrez, J.A. 1992. Performance of small-seeded common bean from the second selection cycle and multiple-cross intra- and interracial populations. *Can. J. Plant Sci.* 72(3):735-741.

Smartt, J. 1969. Evolution of American *Phaseolus* beans under domestication. In P.J. Ucko and G.W. Dimbleby (Eds.), *The domestication and exploitation of plants and animals* (pp. 451-461). Duckworth, London.

Smartt, J. 1988. Morphological, physiological and biochemical changes in *Phaseolus* beans under domestication. In P. Gepts (Ed.), *Genetic resources of Phaseolus beans* (pp. 143-161). Kluwer, Dordrecht, Netherlands.

Stavely, J.R. 1984. Pathogenic specialization in *Uromyces phaseoli* in the United States and rust resistance in beans. *Plant Dis.* 68(2):95-99.

Stavely, J.R. and Grafton, K.F. 1989. Registration of BelDak-Rust Restistant-1 and -2 pinto dry bean germplasm. *Crop Science* 29(3):834-835.

Sullivan, J.G. and Bliss, F.A. 1983. Recurrent mass selection for increased seed yield and seed protein percentage in the common bean (*Phaseolus vulgaris* L.) using a selection index. *J. Am. Soc. Hort. Sci.* 108(1):42-46.

Thung, M. 1990. Phosphorus: A limiting nutrient in bean (*Phaseolus vulgaris* L.) production in Latin America and field screening for efficiency and response. In N. El Bassam, M. Dambroth, and B.G. Loughman (Eds.), *Genetic aspects of plant mineral nutrition* (pp. 501-521). Kluwer, Dordrecht, Netherlands.

Thung, M.D.T. and de Oliveira, I.P. 1998. *Problemas abióticos que afectam a produção do feijeiro é seus métodos de controle.* EMBRAPA. CNPAF, Santo Antônio de Goias, Brazil.

Thung, M.D.T. and Rao, I. 1999. Integrated management of abiotic stresses. In S.P. Singh (Ed.), *Common bean improvement in the twenty-first century* (pp. 331-370). Kluwer, Dordrecht, Netherlands.

Thung, M.T., Ferreira, R.M., Miranda, P., Moda-Cirino, V., Gava Ferrão, M.A., da Silva, L.O., Dourado, V.V., Hemp, S., Souza, B., Serpa S., E., Zimmermann, M.J.O., and Singh, S.P. 1993. Performance in Brazil and Colombia of common bean lines from the second selection cycle. *Rev. Bras. Genet.* 16(1):115-127.

Toro, Ch. O., Tohme, J., and Debouck, D.G. 1990. *Wild bean* (Phaseolus vulgaris L.) *description and distribution.* IBGPR and CIAT, Cali, Colombia.

Urrea, C.A., Miklas, P.N., Beaver, J.S., and Riley, R.H. 1996. A codominant randomly amplified polymorphic DNA (RAPD) marker useful for indirect selection of bean golden mosaic virus resistance in common bean. *J. Am. Soc. Hortic. Sci.* 121(6):1035-1039.

Urrea, C.A. and Singh, S.P. 1994. Comparison of mass, F_2-derived family, and single-seed-descent selection methods in an interracial population of common bean. *Can. J. Plant Sci.* 74(3):461-464.

Urrea, C.A. and Singh, S.P. 1995. Comparison of recurrent and congruity backcrossing for interracial hybridization in common bean. *Euphytica* 81(1):21-26.

van Schoonhoven, A. and Cardona, C. 1980. Insects and other bean pests in Latin America. In H.F. Schwartz and G.E. Gálvez, (Eds.), *Bean production problems: Disease, insect, soil and climatic constraints of Phaseolus vulgaris* (pp. 363-412). CIAT, Cali, Colombia.

van Schoonhoven, A. and Cardona, C. 1982. Low levels of resistance to the Mexican bean weevil in dry beans. *J. Econ. Entomol.* 75(4):567-569.

van Schoonhoven, A. Cardona, C., and Valor, J. 1983. Resistance to the bean weevil and the Mexican bean weevil (Coleopter: Bruchidae) in noncultivated common bean accessions. *J. Econ. Entomol.* 76(6):1255-1259.

Velez, J.J., Bassett, M.J., Beaver, J.S., and Molina, A. 1998. Inheritance of resistance to bean golden mosaic virus in common bean. *J. Amer. Soc. Hort. Sci.* 123(4):628-631.

Vieira, C. 1967. *O Feijoeiro-Comum: Cultura, doencas e molhora-miento.* Universidade Rural do Estado de Minas Gerais, Vicosa, Brasil.

Vieira, C. 1988. *Phaseolus* genetic resources and breeding in Brazil. In P. Gepts (Ed.), *Genetic resources of* Phaseolus *beans* (pp. 467-483). Kluwer, Dordrecht, Netherlands.

Voysest, O. 1983. *Variedades de frijol en América Latina y su origen.* CIAT, Cali, Colombia.

Voysest, O. and Dessert, M. 1991. Bean cultivars: Classes and commercial seed types. In A. van Schoonhoven and O. Voysest (Eds.), *Common beans: Research for crop improvement* (pp. 119-162). C.A.B. Int., Wallingford, U.K. and CIAT, Cali, Colombia.

Voysest, O., Valencia, M.C., and Amézquita, M.C. 1994. Genetic diversity among Latin American Andean and Mesoamerican common bean cultivars. *Crop Sci.* 34(4):1100-1110.

Wall, D. (Ed.) 1973. *Potentials of field beans and other legumes in Latin America.* CIAT, Cali, Colombia.

Wall, J.R. 1970. Experimental introgression in the genus *Phaseolus.* 1. Effect of mating systems on interspecific gene flow. *Evolution* 24(2):356-366.

Weiseth, G. 1954. Una variedad silvestre del poroto común (*Phaseolus vulgaris*), autóctona del Noroeste Argentino y su relación genética con variedades cultivadas. *Rev. Agron. Noroeste Argentino* 1(2):71-81.

Westphal, E. 1974. *Pulses in Ethiopia, their taxonomy and agricultural significance, agricultural research report no. 815.* Center for Agricultural Publishing and Documentation, PUDOC, Wageningen, Netherlands.

White, J., Hoogenboom, G., Ibarra, F., and Singh, S.P. (Eds.) 1988. *Research on drought tolerance in common bean.* CIAT, Cali, Colombia.

White, J. and Laing, D.R. 1989. Photoperiod response of flowering in diverse genotypes of common bean (*Phaseolus vulgaris*). *Field Crops Res.* 22(1):113-128.

Wilkinson, R.E. 1983. Incorporation of *Phaseolus coccineus* germplasm may facilitate production of high yielding *P. vulgaris* lines. *Annu. Rep. Bean Improv. Coop.* 26:28-29.

Wood, D.R. and Keenan, J.G. 1982. Registration of the Olathe bean. *Crop Science* 22(6):1259-1260.

Woolley, J., Lepiz I., R., Portes e Castro, T. De A., and Voss, J. 1991. Bean cropping systems in the tropics and subtropics and their determinants. In A. van Schoonhoven and O. Voysest (Eds.), *Common beans: Research for crop improvement* (pp. 679-706). C.A.B. Int., Wallingford, UK and CIAT, Cali, Colombia.

Wortmann, C.S., Kirkby, R.A., Eledu, C.A., and Allen, D.J. 1998. *Atlas of common bean* (Phaseolus vulgaris L.) *production in Africa.* CIAT, Cali, Colombia.

Yoshii, K., Galvez, G.E., and Alvarez, G. 1976. Estimation of yield losses in beans caused by common blight. *Proc. Am. Phytopathol. Soc.* 3:298-299 (Abstr.).

Zaumeyer, W.J., and Meiners, J.P. 1975. Disease resistance in beans. *Annu. Rev. Phytopath.* 13:313-334.

Zaumeyer, W.J., and Thomas, H.R. 1957. *A monographic study of bean diseases and methods for their control.* USDA Agric. Tech. Bull. No.868.

Zimmermann, M.J.O., Rocha, M., and Yamada, T. (Eds.) 1988. *Cultura do feijoeiro: Fatores que afetam a produtividade.* Associacaõ Brasileira para Pesquisa da Potassa e do Fosfato, Piracicaba, SP, Brasil.

Chapter 7

Sugar Beet Breeding and Improvement

Larry G. Campbell

INTRODUCTION

Beets (*Beta vulgaris* L.) have provided food for humans and animals for centuries, but their use as a sweetener source is relatively recent. Andreas Marggraf reported in 1747 that the sugar from beets had properties similar to that from sugarcane (*Saccharum* spp.). Commercial extraction of sugar (sucrose) from beets began approximately 55 years later with the construction of the first factory, based upon inventions of Franz Karl Achard, a student of Marggraf. Achard also is credited with the development of the progenitor of modern sugarbeet cultivars, the white Silesian beet (Winner, 1993). Louis de Vilmorin's classic breeding work with the sugar beet in the middle 1800s resulted in substantial increases in sugar concentration and subsequently enhanced the competitive position of sugar beet as a sweetener source (Allard, 1960).

Sugar beet can be grown in many environments; however, most production is between 30° and 60° north latitude. It is grown as a summer crop in maritime, prairie, and semicontinental climates and as a winter or summer crop in Mediterranean and some semiarid environments (Draycott, 1972). Thirty-five to 40 percent of the sucrose consumed by humans is extracted from sugar beet, and the remainder from sugarcane. Approximately 255 million tons of sugar beet currently are produced on 7.76 million ha. Russia and Ukraine each produces more than 1 million ha annually. Other leading producers are Poland, Germany, United Kingdom, France, Italy, Turkey, China, and the United States.

Improved cultivars, accompanied by advances in production technology, have resulted in steady gains in the productivity of sugar beet while reducing the labor required for production. Gradual sugar yield increases over a 38-year period have been documented in Western Europe (Bosemark, 1993). Root yields in a major North American production region where irrigation is rare increased approximately 0.5 Mg·ha^{-1} per year from 1955 to 1995 (Campbell, 1995). Among the highest sugar beet yields reported is a root yield of 132 Mg·ha^{-1} (22,042 kg sugar·ha^{-1}) in the Imperial Valley of California in 1998 (Melin, 1999). The extent to which these successes can be attributed to plant breeding efforts is difficult to establish, but most would agree that improved hybrids have played a significant role. Disease-resistant hybrids have increased productivity while reducing dependence upon pesticides.

Sugar beet has a number of unique characteristics that require consideration in the implementation of an applied breeding program. The objective of this chapter is to point out these characteristics and provide examples of adaptations that have been used by breeders. It is assumed that the reader has some familiarity with general plant breeding principles and practices, especially as they apply to cross-pollinated crop species, such as corn (*Zea mays* L.) or sorghum [*Sorghum bicolor* (L.) Moench]. The examples of pest resistance included are not intended to cover all diseases and insects but are offered as a basis for developing specific programs. Most of the maladies affecting sugar beet have been described by Lejealle and d'Aguilar (1982) and Whitney and Duffus (1986).

PLANT CHARACTERISTICS AND INHERITANCE

Beta vulgaris is a herbaceous, allogamous dicotyledon belonging to the Chenopodiaceae family. Cultivated sugar beet is a biennial, producing a succulent tap root the first year and a seed stalk the second. The mature beet has three regions: a broad somewhat cone-shaped crown from which the leaves arise, a smooth thickened hypocotyl that forms the broadest portion of the beet, and the root region that constitutes the bulk of the beet. Leaves are arranged in a close spiral on the crown. The lamina of the leaf is an elongated triangle with a rounded tip and undulate margin (Artschwager, 1926). During bolting (flower stalk formation) the stem elongates, producing a tall angular structure. Shoots that develop in the axils of leaves along the seed stalk produce racemes. Flowers are sessile, subtended by small bracts (Elliot and Weston, 1993).

The small sessile flowers of sugar beet are typically borne in clusters of two to seven. The flowers are perfect, containing a three-lobed (sometimes four or five) stigma and five stamens. Pollen germinates soon after dispersal and fertilization occurs approximately one day after anthesis. Under field conditions, pollen usually does not remain viable for more than a day. Stigmas begin to open in midafternoon, approximately five to seven hours after anthesis, and expansion is completed within 24 to 36 hours. Newly formed embryos require 12 to 14 days to mature (Artschwager and Starrett, 1933).

Hand emasculation of sugar beet generally is utilized only when male sterility or genetic markers cannot insure or allow recognition of hybrid progeny. Sepals of flowers ready to open may be teased back to expose anthers. Anthers are then removed with forceps, taking care not to damage the stigma. Unemasculated flowers are removed from the branch prior to bagging to prevent contamination. Pollinations can be made immediately or up to 12 days after normal anthesis would have occurred (Smith, 1980). Sugar beet is indeterminate, flowering over three to 10 weeks, depending on environmental conditions (Elliot and Weston, 1993).

Hypocotyl color frequently is utilized to identify hybrid progeny. Root, hypocotyl, and foliage color are determined by two linked loci, Y (sometimes designated G) and R, with multiple alleles at each locus (Keller, 1936). Plants with white roots, hence all sugar beets, are homozygous for the recessive y allele *(yy)*. Sugar beets homozygous recessive at the R locus *(rr)* will have green hypocotyls. All of the other R alleles are dominant to r and produce coloration in the hypocotyl. F_1 progeny of green hypocotyl plants *(rr)* pollinated with pollen from plants with colored hypocotyls *(RR)* will have colored hypocotyls *(Rr)*, whereas selfs or sibs will have green hypocotyls *(rr)* and can be rogued as seedlings. If some of the pollinator plants are not homozygous at the R locus, some F_1 progeny may have green hypocotyls and be discarded. In addition to the R and Y loci, another locus, P, controls color. In the homozygous recessive condition *(pp)*, expression of any dominant alleles at the R and Y loci is suppressed, resulting in white roots and green hypocotyls. The frequency of the recessive allele *(p)* is low, so it is not a factor in most sugar beet crosses. The P locus is closely linked to the R and Y loci (Linde-Laursen, 1972). Other simply inherited characters (Theurer, 1968; Smith, 1987, pp. 587-589), resistance to specific herbicides, isozyme markers (Smed et al., 1989), or molecular markers (Nilsson et al., 1997) that can be identified easily prior to flowering also could be used to identify F_1 progeny.

Sugar beet, a biennial, requires a period of low temperature for induction of the reproductive stage. In contrast, many of the wild forms of *Beta* are annuals. The presence of a single dominant gene, designated B, produces the annual growth habit. The B locus is linked with the R locus controlling hypocotyl color (Aberg, 1936). Restricted daylengths may suppress bolting in some plants possessing the dominant B allele. This daylength response appears to be controlled by a number of genes, one of which *(Lr)* is closely linked to B (Abe et al., 1997a). Among biennials, the length of the cold requirement for induction of bolting is genotype dependent (McFarlane et al., 1948; Marcum, 1948). Efficiency in sugar beet breeding programs is enhanced by producing seed in the greenhouse. Greenhouse seed production includes a preinduction phase, a photothermal induction treatment, and a postinduction period (Gaskill, 1952a). During the preinduction phase, seedlings are grown in small individual pots (5 cm) for two to eight weeks in a warm (20 to 23 º C) greenhouse with a long (18 to 24 hour) photoperiod. Larger preinduction seedlings favor the development of larger plants with greater seed yields (Gaskill, 1952b). Following this preinduction phase, seedlings are transferred to a cold chamber with a constant temperature between 4° and 7° C and long daylengths or continuous light. Duration of the induction treatment ranges from 90 to 120 days depending upon the genetic resistance to bolting of the particular lines or populations. Upon completion of the induction treatment, the seedlings are transplanted to larger pots or greenhouse benches. Successful postinduction requires long photoperiods with day/night temperatures near 25/15 º C. Photothermal induction may be reversed by high temperatures or short photoperiods. In some situations, it is advantageous to obtain seed from beets selected in the field or to use plants selected from field plots as parental material. To accomplish this, beets are dug and the leaves carefully trimmed to avoid injury to crown tissue. The beets are then stored at 6° to 9° C in the dark with high relative humidity to prevent dehydration. The length of the storage period is similar to that required to induce seed-stalk formation in seedlings. After storage, the beets are immediately planted in the greenhouse under conditions similar to those utilized in the postinduction period for seedlings (Stout, 1946). Field seed increases and crossing blocks are also used in breeding programs. Seed can be planted in the late summer or early autumn, allowed to overwinter, and harvested the next summer in areas where winters are not too severe. An alternative is the production of stecklings (small plants). With this system, seed frequently will be planted in a separate steckling nursery while the line is also being evaluated in agronomic trials at other sites. Stecklings are dug, roots are stored in cold chambers to induce seed stalk formation, and selected lines are transplanted to crossing blocks or for increase. In areas where winters are cold enough to induce flowering and mild enough to allow survival, stecklings can overwinter in the nursery (Bornscheuer et al., 1993). Selected lines can then be dug and transplanted as desired. Field seed production blocks must be isolated by distance (1,000 m) or covered with tents or bags. A forced-air ventilation system with filtered air (to avoid entry of pollen from adjacent tents) will reduce the incidence of some diseases and enhance pollen dispersal in tents.

Typically, two to seven adjoining flowers within a cluster cohere, producing a single "multigerm" fruit that is not easily divided. This so-called multigerm seed will give rise to several seedlings, and careful hand thinning is required to remove all but one of the closely grouped seedlings. The discovery of plants that produced flowers singly (not in clusters) and, therefore, monogerm fruits was a major milestone in sugar beet breeding and production. The use of monogerm seed has greatly reduced or eliminated the need for hand thinning of sugar beet seedlings in commercial fields. The most widely used source of the monogerm character can be traced to a single plant found by V. F. Savitsky in 1948. The trait is controlled by a single locus with monogerm being recessive (Savitsky, 1952). Heterozygous *(Mm)* plants are multigerm but produce fewer flowers per cluster than homozygous multigerm plants *(MM)*. In some crosses, modifying genes result in the formation of a few double-germ fruits on homozygous *(mm)* predominately monogerm plants. An association between the monogerm trait and a tendency for fasciation (abnormal broadening and flattening of the seed stalks) in the original source (SLC 101) exists in some current monogerm lines. The monogerm trait found in some European lines (Shavrukov, 1997) is not controlled by the single monogerm gene (Savitsky, 1952) used in the United States and Western European breeding programs.

Outcrossing in sugar beet is promoted by a gametophytic self-incompatibility system controlled by at least four linked S loci $(S_a, S_b, S_c,$ and $S_d)$ with complementary interaction. Incompatibility occurs when the four S genes in the pollen match those in the pistil (Larsen, 1977a). Due to the high number of possible S genotypes, this system permits mating of close relatives (Bosemark, 1993). Self-incompatible plants often set some seed upon selfing. This pseudocompatibility or pseudoself-fertility is influenced by both genotype and environment, especially temperature. Larsen (1977b) hypothesized that increased pseudocompatibility was associated with increased heterozygosity in the style. Genotypes with almost obligate self-fertility also have been identified and utilized in breeding programs (Owen, 1942). Self-fertile plants carry a dominant gene (Savitsky, 1954) designated S^F. Both homozygous *(S^F S^F)* and heterozygous *(S^F S^f)* self-fertile plants will produce predominately selfed seed, even when pollen from other plants is abundant.

Genetic (also referred to as nuclear or Mendelian) male sterility may be used to facilitate crossing, especially in self-fertile breeding populations (Fehr, 1987, pp. 199-218). Genetic male sterility is controlled by a single locus, the homozygous recessive $(a_1 a_1)$ condition producing male-sterile plants (Owen, 1952). Since genetic male sterility cannot be maintained in lines or populations producing all male-sterile plants, its use is somewhat restricted.

The cytoplasmic male-sterility (CMS) system, used almost exclusively for production of commercial sugar beet hybrids, was first described by Owen in 1945. It involves the interaction between nuclear genes at two loci *(X and Z)* and two types of cytoplasm, designated N for normal and S for sterile. Plants with the N cytoplasm produce abundant viable pollen regardless of the alleles at the X and Z loci. Plants with the S cytoplasm that are homozygous recessive at both the X and Z loci *(S; xx zz)* are completely male sterile. Owen described two types of semimale-sterile offspring from male-sterile *(S; xx zz)* females. Type 1 beets *(S; Xx zz or S: xx Zz)* produce yellow anthers that usually do not dehisce, or, if they do, most of the pollen grains adhere to the anthers and are nonviable. Offspring producing too much pollen to be classified as Type 1 are assumed to be heterozygous at both loci *(S: Xx Zz)* and classified as Type 2. Classification may be muddled by environmental effects upon anther development. Homozygous recessive plants with the *N* cytoplasm *(N; xx zz)* are referred to as O types (commonly referred to as B lines in other crops) and are used as maintainers for the corresponding CMS lines. Other unique CMS sources have been identified in *Beta* (Oldemeyer, 1957; Coe and Stewart, 1977; Halldén et al., 1990; Dalke and Szota, 1993) but have not been used in commercial hybrid production. Rearrangements within the mitochondrial DNA appear to be the cytoplasmic determinants of the CMS character (Mikami et al., 1985; Halldén et al., 1990).

Genes that partially restore pollen fertility are common; however, strong pollen fertility restoration genes are rare. Hogaboam (1957) identified a gene *(Sh)* that, when present in the dominant state, enhanced pollen production when combined with pollen restoration genes in plants with male-sterile *(S)* cytoplasm. Strong pollen restorer genes have been found in sugar beet lines (Theurer and Ryser, 1969; Roundy and Theurer, 1974; Theurer, 1978) and table beets (Theurer, 1971). Interactions among modifying genes, cytoplasms, and environments complicate the classification of pollen fertility in inheritance studies.

HYBRID PRODUCTION

Commercial sugar beet hybrids were introduced in the mid-1950s. Their utilization has been favored because of the general occurrence of non-additive genetic effects for root yield. Although much of the breeding and production methodology used for other cross-pollinated crops (Hallauer and Miranda, 1981) can be readily applied to sugar beet, some characteristics of sugar beet require unique adaptations. Commercial sugar beet seed-production practices have been described by Bornscheuer and colleagues (1993).

Development of female parental (CMS) lines in sugar beet is typical of that utilized in a number of other crops. O-type plants *(N; xx zz)* are infrequent in most populations. The standard method for identifying O-type individuals is to cross single plants of interest with a CMS tester and observe the progeny for the presence or absence of viable pollen. In some situations, it may be efficient to use an annual CMS line (Panella and Hecker, 1995) as a tester. Breeding populations developed for the purpose of producing female parental lines are generally derived from self-fertile materials. Self-fertile plants can be self-pollinated and crossed to a CMS tester simultaneously. Selfed seed from plants producing all male-sterile *(S; xx zz)* testcross progeny with white empty anthers is retained for further development as female parents. Seed from plants whose progeny *(S; Xx Zz)* produced abundant pollen are discarded as candidates for female parents. Selfed seed from plants whose testcross progeny segregated for sterile and semisterile may be planted for retesting on an individual plant basis. Several backcrosses with the O-type line as the recurrent parent and any CMS line as the donor parent will produce the CMS equivalent of the O-type parent. As an alternative to this approach, one may use a combination of selfing with concurrent backcrossing (Mackay et al., 1999). Initially, O-type self-fertile plants from a population are simultaneously selfed and crossed to any CMS plant. In the following generation, one of the selfed, fertile progeny is paired and isolated with one of its corresponding crossed CMS progeny. A series of such crosses will produce a new inbred line and its CMS equivalent. Selection among lines (testcrosses) may take place at one or more stages of the backcrossing. Selection should begin after three generations of backcrossing if the initial plant was not inbred. Any CMS line, and hence its O-type maintainer line, that is to be used as a female parental line for a commercial hybrid must be monogerm *(mm)*. The low frequency of O-type plants, the desirability of self-fertility, the need for monogerm seed, and the time required for progeny testing for O-type genotypes complicate development of female parental lines with desirable yield and quality characteristics and useful levels of resistance to predominant pests. Seed yield, a trait easily overlooked when selecting for root characteristics, also must be considered when developing CMS parental lines for commercial seed production.

Parental lines to be used as pollinators in the production of commercial hybrids are almost always multigerm. Plants with clustered flowers that produce multigerm seed generally produce more pollen than comparable monogerm plants, a desirable character in a pollinator. The genotype of the commercial hybrid, with regard to the monogerm-multigerm trait, is not important. Only the phenotype of the female parent is important in the production of monogerm commercial hybrid seed. It is not necessary that the pollinator parent possess any pollen fertility-restoration genes, as is the case when seed production is required in the commercial crop.

Most of the commercial sugar beet hybrids are best characterized as three-way topcross hybrids (Figure 7.1). The female (CMS) parent in the final cross is a monogerm F_1 hybrid between an inbred monogerm CMS line and an unrelated monogerm O-type pollinator. The pollinator parent frequently is a heterogeneous multigerm line. Although the resulting heterogeneity within the commercial hybrid may cause some problems related to lack of plant uniformity, it provides stability across environments. Double-cross hybrids can be produced by crossing a monogerm CMS F_1 hybrid produced in the manner described for a three-way cross with an unrelated pollen-fertile F_1 hybrid produced by crossing a CMS line with a line (R line)

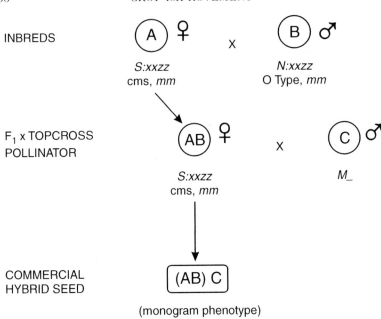

FIGURE 7.1. Synthesis of a three-way topcross sugar beet hybrid. Triploid hybrids can be produced by using a tetraploid line as the topcross parent (C).

that carries a gene or genes for fertility restoration (Theurer and Ryser, 1969). Alternatively, the F_1 hybrid used as the final pollinator for a double-cross hybrid can be obtained by pollinating the male-sterile progeny of a line segregating for genetic male-sterility with any unrelated pollen-fertile line (Owen, 1954). With this method, all male-fertile plants (approximately 50 percent of plants) must be rogued from the line segregating for male sterility prior to pollen shedding (during the bud stage). Single-cross hybrids between an inbred monogerm CMS parent and an inbred pollinator would produce a uniform commercial crop and allow the maximum expression of heterosis. The low seed yield of most inbred CMS lines has limited the use of single-cross hybrids.

A tetraploid pollinator and a diploid CMS line or F_1 (for a three-way hybrid) are used routinely as parents of commercial triploid sugar beet hybrids. An advantage of triploid hybrids is that only one of the three genomes of the hybrid originate from the CMS parent, the parent that often is the most difficult to improve because of constraints noted in the earlier discussion of CMS line development. The continuing availability of commercial diploid hybrids supports published results indicating that triploids are not clearly superior to diploids. Neither a general beneficial nor a detrimental effect on root yield or sucrose concentration appears to be associated with the addition of a genome; however, some genotypes respond favorably to the additional genome in some environments (Hecker et al., 1970; McFarlane et al., 1972; Smith et al., 1979). In cases where triploids enhanced sugar production, the benefit almost always was related to an increase in root yield, with little effect upon sucrose concentration. The sucrose yield of triploid hybrids may be more stable across environments than that of diploids (Lasa et al., 1989). There is evidence to suggest that triploid hybrids produced using tetraploid CMS lines and diploid pollinators will outyield hybrids produced with a diploid CMS parent and a tetraploid pollinator (Fitzgerald, 1977; Smith et al., 1979). The characteristic low seed set of tetraploid CMS lines and the low germination and emergence of triploid seed harvested from tetraploid male-steriles has restricted the use of tetraploid CMS lines in commercial production (Bosemark, 1993). Tetraploid sugar beet hybrids do not yield as well as diploid or triploid hybrids (Smith et al., 1979). Tetraploid pollinators release less pollen, require drier air for pollen release, and release their pollen later in the day than diploid pollinators (Scott and Longden, 1970). These factors should be considered in establishing isolation distances between seed production fields and in determining the ratio of CMS to pollinator.

An established method for producing tetraploid sugar beet is to treat pregerminated or germinated seed with colchicine (Kloen and Speckmann, 1953; Savitsky, 1966). Soaking germinated seed in a colchicine solution has been more efficient than treating seedlings or plant parts, and dry seed requires higher colchicine concentrations than germinated seed. Leaves of doubled plants often have irregular margins and are thicker and darker green than their diploid counterparts (Kloen and Speckmann, 1953). Tetraploids also have larger stomata (Artschwager, 1942) with more chloroplasts in the guard cells (Dudley, 1958). Tetraploid plants produce larger but fewer seeds than diploid plants (Peto and Hill, 1942). While changes in some morphological characteristics may be associated with polyploidy, confirmation should be based upon the larger pollen grains produced by tetraploids (Savitsky, 1966). Because of considerable irregularity in meiosis of sugar beet tetraploids, some aneuploids are expected among their progeny. However, most of the progeny of tetraploid parents will be tetraploid (Savitsky, 1952). Segregation ratios in tetraploid populations are more complicated than in diploids (Fehr, 1987, pp. 59-65). Hence, as much selection as is possible should be completed before a diploid line is converted to a tetraploid line. Adaptation of in vitro vegetative propagation methods (Saunders and Shin, 1986) could facilitate selection within tetraploid populations and aid in the development and maintenance of tetraploid lines in some situations.

SELECTION TECHNIQUES

Although the goals of a hybrid sugar beet-improvement program have much in common with those of most other cross-pollinated crops, some of the characteristics of the sugar beet plant and crop require unique adaptations of traditional breeding methods. Mass selection based upon phenotype is effective for traits controlled by additive gene action with at least intermediate heritability. It has been effective in developing bolting-resistant populations, enhancing the resistance to some diseases, increasing sucrose concentration, and selecting for desirable root shape. Most of the economically important traits of sugar beet are expressed long before flowering and undesirable individuals can be discarded prior to intermating of selected plants; enhancing the progress obtained with mass selection. In some cases, imposing a grid system will assist in reducing the effects of environmental variation within the experimental area. A few cycles of mass selection in a newly formed population may eliminate the most undesirable individuals prior to the utilization of more intensive selection techniques. After a few cycles of mass selection, some form of progeny selection often should hasten progress.

With half-sib selection, plants selected from a source population are induced to flower and allowed to interpollinate as a group. Seed is harvested from each plant and evaluated for traits of interest as a half-sib family. Stecklings representing each family are produced simultaneously. Once the desired families have been identified, the corresponding stecklings are induced to flower and allowed to interpollinate. This improved population may then be subjected to another cycle of half-sib family selection. Individual half-sib families may need to be increased separately at some stage to allow for multisite replicated evaluations of traits, such as root yield. As an alternative to producing stecklings, roots from selected families can be harvested from field evaluations, induced to flower, and allowed to interpollinate. This approach allows for selection based upon family performance and selection of individuals within the better half-sib families (Campbell, 1989).

Pair crosses (or full-sib progeny selection) involving two selected roots from within a population are a form of progeny testing frequently used by sugar beet breeders, especially when self-incompatibility precludes selfing. Including a CMS tester in a small tent isolator with the two plants being mated will provide seed for initial combining-ability tests. It is often necessary to increase promising individual pair crosses to have sufficient seed for final evaluations. Pair crossing within a line derived from a pair cross will increase the degree of inbreeding in the subsequent progeny. Inbred lines of sugar beet generally yield poorly and lack vigor.

Reciprocal recurrent selection is a technique that capitalizes on both additive and nonadditive gene action (Wricke and Weber, 1986). This technique is not adapted readily to sugar beet because it requires that plants be selfed as well as crossed. Doney and Theurer (1978) outlined a method that overcomes these obstacles by incorporating the self-fertile (S^F) gene and the genetic male-sterile gene *(a)* into source populations. The implementation and utility of these adaptations has been examined (Hecker, 1985) and compared to other selection procedures (Hecker, 1978).

Gametophytic (pollen) screening has been suggested as a method of selecting superior sugar beet genotypes (Smith and Moser, 1985). The proposed procedure involves placing pollen from an individual plant on pollen germination media that contains a challenging agent, such as a herbicide or pathotoxin. Surviving pollen could then be used to pollinate a flowering plant. Probably a more useful approach is to use the pollen response to identify resistant plants or to confirm observations from established screening tests. Selected plants would then be crossed in a traditional manner.

Vegetative propagation or cloning is sometimes used to perpetuate genotypes that are difficult or impossible to maintain with seed production. Stem cuttings from bolting plants (Hogaboam, 1962) and longitudinal splitting of roots so that each root section contains one or more crown buds are time-tested methods of cloning sugar beet (Smith, 1980). Saunders (1982) and Mezei and colleagues (1990) described methods for propagating lines using excised axillary buds. Plants derived from explants were identical to the original plant, phenotypically homogeneous, and cytologically stable (Mezei et al., 1990). In some situations, somatic embryos from tissue culture could facilitate propagation and gene-transfer efforts. Methods for plantlet initiation and development that are not genotype specific need to be perfected (Zhong et al., 1993; Tsai and Saunders, 1995). Cell cultures could then be grown on media that is selective for some agronomically desirable traits (Saunders et al., 1992). Methods of producing doubled haploid sugar beet plants from unfertilized ovule cultures have been described (Lux et al., 1990; Galatowitsch and Smith, 1990) and likely will be utilized by breeders. Successful protoplast isolation and culture techniques may provide a method for transferring cytoplasmically inherited traits, such as CMS, to sugar beet (Kerns et al., 1990).

Genetic linkage maps utilizing molecular markers are becoming more available in sugar beet (Pillen et al., 1992; Pillen et al., 1993; Barzen et al., 1995; Nilsson et al., 1997; Hansen et al., 1999). This technology allows breeders to identify accurately and efficiently desirable plants without the use of expensive and time-consuming field or greenhouse tests. Marker-assisted selection is especially adapted to simply inherited traits. Identification of quantitative trait loci is not as straightforward (Hyne et al. 1995; Bernardo, 1998) and results may be less definitive than with single-locus inheritance. Molecular markers can be used to examine phylogenetic relationships among related species (Raybould et al., 1996) to show the extent of differences or relationships among sugar beet populations, to verify crosses, or to verify that specific chromosome segments have been transferred in interspecific crosses.

In spite of some difficultly in obtaining genetically transformed sugar beet (Hall et al., 1996), productive transformed hybrids with resistance to broad-spectrum herbicides are currently a reality and transformations affecting other traits will likely follow. A review by Wozniak (1999) summarizes much of the relevant literature related to sugar beet transformation and is recommended to those interested in initiating a genetic engineering program in sugar beet.

AGRONOMIC TRAITS

A negative correlation between root yield and sucrose concentration (Wyse, 1979; Campbell and Kern, 1983; Campbell and Cole, 1986) has perplexed generations of sugar beet breeders. Some sugar processors pay premiums for higher sucrose concentrations, hence, hybrids producing the most sugar per hectare are not necessarily the most profitable for the grower, and breeding for maximum economic return becomes more complicated. Also, the effectiveness of breeding for combined high sucrose concentration and weight per root may depend upon environmental conditions (Powers, 1963). In general, the genetic variance for sucrose concentration appears to be predominantly additive. In contrast, nonadditive genetic variance and specific combining ability components are significant in determining root yield, and, therefore, have a role in determining sucrose yield (MacLachlan, 1972; Smith et al., 1973; Hecker, 1991). Increasing sucrose yields of elite populations by increasing root yield is often more fruitful than attempting to increase sucrose concentration. Perhaps this is because a large portion (75 percent) of the dry weight of current commercial sugar beets is sucrose (Bohn and Clarke, 1998) and any substantial increase in sucrose concentration might negatively affect the plant's well-being. Certain soluble nonsucrose constituents in extracted sugar beet juice, referred to as impurities, may impede crystallization and lower sucrose extraction rates. Sodium, potassium, amino-nitrogen, and betaine are considered the most important impurities (Smith et al., 1977). Sodium, potassium, and amino-nitrogen levels can be shifted dramatically with only a few cycles of selection (Coe, 1987, Smith and Martin, 1989), confirming the importance of additive genetic variance in determining relative levels of these traits (Smith et al., 1973). Interactions among the impurity components and between the impurity components, sucrose concentration, and root yield complicate selection for optimum levels of yield and quality traits. Snyder and Carlson (1978) presented evidence that concurrent selection for high taproot weight and high taproot-leaf weight ratio in seedlings increased yield. Cultivar × environment interactions (Campbell and Kern, 1982) must be considered in any plant breeding program; however, since commercial sugar beet production does not require completion of the plant's reproductive cycle, these interactions may be less important in sugar beet than in some other crops. Cultivar × herbicide interactions have been noted (Smith and Schweizer, 1983), making clear the necessity for testing under conditions duplicating those used in commercial production.

Extended periods of cool weather combined with long daylengths may induce bolting (flowering) in commercial fields. Early bolting plants that form seed stalks produce about half the root yield of vegetative plants (Nelson and Deming, 1952). In addition, any seed produced may result in future weed beet infestations (Longden, 1993). Bolting resistance is required in Northwestern Europe, Southwestern United States, and anywhere mild winters allow autumn planting. The inheritance of bolting is complex with mainly additive effects and some nonadditive effects in certain crosses (Sadeghian and Johansson, 1993; Sadeghian et al., 1993). Some genes may control response to photoperiod while others regulate response to low temperature. In breeding programs, early planting will allow identification of lines or individuals that bolt easily. A few generations of selection generally will increase bolting resistance significantly (Bosemark, 1993). Lewellen (1989) proposed that the combination of nonbolting tendency and male sterility in C600 CMS, an annual requiring an exceptionally long period of exposure to long-day conditions, make it useful as a tester for sorting biennial *(bb)* genotypes for bolting tendency in greenhouse or field tests under warm long-day conditions. Isozyme loci linked to genes controlling bolting tendency may facilitate marker-assisted selection for bolting resistance (Abe et al., 1997b).

Cercospora leaf spot caused by the fungus *Cercospora beticola* Sacc. is one of the most widespread and destructive foliar diseases of sugar beet. Smith and Gaskill (1970) reported that *Cercospora* resistance was quantitatively inherited, principally controlled by four or five major genes. Realized heritabilities of approximately 0.25 were obtained by Smith and Ruppel (1974). Bilgen and colleagues (1969) indicated that *Beta maritima* possessed high levels of resistance to *Cercospora* leaf spot, but low heritability for resistance-complicated utilization of interspecific hybrids. Some of the earliest known *Cercospora*-resistant varieties can be traced to interspecific hybrids developed in Italy (McFarlane, 1971). Resistance-breeding efforts continue, and highly resistant, monogerm, O-type lines are available to breeders (Smith and Ruppel, 1988). Parental lines selected for resistance in the United States provided protection against *Cercospora* leaf spot in Southern Europe and vice versa (Smith, 1985). In areas where effective fungicides are available, many sugar beet breeders have chosen to concentrate on resistance to other pests and yield at the expense of *Cercospora* resistance. This is, in part, because resistance is not simply inherited and, perhaps more important, because breeders have experienced difficulty in producing highly resistant lines with competitive yield potential (Miller et al., 1994; Smith and Campbell, 1996). This approach is ineffective because the fungus can readily produce fungicide-resistant strains (Campbell et al., 1998).

Rhizoctonia root and crown rots (*Rhizoctonia solani* Kühn) are endemic to many sugar beet production areas. *Rhizoctonia* survives in the soil as a saprophyte for many years, making cultural controls ineffective. Most of the resistance-breeding efforts have utilized field screening and testing under artificially induced epiphytotics. Individual plant ratings are required to obtain meaningful progress with most selection schemes (Hecker and Ruppel, 1977). Timing of inoculation is a critical factor for regulating disease intensity and selection pressure (Ruppel and Hecker, 1988). Greenhouse screening procedures may provide preliminary information in a short time (Campbell and Altman, 1976) but cannot be substituted for field testing. Hecker and Ruppel (1975) reported that resistance was conditioned by two or more loci with some additive gene action. The partial dominance for resistance observed in F_1s was sufficient for the production of useful commercial hybrids. Hecker and Ruppel (1976) found no difference in resistance between diploid and tetraploid lines but observed a dosage effect in triploid hybrids. No cytoplasmic by ploidy interactions were observed. They recommended using resistant tetraploid pollinators for the production of triploid hybrids for regions where *Rhizoctonia* resistance would be beneficial. Immunity to *R. solani* has not been observed in sugar beet.

Black root or *Aphanomyces* root rot (*Aphanomyces cochlioides* Drechsler) development is favored by high soil moisture and warm temperatures. Resistant cultivars are available for some regions (Hogaboam et al., 1982). Schneider (1954) described optimum conditions for disease development in greenhouse tests. In areas where *Aphanomyces* root rot causes frequent damage, annual selection for resistance within most breeding populations may be necessary (Schneider and Hogaboam, 1983). Coe and Schneider (1966) warned that caution should be exercised when comparing hybrids to inbreds as heterosis may partially compensate for lack of resistance. Some inheritance studies have indicated that resistance is dominant (Coons et al., 1946; Bockstahler et al., 1950); however, others (Coe and Schneider, 1966) point out that high levels of resistance are obtained only after numerous selection cycles and conclude that *Aphanomyces* resistance is not simply inherited. Immunity to *Aphanomyces* has not been found.

Fusarium oxysporum Schlecht is a soilborne fungus that invades the vascular system. It frequently occurs as a stalk blight in seed-production fields but can also cause losses in growers' fields. McFarlane (1981) observed responses from near-immunity to plant death among inbred lines screened for resistance to *Fusarium* stalk blight. Selection for resistance was effective and relatively easy. Resistance appeared to be dominant.

The extreme severity of *Sclerotium* rot (*Sclerotium rolfsii* Sacc.) has discouraged the expansion of sugar beet into more humid areas and is a problem in some irrigated areas with relatively high temperatures. Lawlor and Doxtator (1950) observed differences among genotypes for survival rate, suggesting that *Sclerotium*-resistant cultivars could be produced. Coe and O'Neil (1983) developed a greenhouse screening procedure but did not find sufficient *Sclerotium* resistance to allow sugar beet production in disease-prone areas.

Bacterial vascular necrosis and rot or *Erwinia* root rot [*Erwinia carotova* (Jones) Holland] is the only sugar beet root rot caused by a bacterium. Resistance is simply inherited with a large dominance component. A second, primarily additive, component determines the amount of rot in susceptible individuals. This additive component may confer useful levels of resistance in the absence of the major resistance gene (Lewellen et al., 1978).

Storage rots caused by three fungi are major contributors to deterioration during sugar beet storage. *Phoma betae* (Oud.) Frank is potentially the most devastating pathogen because its disease cycle is closely associated with the life cycle of the sugar beet. Although many species of *Penicillium* cause storage rot, *P. claviforme* Brainier is the most damaging in some regions. *Botrytis cinerea* Pers. ex Fr. is more aggressive than *Phoma* or *Penicillium* and is able to rot tissue quickly over a wide range of temperatures. Bugbee (1979) described methods for evaluating individual roots for response to storage-rot fungi and demonstrated that selection for combined resistance was possible. Comparisons of rot-resistant lines and commercial hybrids demonstrated that genetic resistance could reduce sucrose losses comparable to the application of a fungicide (Bugbee and Cole, 1979). Germplasm lines with resistance to storage-rot fungi have been released (Bugbee, 1978; Campbell and Bugbee, 1985).

Rhizomania, caused by the beet necrotic yellow vein virus, is spread by a soil fungus, *Polymyxa betae* Keskin, that is capable of surviving in the soil for 15 years or more. Chemical control measures that might reduce fungal populations are cost prohibitive. Diseased plants produce stunted taproots that take on a characteristic "wineglass" shape and produce masses of hairy secondary roots. Aboveground symptoms include poor growth, yellow-green foliage, and narrow leaves with long erect petioles. Differential cultivar response to rhizomania appears to be identical in Europe and the United States. Both quantitatively and qualitatively inherited resistances (to virus or vector) have been identified. The simply inherited resistance, originally noted in a commercial hybrid, is conditioned by a single dominant gene that greatly reduces symptom expression and damage but does not confer immunity (Lewellen et al., 1987). High levels of resistance also have been observed in *B. maritima* accessions and sugar beet lines with *B. maritima* in their parentage (Lewellen, 1999). These also appear to be simply inherited and dominant. Resistance in more distantly related *Beta* species (Paul et al., 1994; Mesbah et al., 1997) probably will not be used by commercial breeders as long as the resistance found in *B. vulgaris* and *B. maritima* is effective. The coat-protein gene of the virus causing rhizomania has been introduced into sugar beet through *Agrobacterium*-mediated transformation. Transgenic plants had reduced virus multiplication rates, compared to nontransgenic plants in greenhouse and field trials (Mannerlöf et al., 1996).

Nematodes that attack sugar beet roots can be devastating and are difficult and expensive to control. Nematode-resistant hybrids would provide a desirable control alternative; however, useful levels of resistance have not been found in the sugar beet germplasm that has been screened. Yu (1995) found root-knot nematode (*Meloidogyne* spp.)-resistant plants within a wild sea beet accession (PI 546387). Selection within this accession and another (PI 546426) that also carried resistance to rhizomania produced two *B. maritima* populations (M66 and Mi-1) with a relatively high frequency of resistant plants (Yu, 1996; Yu, 1997). The close phylogenetic relationship between *B. maritima* and cultivated sugar beet should facilitate the introduction of root-knot nematode resistance into elite breeding populations. Except for some partial resistance reported in a *B. maritima* accession (Lang and deBock, 1994), the recognized sources of resistance to the beet cyst nematode (*Heterodera schachtii* Schmidt) are found in more distantly related *Beta* species, *B. procumbens*, *B. patellaris*, and *B. webbiana*. Mesbah and colleagues (1997) found that a single chromosome of *B. patellaris* carried the genes, or gene, conferring full resistance to the beet cyst nematode. Paul and colleagues (1990) demonstrated the potential of hairy root culture [induced by *Agrobacterium rhizogenes* (Riker et al.) Conn.] as an in vitro technique for assessing nematode resistance. Cai and colleagues (1997) have cloned a gene (*Hs1$^{pro=1}$*) from *B. procumbens* that confers resistance to the beet cyst nematode. Isolation of this gene enhances the possibility of transferring resistance to agronomically acceptable sugar beet.

The sugar beet root aphid, *Pemphigus populivenae* Fitch, is one of the most widespread insect pests of sugar beet. The root aphid reduces both size and quality of the roots by sucking sap from them. Root aphid populations can increase rapidly and significant damage may occur before they are noticed. A root aphid damage index has been developed (Hutchinson and Campbell, 1994) and greenhouse and field inoculation techniques described (Campbell and Hutchinson, 1995). Lines and hybrids differ substantially in their resistance to root aphid (Wallis and Turner, 1968). Both antibiosis and antixenosis (nonpreference) appeared to be important in determining resistance (Campbell and Hutchinson, 1995).

The sugar beet root maggot (*Tetanops myopaeformis* Röder) is a serious insect pest in some of the major U.S. production areas (Campbell et al., 1998). Developing larvae feed on the root by tunneling along the surface. Yield reductions may be the result of stand loss early in the season but occur primarily from feeding throughout the growing season. Selection for high- and low-maggot damage showed a linear trend in increasing and decreasing maggot damage, respectively. Both tolerance and antibiosis appeared to contribute to resistance (Theurer et al., 1982). Two root maggot-resistant sugar beet germplasm lines, F1015 and F1016, have been developed and released (Campbell et al., 1997) and two red beet accessions (PI 179180 and PI 181718) with resistance have been identified (Campbell et al., 1993). All evaluations have been dependent upon natural infestations at a site where sugar beet root maggot populations are consistently high. Resistance is not simply inherited, and it is probable that more than one parental line would need to be resistant to the maggot for the resultant hybrid to have sufficient resistance to be grown commercially without insecticides.

Transgenic sugar beet hybrids tolerant to broad-spectrum herbicides promise to simplify weed control while reducing the amount of herbicide applied to the crop. A gene, designated *epsps*, from *Agrobacterium* ssp. strain CP4 confers a high level of glyphosate (Roundup Ready) tolerance. This gene can be combined with a gene *(gox)* from *Achromobacter* ssp. to provide an additional resistance source and potentially provide higher levels of glyphosate resistance (Mannerlöf et al., 1997). Unfortunately, the genotypes most amenable to transformation are not elite parental lines. A wide range of tolerance to glyphosate among different transformed plants was assumed to be due to positional effects (Pedersen and Steen, 1995; Mannerlöf et al., 1997). A gene *(bar)* conditioning resistance to glufosinate ammonium (Liberty) has been incorporated into sugar beet through *Agrobacterium*-mediated transformation (D'Halluin et al., 1992), and resistant hybrids are now a commercial reality. The influence of different promoters on expression of the resistance gene was apparent in the development of the glufosinate ammonium-resistant lines. Hence, it is necessary to produce large quantities of different transgenic plants for field evaluations. The herbicide-resistance trait must then be transferred to elite lines using conventional breeding techniques. A dominant resistance gene can be introduced into either the female parent, of a hybrid or the pollinator. In a three-way cross, for the resistance in the hybrid to come from the female parent, it would need to be homozygous in both lines that produce the F_1 (CMS), i.e. the female parent. Introducing resistance into a pollinator may be slowed if the line is not self-fertile, as is often the case, and introduction into a tetraploid pollinator line would be complicated. The preferred approach will depend upon conditions unique to individual situations (Steen and Pedersen, 1995). Large-scale production of the herbicide-tolerant hybrids may create new weed beet problems (Boudry et al., 1993) that researchers should anticipate.

The yield and quality traits, pest resistance, and herbicide tolerance discussed previously are of obvious importance. In addition to these, some unique characteristics of sugar beet that may be altered in breeding programs are important under certain conditions. In many areas, sugar beets are stored in large exposed piles while awaiting processing. Selecting cultivars with reduced storage respiration rates could reduce losses during the storage period (Akeson and Widner, 1981). Soil adhering to harvested roots causes problems during harvesting, increases losses during storage, and increases processing costs. Storage roots devoid of the two vertical grooves that occur on standard cultivars (smooth-root cultivars) would reduce the quantity of soil remaining on harvested roots (Theurer, 1993). Sugar beet, fodder beet, and sugar-beet × fodder beet hybrids have been considered as sources of ethanol fuel that would provide an alternative to petroleum-based products. In an extensive U.S. study, the ethanol production potential of sugar beet was equal to or better than fodder beet or sugar-beet × fodder beet hybrids (Theurer et al., 1987). A gene from *Helianthus tuberosus* L. that causes the sucrose normally stored in sugar beet roots to be converted into low molecular weight fructans has been introduced into sugar beet (Sévenier et al., 1998). This transformed sugar beet produces a low-calorie sweetener, a potential new market for sugar beet growers and processors.

GERMPLASM RESOURCES

The genetic base of the commercial sugar beet crop is generally considered to be quite narrow (Lewellen, 1992; Doney, 1995b; Stander, 1993; McGrath et al., 1999). As suggested in the introduction, the early beet-sugar industry was dependent upon a few open-pollinated lines that had been selected for high sugar concentration. With this limited base as a beginning and the utilization of a single, or at best a few, sources for monogerm seed, cytoplasmic male sterility, and resistance to some of the major diseases, the current crop should be considered vulnerable to disease epidemics or insect infestations. Even in the absence of new insect or disease problems, success of any long-term breeding program is dependent upon genetic variability. Because of the difficulties and uncertainty of progress in a short time, the incorporation of exotic germplasm into agronomically useful populations (prebreeding) is primarily the domain of public breeders (Oldemeyer, 1975; Frey, 1996; Frey, 1998).

Taxonomic classification schemes frequently are subjected to revisions and the genus *Beta* is no exception (Krasochkin, 1959; Letschert et al., 1994; Winner, 1993); however, four sections are recognized in many schemes (Jassem, 1991). *Beta vulgaris* is included in the section *Beta* along with another primarily self-sterile species, *B. maritima,* and three predominately self-fertile species, *B. atriplicifolia, B. orientalis,* and *B. macrocarpa.* Species within this section readily cross with each other, producing fertile hybrids. *Beta maritima* has provided *Cercospora* leaf spot resistance (McFarlane, 1971), been considered as a source for male-sterility (Coe and Stewart, 1977; Dalke and Szota, 1993), and been used to enhance the genetic variability within the crop (Doney, 1995a). The section Corollinae includes diploid *(B. lomatogona* and *B. macrorhiza)* and polyploid *(B. corolliflora* and *B. trigyna)* species. Some of the Corollinae are apomictic, a trait that could be utilized to produce pure-breeding triploid lines equivalent to current hybrids. Fertile hybrids between *B. macrorhiza* and sugar beet have been reported; however, success in interspecific crosses involving other Corollinae species has been limited. The *Procumbentes* (or *Patellares*) section is comprised of three perennial species, *B. webbiana, B. procumbens,* and *B. patellaris.* The desire to transfer nematode resistance from Procumbentes species has stimulated interest in crossing species representing this section with sugar beet; however, lack of homology between chromosomes of the species has made interspecific hybrids difficult to obtain. The section *Nanae* contains a single species, *B. nana.* Hybrids between *B. nana* and sugar beet have not been reported (Smith, 1987). Although the species related to sugar beet contain large amounts of genetic variability that has not been utilized, the variability within *B. vulgaris* is considerable and generally more easily incorporated into elite populations (Ford-Lloyd, 1983; Dale et al., 1985). *Beta vulgaris* includes diverse plant types, such as Swiss chard, table beet, fodder beet, and old landraces and obsolete cultivars of sugar beet, and has a wide geographic distribution.

In many locations, the wild *Beta* populations are threatened by human activities. Recognition of this has prompted a number of collection expeditions (Doney, 1995b). In the United States, this germplasm is maintained under the auspices of the U.S. Department of Agriculture National Plant Germplasm System. A working collection, including wild accessions and public germplasm releases (Doney, 1995c), is maintained at Pullman, Washington. Information regarding individual accessions can be obtained from the GRIN (Germplasm Resources Information Network) database, and small quantities of seed are supplied to researchers when requested. A near-duplicate collection is stored in the National Seed Storage Laboratory in Fort Collins, Colorado. This collection is less accessible to the public and is considered a reserve or back-up source. The World *Beta* Network (WBN) was organized in 1989 (Doney, 1995b). This organization will facilitate the development of an international database and coordinate seed regeneration and collection activities.

REFERENCES

Abe, J., Guan, G.-P., and Shimamoto, Y. 1997a. A gene complex for annual habit in sugar beet (*Beta vulgaris* L.). Euphytica 94(2):129-135.

Abe, J., Guan, G.-P., and Shimamoto, Y. 1997b. A marker-assisted analysis of bolting tendency in sugar beet (*Beta vulgaris* L.). *Euphytica* 94(2):137-144.

Aberg, F.A. 1936. A genetic factor for the annual habit in beets and linkage relationships. *J Agric. Res.* 53(7):493-511.

Akeson, W.R. and Widner, J.N. 1981. Differences among sugar beet cultivars in sucrose loss during storage. *J. Am. Soc. Sugar Beet Technol.* 21(1):80-91.

Allard, R.W. 1960. *Principles of plant breeding.* John Wiley and Sons, Inc., New York.

Artschwager, E. 1926. Anatomy of the vegetative organs of the sugar beet. *J. Agric. Res.* 33(2):143-176.

Artschwager, E. 1942. Colchicine-induced tetraploidy in sugar beets: Morphological effects shown in selection of progenies of a number of selections. *Proc. Am. Soc. Sugar Beet Technol.* 3: 296-303.

Artschwager, E. and Starrett, R. C. 1933. The time factor in fertilization and embryo development in the sugar beet. *J. Agric. Res.* 47(11):823-843.

Barzen, E., Mechelke, W., Ritter, E., and Schulte-Kappert, E. 1995. An extended map of the sugar beet genome containing RFLP and RAPD loci. *Theor. Appl. Genet.* 90(2):189-193.

Bilgen, T., Gaskill, J.O., Hecker, R.J., and Wood, D.R. 1969. Transferring Cercospora leaf spot resistance from *Beta maritima* to sugar beet by backcrossing. *J. Am. Soc. Sugar Beet Technol.* 15(5):444-449.

Bernardo, R. 1998. A model for marker-assisted selection among single crosses with multiple genetic markers. *Theor. Appl. Genet.* 97(3):473-478.

Bockstahler, H.W., Hogaboam, G.J., and Schneider, C.L. 1950. Further studies on the inheritance of black rot resistance in sugar beet. *Proc. Am. Soc. Sugar Beet Technol.* 6:104-107.

Bohn, K. and Clarke, M.A. 1998. Composition of sugar beet and sugarcane and chemical behavior of constituents in processing. In P. W. van der Poel, H. Schiweck, and T. Schwartz (Eds.), *Sugar Technology* (pp. 115-208).Verlag, Dr. Albert Bartens KG, Berlin.

Bornscheuer, E., Meyerholz, K., and Wunderlich, K. H. 1993. Seed production and quality. In D.A. Cooke and R.K. Scott (Eds.), *The sugar beet crop* (pp. 121-155). Chapman and Hall, London.

Bosemark, N. O. 1993. Genetics and Breeding. In D.A. Cooke and R.K. Scott (Eds.), *The sugar beet crop* (pp. 67-119). Chapman and Hall, London.

Boudry, P., Mörchen, M., Saumitou-Laprade, P., Vernet, Ph., and Van Dijk, H. 1993. The origin and evolution of weed beets: Consequences for the breeding and release of herbicide-resistant transgenic sugar beets. *Theor. Appl. Genet.* 87:471-478.

Bugbee, W.M. 1978. Registration of F1001 and F1002 sugar beet germplasm. *Crop Sci.* 18(2):358.

Bugbee, W.M. 1979. Resistance to sugar beet storage rot pathogens. *Phytopathology* 69(12):1250-1252.

Bugbee, W.M. and Cole, D.F. 1979. Comparison of thiabendazole and genetic resistance for control of sugar beet storage rot. *Phytopathology* 69(12):1230-1232.

Cai, D., Kleine, M., Kifle, S., Harloff, H.-J., Sandal, N.N., Marcker, K.A., Klein-Lankorhorst, R.M., Salentijn, E.M.J., Lang, W., Stiekema, W.J., Wyss, U., Grundler, F.M.W., Jung, C. 1997. Positional cloning of a gene for nematode resistance in sugar beet. *Science* 275(5301):832-834.

Campbell, C.D. and Hutchinson, W.D. 1995. sugar beet resistance to Minnesota populations of sugar beet root aphid (Homoptera: Aphididae). *J. Sugar Beet Res.* 32(1):37-46.

Campbell, C.L. and Altman, J. 1976. Rapid laboratory screening of sugar beet cultivars for resistance to *Rhizoctonia solani. Phytopathology* 66(11):1373-1374.

Campbell, L.G. 1989. *Beta vulgaris* NC-7 collection as source of high sucrose germplasm. *J. Sugar Beet Res.* 26(1):1-9.

Campbell, L.G. 1995. Long-term yield patterns of sugar beet in Minnesota and Eastern North Dakota. *J. Sugar Beet Res.* 32(1):9-22.

Campbell, L.G., Anderson, A.W., Dregseth, R., and Smith, L.J. 1998. Association between sugar beet root yield and sugar beet root maggot (Diptera: Otitidae) damage. *J. Econ. Entomol.* 91(2):522-527.

Campbell, L.G., Anderson, A.W., and Prodoehl, K.A. 1993. The use of exotic and domestic germplasm for resistance to the sugar beet root maggot. *J. Sugar Beet Res.* 30(1-2):84.

Campbell, L.G. and Bugbee, W.M. 1985. Registration of storage rot resistant sugar beet germplasms. *Crop Sci.* 25(3):577.

Campbell, L.G. and Cole, D.F. 1986. Relationships between taproot and crown characteristics and yield and quality traits in sugar beets. *Agron. J.* 78(6):971-973.

Campbell, L.G. and Kern, J.J. 1982. Cultivar × environment interactions in sugar beet yield trials. *Crop Sci.* 22(5):932-935

Campbell, L.G. and Kern, J.J. 1983. Relationships among components of yield and quality of sugar beet. *J. Am. Soc. Sugar Beet Technol.* 22(2):135-145.

Campbell, L.G., Smith, G.A., Eide, J.D., Anderson, A.W., and Smith, L.J. 1997. Alternatives to insecticides for control of the sugar beet root maggot. *Proc. 60th Congress of Inst. for Beet Res.*, 419-421.

Campbell, L.G., Smith, G.A., Lamey, H.A., and Cattanach, A.W. 1998. *Cercospora beticola* tolerant to triphenyltin hydroxide and resistant to thiophanate methyl in North Dakota and Minnesota. *J. Sugar Beet Res.* 35(1/2):29-41.

Coe, G.E. 1987. Selecting sugar beets for low content of nonsucrose solubles. *J. Am. Soc. Sugar Beet Technol.* 24(1):41-48.

Coe, G.E. and O'Neil, N.R. 1983. Selecting sugar beet seedlings in the greenhouse for resistance to *Sclerotium rolfsii*. *J. Am. Soc. Sugar Beet. Technol.* 22(1):35-45.

Coe, G.E. and Schneider, C.L. 1966. Selecting sugar beet seedlings for resistance to *Aphanomyces cochlioides. J. Am. Soc. Sugar Beet. Technol.* 14(2):164-167.

Coe, G.E. and Stewart, D. 1977. Cytoplasmic male sterility, self-fertility, and monogermness in *Beta maritima* L. *J. Am. Soc. Sugar Beet Technol.* 19(3): 257-261.

Coons, G.H., Kotila, J.E., and Bockstahler, H.W. 1946. Black rot of sugar beet and possibilities of its control. *Proc. Am. Soc. Sugar Beet Technol.* 4:365-379.

Dale, M.F.B., Ford-Lloyd, B.V., and Arnold, M.H. 1985. Variation in some agronomically important characters in a germplasm collection of beet (*Beta vulgaris* L.). *Euphytica* 34(2):449-455.

Dalke, L. and Szota, M. 1993. Utilizing male sterility from *Beta maritima* in sugar beet breeding. *J. Sugar Beet Res.* 30(4):253-260.

D'Halluin, K., Bossut, M., Bonne, E., Mazur, B., Leemans, J., and Botterman, J. 1992. Transformation of sugar beet (*Beta vulgaris* L.) and evaluation of herbicide resistance in transgenic plants. *Biotechnology* 10(3):309-314.

Doney, D.L. 1995a. Registration of four sugar beet germplasms: y317, y318, y322, y387. *Crop Sci.* 35(3):947.

Doney, D.L. 1995b. International activities in *Beta* germplasm. In R. R. Duncan (Ed.), *International germplasm transfer: Past and present* (pp. 183-191). CSSA Spec. Publication 23. Crop Science Society of America. Madison, WI.

Doney, D.L. 1995c. USDA-ARS sugar beet releases. *J. Sugar Beet Res.* 32(4):229-257.

Doney, D.L. and Theurer, J.C. 1978. Reciprocal recurrent selection in sugar beet. *Field Crops Res.* 1(2):173-181.

Draycott, A.P. 1972. *Sugar-beet nutrition.* John Wiley and Sons, Inc., New York.

Dudley, J.W. 1958. Number of chloroplasts in the guard cells of inbred lines of tetraploid and diploid sugar beets. *Agron. J.* 50(3):169-170.

Elliott, M.C. and Weston, G.D. 1993. Biology and physiology of the sugar-beet plant. In D. A. Cooke and R. K. Scott (Eds.), *The sugar beet crop* (pp. 37-66). Chapman and Hall, London.

Fehr, W.R. 1987. *Principles of cultivar development, Vol 1: Theory and technique.* Macmillan, New York.

Fitzgerald, P. 1977. Influence of crossing direction on the agronomic performance of sugar-beet triploids. *Irish J. Agric. Res.* 16(2):149-153.

Ford-Lloyd, B.V. 1983. Progress in beet germplasm utilisation. *Genetika* 15(2):269-272.

Frey, K.J. 1996. *National plant breeding study I: Human and financial resources devoted to plant breeding research and development in the United States in 1994.* Special Report 98. Iowa Agricultural and Home Economics Experiment station, Ames, Iowa.

Frey, K.J. 1998. *National plant breeding study III: National plan for genepool enrichment of U.S. crops.* Special Report 101. Iowa Agricultural and Home economics Experiment station, Ames, Iowa.

Galatowitsch, M.W. and Smith, G.A. 1990. Regeneration from unfertilized ovule callus of sugar beet. *Can. J. Plant Sci.* 70(1):83-89.

Gaskill, J.O. 1952a. A new sugar-beet breeding tool—two seed generations in one year. *Agron. J.* 44(6):338.

Gaskill, J.O. 1952b. Induction of reproductive development in sugar beets by photo-thermal treatment of young seedlings. *Proc. Am. Soc. Sugar Beet Technol.* 7: 112-120.

Hall, R.D., Riksen-Bruinsam, T., Weyens, G.J., Rosquin, I.J., Denys, P.N., Evans, I.J., Lathouwers, J.E., Lefébvre, M.P., Dunwell, J.M., van Tunen, A., and Kerns, F.A. 1996. A high efficiency technique for the generation of transgenic sugar beets from stomatal guard cells. *Nature Biotechnology* 14(9):1133-1138.

Hallauer, A.R. and Miranda, J.B. 1981. *Quantitative genetics in maize breeding.* Iowa State Univ. Press., Ames, Iowa.

Halldén, C., Lind, C., Säll, T., Bosemark, N.O., and Bengtsson, B.O. 1990. Cytoplasmic male sterility in *Beta* is associated with structural rearrangements of the mitochondrial DNA and is not due to interspecific organelle transfer. *J. Mol. Evolution* 31(3):365-373.

Hansen, M., Kraft, T., Christiansson, M., and Nilsson, N.-O. 1999. Evaluation of AFLP in *Beta. Theor. Appl. Genet.* 98(6/7):845-852.

Hecker, R.J. 1978. Recurrent and reciprocal recurrent selection in sugar beet. *Crop Sci.* 18(5):805-809.

Hecker, R.J. 1985. Reciprocal recurrent selection for the development of improved sugar beet hybrids. *J. Am. Soc. Sugar Beet Technol.* 23(1/2):47-58.

Hecker, R.J. 1991. Effect of sugar beet root size on combining ability of sucrose yield components. *J. Sugar Beet Res.* 28(1-2):41-48.

Hecker, R.J. and Ruppel, E.G. 1975. Inheritance of Rhizoctonia root rot in sugar beet. *Crop Sci.* 15(4):487-490.

Hecker, R.J. and Ruppel, E.G. 1976. Polyploid and maternal effects on Rhizoctonia root rot resistance in sugar beet. *Euphytica* 25(2):419-423.

Hecker, R.J. and Ruppel, E.G. 1977. Rhizoctonia root-rot resistance in sugar beet: Breeding and related research. *J. Am. Soc. Sugar Beet Technol.* 19(3):246-256.

Hecker, R.J., Stafford, R.E., Helmerick, R.H., and Maag, G.W. 1970. Comparison of the same sugar beet F_1 hybrids as diploids, triploids, and tetraploids. *J. Am. Soc. Sugar Beet Technol.* 16(2):106-116.

Hogaboam, G.J. 1957. Factors influencing phenotypic expression of cytoplasmic male sterility in the sugar beet (*Beta vulgaris* L.). *J. Am. Soc. Sugar Beet Technol.* 9(5):457-465.

Hogaboam, G.J. 1962. Plastic chambers for humidity and temperature control in vegetative propagation and growth of sugar beets. *J. Am. Soc. Sugar Beet Technol.* 11(7):661-667.

Hogaboam, G.J., Zielke, R.C., and Schneider, C.L. 1982. Registration of EL40 sugar beet parental line. *Crop Sci.* 22(3):700.

Hutchinson, W.D. and Campbell, C.D. 1994. Economic impact of sugar beet root aphid (Homoptera: Aphididae) on sugar beet yield and quality in Southern Minnesota. *J. Econ. Entomol.* 87(2):465-475.

Hyne, V., Kearsey, M.J., Pike, D.J., and Snape, J.W. 1995. QTL analysis: Unreliability and bias in estimation procedures. *Mol. Breed.* 1(3):273-282.

Jassem, B. 1991. *Species relationship in the genus Beta as revealed by crossing experiments. Report of 2nd International* Beta *Genetics Workshop, Braunschweig, Germany.* IBPGR, International Crop Network Series 7: 55-61.

Keller, W. 1936. Inheritance of some major color types in beets. *J. Agric. Res.* 52(1):27-38.

Kerns, F.A., Jamar, D., Rouwendal, G.J.A., and Hall, R.D. 1990. Transfer of cytoplasm from new *Beta* CMS sources to sugar beet by asymmetric fusion. *Theor. Appl. Genet.* 79(3):390-396.

Kloen, D. and Speckmann, G.J. 1953. The creation of tetraploid beets. *Euphytica* 2(2):187-196.

Krasochkin, V.G. 1959. Review of the species of the genus *Beta. Genetike I Selektsii* 32(1):3-35 (In Russian; English translation published by Amerind Publishing Co., New Delhi, 1973).

Lang, W. and deBock, Th.S.M. 1994. Pre-breeding for nematode resistance in beet. *J. Sugar Beet Res.* 31(1):13-26.

Larsen, K. 1977a. Self-incompatibility in *Beta vulgaris* L.: 1. Four gametophytic, complementary S-loci in sugar beet. *Hereditas* 85(1):227-248.

Larsen, K. 1977b. Pseudo-compatibility in *Beta vulgaris* L.: A quantitative character, dependent on the degree of S-gene heterozygosity. *Incompatibility Newsletter* 8:48-51.

Lasa, J.M., Romagosa, I., Hecker, R.J., and Sanz, J.M. 1989. Combining ability in diploid and triploid sugar beet hybrids from diverse parents. *J. Sugar Beet Res.* 26(1):10-18.

Lawlor, N. Jr. and Doxtator, C.W. 1950. Breeding for resistance to root rot caused by *Sclerotium rolfsii. Proc. Am. Soc. Sugar Beet Technol.* 6:108-110.

Lejealle, F. and d'Aguilar, J. 1982. *Pests, diseases, and disorders of sugar beet.* Deleplanque, Maisons Lafitte, Paris (English version text by Andrew Dunning and William Byford).

Letschert, J.P.W., Lange, W., Frese, L., and Van Den Berg, R.G. 1994. Taxonomy of the section *Beta. J. Sugar Beet Res.* 31(1/2):69-85.

Lewellen, R.T. 1989. Registration of cytoplasmic male-sterile sugar beet germplasm C600CMS. *Crop Sci.* 29(1): 246.

Lewellen, R.T. 1992. Use of plant introductions to improve populations and hybrids of sugar beet. In H.L. Shands and L.E. Wiesner (Eds.), *Use of plant introductions in cultivar development. Part 2* (pp. 117-136). CSSA Spec. Publ. 20. Crop Science Society of America. Madison, WI.

Lewellen, R.T. 1999. Comparison of two sources of resistance to rhizomania and associated high temperature root rots in sugar beet. *J. Sugar Beet Res.* 36(1):83.

Lewellen, R.T., Skoyen, I.O., and Erichsen, A.W. 1987. Breeding sugar beet for resistance to rhizomania: Evaluation of host plant reactions and selection for and inheritance of resistance. *Proceedings of the 50th winter congress of the Int. Inst. Sugar Beet Res. II:*139-156.

Lewellen, R.T., Whitney, E.D., and Goulas, C.K. 1978. Inheritance of resistance to Erwinia root rot in sugar beet. *Phytopathology* 68(6): 947-950.

Linde-Laursen, I. 1972. A new locus for colour formation in beet, *Beta vulgaris* L. *Hereditas* 70(10):105-112.

Longden, P.C. 1993. Weed beet: A review. *Aspects of Applied Biology* 35:185-194.

Lux, H., Herrmann, L., and Wetzel, C. 1990. Production of haploid sugar beet (*Beta vulgaris* L.) by culturing unpollinated ovules. *Plant Breed.* 104(3):177-183.

Mackay, I.J., Gibson, J.P., and Caligari, P.D.S. 1999. The genetics of selfing with concurrent backcrossing in breeding hybrid sugar beet (*Beta vulgaris altissima* L.). *Theor. Appl. Genet.* 98(6/7):1156-1162.

MacLachlan, J.B. 1972. Estimation of genetic parameters in a population of monogerm sugar beet (*Beta vulgaris*). 1. Sib-analysis of mother-line progenies. 2. Offspring/parent regression analysis of mother-line progenies. 3. Analysis of a diallel set of crosses among heterozygous populations. *Irish J. Agric. Res.* 11(2-3): 237-246, 319-325, 327-338.

Mannerlöf, M., Lennerfors, B.-L., and Tenning, P. 1996. Reduced titer of BNYVV in transgenic sugar beets expressing the BNYVV coat protein. *Euphytica* 90:293-299.

Mannerlöf, M., Tuvesson, S., Steen, P., and Tenning, P. 1997. Transgenic sugar beet tolerant to glyphosate. *Euphytica* 94(4):83-91.

Marcum, W.B. 1948. Inheritance of bolting resistance. *Proc. Am. Soc. Sugar Beet Technol.* 5: 154-155.

McFarlane, J.S. 1971. Variety development. In R. T. Johnson, J. T. Alexander, G. E. Rush, and G. R. Hawkes (Eds.), *Advances in sugar beet production: Principles and practices* (pp. 401-435). Iowa State Univ. Press, Ames, Iowa.

McFarlane, J.S. 1981. Fusarium stalk blight resistance in sugar beet. *J. Am. Soc. Sugar Beet Technol.* 21(2):175-183.

McFarlane, J.S., Price, C., and Owen, F. V. 1948. Strains of sugar beets extremely resistant to bolting. *Proc. Am. Soc. Sugar Beet Technol.* 5:151-153.

McFarlane, J.S., Skoyen, I.O., and Lowellen, R.T. 1972. Performance of sugar beet hybrids as diploids and triploids. *Crop Sci.* 12(1):118-119.

McGrath, J.M., Derrico, C.A., and Yu, Y. 1999. Genetic diversity in selected, historical US sugar beet germplasm and *Beta vulgaris* spp. *maritima. Theor. Appl. Genet.* 98(6/7):968-976.

Melin, D. 1999. 1998 crop year in review: Spreckels Sugar Co., Brawley, Calif. *The sugar beet Grower* 37(1):20-22.

Mesbah, M., Scholten, O.E., deBock, T.S.M., and Lange, W. 1997. Chromosome localisation of genes for resistance to *Heterodera schachtii, Cercospora beticola* and *Polymyxa betae* using sets of *Beta procumbens-* and *B. patellaris-*derived monosomic additions in *B. vulgaris. Euphytica* 97(1):117-127.

Mezei, S., Jelaska, S., and Kovacev, L. 1990. Vegetative propagation of sugar beet from floral ramets. *J. Sugar Beet Res.* 27(3-4):90-96.

Mikami, T., Kishima, Y., Sugiura, M., and Kinoshita, T. 1985. Organelle genome diversity in sugar beet with normal and different sources of male sterile cytoplasms. *Theor. Appl. Genet.* 71(2):166-171.

Miller, J., Rekoske, M., and Quinn, A. 1994. Genetic resistance, fungicide protection, and varietal approval policies for controlling yield losses from Cercospora leaf spot infections. *J. Sugar Beet Res.* 31(1/2):7-12.

Nelson, R.T., and Deming, G.W. 1952. Effect of bolters on yield and sucrose content of sugar beets. *Proc. Am. Soc. Sugar Beet Technol.* 7:441-444.

Nilsson, N.-O., Halldén, C., Hansen, M., Hjerdin, A., and Säll, T. 1997. Comparing the distribution of RAPD and RFLP markers in a high density linkage map of sugar beet. *Genome* 40(5):644-651.

Oldemeyer, R.K. 1957. Sugar beet male sterility. *J. Am. Soc. Sugar Beet Technol.* 9(5):381-386.

Oldemeyer, R.K. 1975. Introgressive hybridization as a breeding method in *Beta vulgaris. J. Am. Soc. Sugar Beet Technol.* 18(3):269-273.

Owen, F.W. 1942. Inheritance of cross- and self-sterility in *Beta vulgaris. J. Agric. Res.* 64(12):679-698.

Owen, F.W. 1945. Cytoplasmically inherited male-sterility in sugar beets. *J. Agric. Res.* 71(10):423-440.

Owen, F.W. 1952. Mendelian male sterility in sugar beets. *Proc. Am. Soc. Sugar Beet Technol.* 7:371-376.

Owen, F.W. 1954. Hybrid sugar beets made by utilizing both cytoplasmic and Mendelian male sterility. *Proc. Am. Soc. Sugar Beet Technol.* 8:66.

Panella, L.W. and Hecker, R.J. 1995. Registration of annual O-type and CMS sugar beet germplasm lines FC404 and FC404CMS. *Crop Sci.* 35(6):1721.

Paul, H., Henken, B., Scholten, O.E., deBock, Th.S.M., and Lang, W. 1994. Resistance to Polymyxa beta and beet necrotic yellow vein virus in *Beta* species of the section *Corollinae. J. Sugar Beet Res.* 31(1):1-6.

Paul, H., van Deelen, J.E.M., Henken, B., deBock, Th.S.M., Lang, W., and Kerns, F.A. 1990. Expression of in vitro resistance to *Heterodera schachtii* in hairy roots of an alien monotelosomic addition plant of *Beta vulgaris,* transformed by *Agrobacterium rhizogenes. Euphytica* 48(2):153-157.

Pedersen, H.C., and Steen, P. 1995. The stability of transgenes inserted into sugar beet (*Beta vulgaris* L.). *Proc. 58th Congress of Int. Inst. for Beet Res.,* 197-200.

Peto, F.H., and Hill, K.W. 1942. Colchicine treatment of sugar beets and the yielding capacity of the resulting polyploids. *Proc. Am. Soc. Sugar Beet Technol.* 3: 287-295.

Pillen, K., Steinrücken, G., Herrmann, R.G., and Jung, C. 1993. An extended linkage map of sugar beet (*Beta vulgaris* L.) including nine putative lethal genes and the restorer X. *Plant Breed.* 111(4):265-272.

Pillen, K., Steinrücken, G., Wricke, G., Herrmann, R.G., and Jung, C. 1992. A linkage map of sugar beet (*Beta vulgaris* L.) *Theor. Appl. Genet.* 84(1-2):129-135.

Powers, L. 1963. The partitioning method of genetic analysis and some aspects of its application to plant breeding. In W.D. Hanson and H.F. Robinson (Eds.), *Statistical Genetics and Plant Breeding*. National Acad. Sci—National Res. Council. Pub. 982: 280-318.

Raybould, A.F., Mogg, R.J., and Clarke, R.T. 1996. The genetic structure of *Beta vulgaris* ssp. *maritima* (sea beet) populations: RFLPs and isozymes show different patterns of gene flow. *Heredity* 77(Pt. 3):245-250.

Roundy, T.E. and Theurer, J.C. 1974. Inheritance of a yellow-leaf mutant and a pollen fertility restorer in sugar beet. *Crop Sci.* 14(1):62-63.

Ruppel, E.G. and Hecker, R.J. 1988. Variable selection pressure for different levels of resistance to Rhizoctonia root rot in sugar beet. *J. Sugar Beet Res.* 25(1):63-69.

Sadeghian, S.Y., Becker, H.C., and Johansson, E. 1993. Inheritance of bolting in three sugar beet crosses with different periods of vernalization. *Plant Breed.* 110(4):328-333.

Sadeghian, S.Y. and Johansson, E. 1993. Genetic studies of bolting and stem length in sugar beet (*Beta vulgaris* L.) using a factorial cross design. *Euphytica* 65(3):177-185.

Saunders, J.W. 1982. A flexible in-vitro shoot culture propagation system for sugar beet that includes rapid floral induction of ramets. *Crop Sci.* 22(6):1102-1105.

Saunders, J.W., Acquaah, G., Renner, K.A., and Doley, W.P. 1992. Monogenic dominant sulfonylurea resistance in sugar beet from somatic cell selection. *Crop Sci.* 32(6):1357-1360.

Saunders, J.W. and Shin, K. 1986. Germplasm and physiologic effects on high-frequency hormone autonomous callus and subsequent shoot regeneration in sugar beet. *Crop Sci.* 26(6):1240-1245.

Savitsky, H. 1952. Polyploid sugar beets—cytological study and methods of production. *Proc. Am. Soc. Sugar Beet Technol.* 7:470-476.

Savitsky, H. 1954. Self-sterility and self-fertility in monogerm beets. *Proc. Am. Soc. Sugar Beet Technol.* 8(2): 29-33.

Savitsky, H. 1966. A method for inducing autopolyploidy in sugar beets by seed treatment. *J. Am. Soc. Sugar Beet Technol.* 14(1):26-47.

Savitsky, V.F. 1952. A genetic study of monogerm and multigerm characters in beets. *Proc. Am. Soc. Sugar Beet. Technol.* 7:331-338.

Schneider, C.L. 1954. Methods of inoculating sugar beets with *Aphanomyces cochlioides* Drechsl. *Proc. Am. Soc. Sugar Beet Technol.* 8:247-251.

Schneider, C.L. and Hogaboam, G.J. 1983. Evaluation of sugar beet breeding lines in greenhouse tests for resistance to *Aphanomyces cochlioides. J. Am. Soc. Sugar Beet Technol.* 22(2):101-107.

Scott, R.K., and Longden, P.C. 1970. Pollen release by diploid and tetraploid sugar-beet plants. *Ann. Appl. Biol.* 66(2):129-135.

Sévenier, R., Hall, R.D., Van der Meer, I.M., Hakkert, H.J.C., Van Tunen, A.J., and Koops, A.J. 1998. High level fructan accumulation in a transgenic sugar beet. *Nature Biotechnology* 16(9):843-846.

Shavrukov, Y.N. 1997. Multiple allelism at the M-m locus controlling monofloret (monogerm) in sugar beet. *Genetika* 33(1):46-52 (translated from Russian).

Smed, E., Van Geyt, J.P.C., and Oleo, M. 1989. Genetical control of linkage relationships of isozyme markers: 1. Isocitrate dehydrogenase, adenylate kinase, phosphoglucomutase phosphate isomerase, and cathodal peroxidase. *Theor. Appl. Genet.* 78(1):97-104.

Smith, G.A. 1980. Sugarbeet. In W. R. Fehr and H. H. Hadley (Eds.), *Hybridization of crop plants* (pp. 601-616). American Society of Agronomy, Madison, Wisconsin.

Smith, G.A. 1985. Response of sugar beet in Europe and the USA to *Cercospora beticola* infection. *Agron. J.* 77(1):126-129.

Smith, G.A. 1987. Sugar beet. In W. R. Fehr (Ed.), *Principles of cultivar development Vol 2: Crop species* (pp. 577-625). Macmillan, New York.

Smith, G.A. and Campbell, L.G. 1996. Association between resistance to *Cercospora* and yield in commercial sugar beet hybrids. *Plant Breed.* 115(1):28-32.

Smith, G.A. and Gaskill, J.O. 1970. Inheritance of resistance to Cercospora leaf spot in sugar beet. *J. Am. Soc. Sugar Beet Technol.* 16(2):172-180.

Smith, G.A., Hecker, R.J., Maag, G.W., and Rasmuson, D.M. 1973. Combining ability and gene action estimates in an eight parent diallel cross of sugar beet. *Crop Sci.* 13(3):312-316.

Smith, G.A., Hecker, R.J., and Martin, S.S. 1979. Effects of polyploidy level on the components of sucrose yield and quality in sugar beet. *Crop Sci.* 19(3):319-323.

Smith, G.A. and Martin S.S. 1989. Effect of selection for sugar beet purity components on quality and sucrose extractions. *Crop Sci.* 29(2):294-298.

Smith, G.A., Martin, S.S., and Ash, K.A. 1977. Path coefficient analysis of sugar beet purity components. *Crop Sci.* 17(2):249-253.

Smith, G.A. and Moser, H.S. 1985. Sporophytic-gametophytic herbicide tolerance in sugar beet. *Theor. Appl. Genet.* 71(2):231-237.

Smith, G.A. and Ruppel, E.G. 1974. Heritability of resistance to Cercospora leaf spot in sugar beet. *Crop Sci.* 14(1):113-115.

Smith, G.A. and Ruppel, E.G. 1988. Registration of FC609 and FC609 CMS sugar beet germplasm. *Crop Sci.* 28(6):1039.

Smith, G.A. and Schweizer, E.E. 1983. Cultivar × herbicide interaction in sugar beet. *Crop Sci.* 23(2):325-328.

Snyder, F.W. and Carlson, G.E. 1978. Photosynthate partitioning in sugar beet. *Crop Sci.* 18(4):657-661.

Stander, J.R. 1993. Pre-breeding from the perspective of the private plant breeder. *J. Sugar Beet Res.* 30(4):197-207.

Steen, P. and Pedersen, H.C. 1995. Strategies in creating transgenic herbicide tolerant sugar beet (*Beta vulgaris* L.) varieties. *Proc. 58th Congress of Int. Inst. for Beet Res.*, 189-192.

Stout, M. 1946. Relation of temperature to reproduction in sugar beets. *J. Agric. Res.* 72(2):49-68.

Theurer, C.J. 1968. Linkage tests of Mendelian male sterility and other genetic characters in sugar beet, *Beta vulgaris* L. *Crop Sci.* 8(6):698-701.

Theurer, C.J. 1971. Inheritance studies of a pollen restorer from Ruby Queen table beet. *J. Am. Soc. Sugar Beet Technol.* 16(4):354-358.

Theurer, C.J. 1978. Registration of two germplasm lines of sugar beet. *Crop Sci.* 18(6):1101-1102.

Theurer, J.C. 1993. Fibrous root growth and partitioning in smooth root sugar beet versus standard root types. *J. Sugar Beet Res.* 30(3):143-150.

Theurer, J.C., Blickenstaff, C.C., Mahrt, G.C., and Doney, D.L. 1982. Breeding for resistance to the sugar beet root maggot. *Crop Sci.* 22(3):641-645.

Theurer, J.C., Doney, D.L., Smith, G.A., Lewellen, R.T., Hogaboam, G.J., Bugbee, W.M., and Gallian, J.J. 1987. Potential ethanol production from sugar beet and fodder beet. *Crop Sci.* 27(5):1034-1040.

Theurer, C.J. and Ryser, G.K. 1969. Inheritance studies with a pollen fertility restorer sugar beet inbred. *J. Am. Soc. Sugar Beet Technol.* 15(6):538-545.

Tsai, C.J. and Saunders, J.W. 1995. Somatic embryos from callus of sugar beet biotechnology clone REL-1. *J. Sugar Beet Res.* 32(4):215-227.

Wallis, R.L. and Turner, J.E. 1968. Resistance of sugar beets to sugar beet root aphids, *Pemphigus populivenae* Fitch. *J. Am. Soc. Sugar Beet Technol.* 14(8):671-673.

Wricke, G. and Weber, W.E. 1986. *Quantitative genetics and selection in plant breeding.* Walter de Gruyter and Co., Berlin.

Whitney, E.D. and Duffus, J.E. (Eds.). 1986. *Compendium of beet diseases and insects.* Am. Phytopathological Soc. St. Paul, Minnesota.

Winner, C. 1993. History of the crop. In D.A. Cooke and R.K. Scott (Eds.), *The sugar beet crop* (pp. 1-35). Chapman and Hall, London.

Wozniak, C.A. 1999. Transgenic sugar beet: Progress and development. In V.L. Chopra, V.S. Malik, and S.R. Bhat (Eds.), *Applied biotechnology* (pp. 301-324). Science Pub., Inc. Enfield, New Hampshire.

Wyse, R. 1979. Parameters controlling sucrose content and yield of sugar beet roots. *J. Am. Soc. Sugar Beet Technol.* 20(4):368-385.

Yu, H.M. 1995. Identification of a *Beta maritima* source of resistance to root-knot nematode for sugar beet. *Crop Sci.* 35(5):1288-1290.

Yu, H.M. 1996. Registration of root-knot nematode resistant beet germplasm M66. *Crop Sci.* 36(2):469.

Yu, H.M. 1997. Registration of Mi-1 root-knot resistant beet germplasm line. *Crop Sci.* 37(1):295.

Zhong, Z., Smith, H.G., and Thomas, T.H. 1993. *In vitro* culture of petioles and intact leaves of sugar beet (*Beta vulgaris*). *J. Plant Growth Regulation* 12(1):59-66.

Chapter 8

Bananas and Plantains: Future Challenges in *Musa* Breeding

Michael Pillay
A. Tenkouano
John Hartman

INTRODUCTION

History and Taxonomy

Bananas and plantains (*Musa* spp. L.) are giant perennial herbs that grow best in the tropical and subtropical regions of the world. They are the world's largest fruit crop. There appears to be no definite botanical distinction between bananas and plantains. Generally, plantain fruits are thicker, longer, and starchier than banana fruits, which are sweeter in their ripened state. In this chapter, the term *banana* will be used to refer to both bananas and plantains. Bananas belong to the Musaceae, one of the six families of the order Zingiberales. The Musaceae have long been associated with human history. The banana plant is described as the "Tree of Paradise" in the Koran and as the "Tree of Knowledge" in the Book of Genesis in the Old Testament (Rowe and Rosales, 1996). These descriptions are probably responsible for the earlier taxonomic identification of bananas as *Musa paradisiaca* L. or *M. sapientum* L. The cultivation of bananas was first documented in ancient scriptures of India dating back to 500 to 600 B.C. (Reynolds, 1951). Bananas originated in Southeast Asia but are now found in all humid tropical regions of the world where they have become major subsistence and cash crops for both domestic and international trade markets.

The Musaceae comprise two genera, *Musa* and *Ensete* Horan. The genus *Ensete* is only of local importance in the highlands of Ethiopia with one major species, *E. ventricosa* (Welw.) Cheem. [syn. *E. edule* (Gmel) Horan., *Musa ensete* Gmel.]. The corm and pseudostem of this species are processed into an edible starch which is fermented to prepare a local dish in parts of southern Ethiopia (Demeke, 1986). The pseudostem is also a source of fiber that is used to make cordage and sacks.

The genus *Musa* consists of four sections, Australimusa, Callimusa, Rodochlamys, and Eumusa. The sections Australimusa and Callimusa have a basic set of 10 chromosomes in contrast to 11 chromosomes in the other two sections. The sections Callimusa and Rodochlamys contain species that are nonparthenocarpic and are thus of ornamental interest only. The Australimusa section contains some economically important species such as *Musa textilis* Nee, which is the source of a strong and highly resilient fiber. This section also comprises some parthenocarpic edible varieties collectively referred to as the Fe'i bananas, distinguishable from other cultivated bananas by their erect fruit bunches and often red sap. The section Eumusa constitutes the most important group of the genus *Musa* as it is the largest and most geographically widespread group that also comprises virtually all edible, parthenocarpic bananas.

The Eumusa section is commonly divided into dessert bananas, cooking bananas, plantains, and beer bananas. Of these, only the dessert bananas have gained prominence on the international market scene and have recently become subject to commercial disputes among nations. The other types of bananas are virtually absent from international trade circuits and are predominantly ethnic staple crops grown for local consumption.

There are many excellent books covering a wide range of topics on bananas. These include texts by Stover and Simmonds (1987), Gowen (1995), and Robinson (1996). This chapter outlines the biology of banana and production systems necessary for an understanding of what we believe to be the future direction of *Musa* breeding. For a fuller understanding, the reader should consult the previously mentioned references.

Importance of Musa

The annual production of *Musa* is about 88 million tons grown over an area of approximately 10 million ha (Horry, Sharrock, et al., 1998). The three main areas of banana production include Africa (35 percent), Asia and the Pacific (29 percent), and Latin America and the Caribbean (35 percent). Economically, bananas are: (1) a major export crop in some countries, especially in Latin America and the Caribbean, (2) an important staple food for rural and urban consumers, and (3) an important source of income for rural populations in the tropics. Bananas play a vital role in the nutrition, well-being, and cultural life of millions of people in Central and Western Africa and South and Central America (Price, 1995). They are the major staple food in the equatorial belt of Africa. Over 70 million in West and Central Africa get more than 25 percent of their carbohydrates from plantains (Robinson, 1996). Bananas are rich in carbohydrates (about 35 percent) and fiber (6 to 7 percent) and contain a relatively low protein and fat content (1 to 2 percent). The banana is a good source of major elements, such as potassium, magnesium, phosphorus, calcium, iron, and vitamins A, B_6, and C (Marriott and Lancaster, 1983). Relative to other common tropical and subtropical fruits, bananas are high-energy fruits that are richer in carbohydrates, phosphorus, iron, and potassium (Robinson, 1996).

There is considerable overlap in the way bananas are eaten (Robinson, 1996). Bananas are mainly consumed raw as a dessert fruit when they are ripe but are also eaten as a starchy food when unripe. They can be boiled, roasted, or fried either in the ripe or unripe state. Besides the fruit, nonfruit parts of the banana plant, including the corm, shoots, and male buds, are eaten as vegetables in Africa and many parts of Asia (Simmonds, 1962). The heart of the pseudostem is eaten in India. Besides a source of food, other parts of the banana plant are also useful in everyday life: the dry leaves are used for fruit packing whereas green leaves serve as plates or for wrapping or cooking. Bananas that grow in the highlands (altitudes of 1200 to 1700 m) of East Africa are distinctly different from dessert bananas and plantains (Rowe and Rosales, 1996). The East African type can be divided into cooking bananas, which are boiled as unripe fruit, and beer bananas. The brewing of ripe bananas to prepare beer, wine, and other alcoholic products forms an important part of the cultural life of many people in East Africa (Stover and Simmonds, 1987).

Bananas form important sources of income in some locations where small landholders produce them in their compounds or home gardens. These home gardens are dumping sites for the household refuse, mulch from harvested plants, and manure from farm animals. The variations in quality and rate of decomposition of these additions result in high levels of soil organic matter (Kirkby, 1990). These plants are productive throughout the year providing a constant source of food as well as a regular source of income.

Origin and Genome Groups in Musa

Musa spp. are considered to have originated in Indochina and Southeast Asia (Simmonds, 1962). From there, bananas gradually spread to all humid tropical regions. Domesticated bananas are considered to be derived from inter- and intraspecific hybridization between two wild diploid ($x = 11$) species, *M. acuminata* Colla. and *M. balbisiana* Colla., whose genomes are designated as *A* and *B,* respectively. The two genomes contribute different traits to the phenotype of the various *Musa* clones. For example, the genes for hardiness, drought tolerance, greater disease resistance, and increased starchiness were contributed by the *B* genome of *M. balbisiana* (Robinson, 1996). Hybridization between various subspecies of *M. acuminata* produced a range of diploid cultivars (*AA* genomes). Diploid *AA*s produced triploid *AAA* types by chromosome restitution. Hybridization between *AA* diploids and *M. balbisiana (BB)* gave rise to the many *AAB* and *ABB* types of today. The cultivated triploid *AAA* comprises the sweeter dessert cultivars whereas the *AAB*s and *ABB*s are the starchier cooking types. Triploid cultivars are the most numerous, diploids are of local importance, and tetraploid forms are rare (Novak, 1992). Other genomic groups, including *AB, ABBB, AAAB,* and *AABB,* that originated naturally or by artificial hybridization also exist. The basic characteristics of triploid bananas include parthenocarpy and female sterility. Hence the development of edible seedless banana that could be propagated vegetatively by suckers.

Constraints to Production

Pest and disease problems are increasingly important constraints to banana production. Black leaf streak (BLS), caused by *Mycosphaerella fijiensis* (Morelet) is considered the most serious production constraint in most of the important banana production areas of the world. This disease causes from 30 to 50 percent yield reduction in plantain landraces in Africa and tropical America and could wipe out commercial production of Cavendish bananas if it were not for the massive application of synthetic fungicides costing an estimated $200 million annually. A related disease, yellow Sigatoka leaf spot (YLS, caused by *M. musicola* Leach) causes similar damage, primarily in highlands and areas where BLS has not yet penetrated.

Fusarium wilt can cause complete destruction of susceptible plantations and in fact did devastate the old 'Gros Michel'-based dessert banana export industry during the middle of the twentieth century (Simmonds, 1966). The causal agent of Fusarium wilt, *Fusarium oxysporum* Schlecht. f.sp. *cubense (F.O.C.),* is a soilborne filamentous fungus that colonizes and occludes the xylem of banana plants (Ploetz, 1994). Four races of *F.O.C.* have been identified by differential host testing (Ploetz, 1994). The increasing importance of Fusarium wilt on Cavendish clones (race 4) poses a serious threat to today's banana export industry and has refocused research on this disease (Ploetz, 1994).

Yield losses from plant parasitic nematodes can exceed 50 percent in some environments (Speijer et al., 1994; Davide, 1996). Several nematode species cause damage to bananas; however, the most damaging and widespread nematodes on banana are the burrowing nematode, *Radopholus similis* Cobb, the spiral nematode, *Helicotylenchus multicinctus* Cobb, and the root-lesion nematodes, *Pratylenchus coffeae* Zimmerman and *Pratylenchus goodeyi* Sher and Allen (Speijer and De Waele, 1997).

Banana streak virus (BSV) is a badnavirus with worldwide distribution that may cause complete loss of individual plants (Frison and Sharrock,

1998). It is vectored by mealybugs and exhibits pararetrovirus behavior similar to cauliflower mosaic virus. Perhaps the greatest challenge to banana improvement posed by BSV is the presence of activatible BSV sequences integrated in the banana genome (LaFleur et al., 1996). This has led to a proposal to breed for genotypes with low propensity to express integrated sequences rather than breeding for traditional virus resistance (B.E.L. Lockhart, May 11, 1998, e-mail).

The banana weevil (*Cosmopolites sordidus* Germ.) causes damage to bananas throughout the tropics, and is thought to be a particular problem on highland cooking bananas and plantains important to small landholding farmers in Africa (Ortiz et al., 1995). New plantations are particularly susceptible as weevils may kill young shoots and prevent crop establishment (Mitchell, 1980). Variation among banana clones in susceptibility to weevil attack has been observed, but breeding for weevil resistance is a very recent activity (Ortiz et al., 1995).

Other important pests and diseases include Moko disease (causal agent, *Pseudomonas solanacearum* E.F. Sm.), banana mosaic disease (caused by cucumber mosaic virus; CMV), and banana bunchy top virus disease (BBTV). Although chemical and cultural practices may be used to ameliorate many banana pest and disease problems, most of these have been deemed inappropriate for small landholding farmers. The development of cultivars with resistance and/or tolerance to these production constraints is the primary focus of banana improvement programs around the world.

Constraints to Musa *Improvement*

Bananas present several impediments to breeders and geneticists wishing to improve them. Most of the widely grown clones are triploid with low female fertility or male and female sterility (Vuylsteke, Swennen, et al., 1993). Meiosis in triploids can result in formation of gametes that contain uneven chromosome numbers resulting in high sterility. Other mechanisms of sterility result from morphological errors in postmeiotic stages and physiological dysfunction during pollination and fertilization (Simmonds, 1962). Seed set per bunch in many clones is less than one seed, and germination in soil is usually less than 1 percent (Ortiz and Vuylsteke, 1995). It takes over a year to go from seed to seed, and each banana plant requires 4 to 9 m² of field space during that time. Shortening this cycle is indeed a goal of *Musa* breeders. In the early stages of banana breeding, most progeny are obtained from 3x by 2x crosses. The effects of multiploidy and autopolyploid chromosome behavior results in an unpredictable frequency of aneuploids and undesirable hyperpolyploids (>5x) in addition to 2x, 3x, and 4x euploids (Simmonds, 1966). Diploid bananas generally have an unacceptably low yield potential, while tetraploid bananas often suffer from premature senescence, fruit drop, short shelf life, and a weak pseudostem. Additionally, although low fertility is a problem for breeding, very high fertility, as is often the case with 2x and 4x bananas, may result in undesirable seed production in a released cultivar. Banana seeds are large and hard and are not acceptable to most banana consumers. Large scale screening of a number of cultivars has shown that sterility and seedlessness is partial in some plants. These plants have formed the basis for conventional breeding schemes.

OVERCOMING INFERTILITY BARRIERS

Ploidy Manipulation, Fertility, and Seed Handling

As was stated above, seed set and germination can be very low in bananas. Embryo rescue techniques have increased the number of progeny per pollination by three to 10 times (Vuylsteke et al., 1990). Selection for fertility can greatly increase the efficiency of pollination (Ortiz and Vuylsteke, 1995). Modifications of pollinating procedures and methods of seed handling, including improvements in embryo rescue techniques and in vivo hybrid-seed germination, will, in combination with improved fertility in advanced breeding populations, greatly improve the efficiency of banana breeding. This will allow breeders to evaluate larger segregating populations, transfer more of their resources from obtaining progeny to evaluating progeny, and thereby increase the number and quality of cultivars released.

Other Methods for Overcoming Infertility

The recalcitrance of banana to genetic improvement makes biotechnological approaches especially attractive (Dale, 1990; Novak, 1992). In addition to improving fertility, problems of fertility can be bypassed by using genetic transformation techniques. Methods for inserting transgenes, including electroporation (Sagi et al., 1995), particle bombardment (Sagi et al., 1995) and cocultivation with *Agrobacterium* (May et al., 1995) have been devised for *Musa*. Transgenes for antifungal proteins have been successfully inserted and expressed in *Musa* at Katholieke Universiteit Leuven, Belgium (Remy et al., 1998). It is hoped that antifungal proteins will enable *Musa* spp. to overcome the devastating fungal diseases of the Sigatoka complex, which includes black and yellow leaf streak. Once regulatory hurdles have been overcome, these methods promise to bring new alleles for fruit quality and pest and disease resistance to relatively infertile *Musa* clones such as those in the Cavendish subgroup (Crouch, Vuylsteke, et al., 1998). Transgenes are also being sought by various laboratories for: (1) Panama disease or banana wilt, the second major fungal disease of banana, (2) viral diseases including BSV and BBTV (banana bunchy top virus), and (3) nematode resistance. These transgenics may also serve as new sources of genetic variation for more conventional breeding. While conventional breeding of *Musa* is faced with difficulties, it must be understood that currently available transformation methods will not solve all these difficulties. The scarcity of useful genes, factors affecting transgene expression, interactions between transgenes and native genes of the plant, and the quantitative nature of some traits must be considered before accepting that genetic transformation is the obvious choice for *Musa* improvement. We advocate that the aim should also be to insert useful genes into an elite background for further conventional breeding.

NEW METHODS FOR MANIPULATING
A MULTIPLOIDY CROP

Cytogenetics

Despite its importance as a staple crop and export commodity in world trade, cytogenetic studies in *Musa* have been neglected. The small size, poor staining of chromosomes, and difficulties in obtaining metaphase chromosomes are some of the reasons cited for this lag in knowledge in *Musa* karyology (Osuji et al., 1996). However, there appears to be a renewed interest in cytogenetic studies of *Musa*. This interest is probably spurred by: (1) the development of better techniques and optics to study small chromosomes, (2) a renewed interest in *Musa* as an important food crop in tropical and subtropical regions, and (3) new breeding schemes for the improvement of bananas that involve interploidy crosses such as 3x-2x, and 4x-2x. These crosses produce progeny with various euploid and aneuploid chromosome numbers. The ploidy status and genomic constitution of these plants must be ascertained if they are to be used further in a breeding program. Flow cytometry and conventional chromosome analysis are, therefore, becoming increasingly important in *Musa* breeding programs. While flow cytometry provides information on the ploidy status of the plant, conventional and more advanced cytogenetic techniques are required to provide information on its exact genomic makeup. In this regard, chromosome banding and the more recently adopted in situ hybridization (ISH) techniques should facilitate the identification of the morphologically similar chromosomes in *Musa*. Newer techniques such as the computer-based chromosome image-analyzing system (CHIAS) have been used for karyotyping plants with small chromosomes (Fukui, 1986; Iijima and Fukui, 1991). The method is compatible with ISH and chromosome painting techniques and provides the opportunity to map linkage groups to chromosomes and specific marker clusters to chromosome regions (Fukui, 1986). Flow cytometry can be used to classify chromosomes according to DNA content, AT/GC ratio, presence of repetitive DNA sequences, and morphology (Dolezel, 1998). The histograms produced from such manipulations are called flow karyotypes. In a heterogeneous karyotype, flow karyotypes should be able to distinguish individual chromosomes that could be sorted in large quantities. Flow sorting of individual chromosomes lends itself to many DNA manipulations, such as obtaining chromosome-specific painting probes, for construction of chromosome-specific libraries and mapping of agronomically important genes (Pich et al., 1995).

Other cytological techniques, such as primed in-situ DNA labeling, (PRINS), cycling PRINS (C-PRINS) and PRINS fluorescent in situ hybridization (PRINS-FISH), are being applied to plants for localization of DNA sequences (Kubalakova et al., 1997; Kubalakova and Dolezel, 1998; Kubalakova et al., 1998). Similar techniques may be useful in banana cytology.

Establishing Karyotypes in Musa

Although a preliminary karyotype of *Musa* has been reported (Dantas et al., 1993), there is need for more detailed studies of *Musa* chromosomes. The individual chromosomes in bananas have not yet been identified and numbered partly because of the small size of the chromosomes and the absence of cytological markers. Preliminary studies show that *M. acuminata* and *M. balbisiana* have one pair of satellite chromosomes. The other chromosomes appear to be metacentric to submetacentric. In situ hybridization studies showed a single pair of sites for the 18S-25S rDNA in *Musa* (Dolezelova et al., 1998; Osuji et al., 1998). Wang and colleagues (1993) also reported a single pair of chromosomes with satellites in wild *AA* bananas. Dolezelova and colleagues (1998) observed that the labeling intensity for the 45S rDNA was stronger in one of the homologues. This was observed consistently in *AA* and *BB* genotypes. The authors concluded that a difference in copy number exists between the homologues carrying the nucleolar organizer regions in *Musa*. Great variation in the number of 5S rDNA sites has been reported for *Musa* (Dolezelova et al., 1998; Osuji et al., 1998). Dolezelova and colleagues (1998) found that the number of sites ranged from four to six in *AA* genotypes (*M. acuminata* ssp. banksii, Pisang Mas, and Pa (Rayong). In *BB* genotypes, six sites were found, the only exception being *M. balbisiana* type Cameroon. Osuji and colleagues (1998) reported seven sites for the *AA* genotypes, *M. acuminata* ssp. *burmannicoides* (Calcutta 4), and Pisang Lilin and six sites for *M. balbisiana* type Butohan. Osuji and colleagues (1998) also reported six sites in *AAA* dessert bananas (Cavendish and Valery), seven sites in *AAB* plantains (Agbagba and Obino l'Ewai) and eight sites in *ABB* cooking bananas (Bluggoe, Fougamou, and Cardaba). Assuming that

the reported numbers of 5S rDNA sites are correct and not marred by arti-
facts of the ISH technique, it would appear that 5S sites could provide useful
physical landmarks for genome mapping in banana. The mitotic metaphase
chromosomes of bananas appear as small dotted structures without great detail in the fully contracted state. To overcome limitations imposed by chromosome size, either mitotic prometaphase or pachytene chromosomes may
be ideal for karyotyping. Mitotic prometaphase chromosomes are generally
about three to five times longer than those found at the fully condensed
metaphase stage. Our studies show that primary and secondary constric-
tions are more easily distinguished in prometaphase chromosomes than in
overcondensed metaphase chromosomes.

Since the 1970s, our ability to distinguish morphologically between
chromosomes has been enhanced by new staining techniques using Giemsa
and fluorochrome dyes. Under controlled conditions, chromosomes were
found to stain in a consistent banding pattern instead of being uniform in appearance. Banding techniques provide cytological markers along the length
of chromosomes enabling easy identification of homologous pairs and individual chromosomes of a karyotype. Many different banding techniques are
available, each giving a different banding pattern based on the specific biochemical consequences of each method (Stace, 1980). The common types
applied to plants include C-banding, G-banding, Q-banding and Hy-band-
ing. Banding techniques have not been applied to *Musa* species. Preliminary
studies show that *Musa* chromosomes display characteristic C-banding patterns that may play a vital role in chromosome identification and development
of a *Musa* karyotype. C-banding patterns are heritable and remain discern-
ible throughout the mitotic and meiotic cell cycles (Jewel and Islam-Faridi,
1994) and are a useful technique to identify parental chromosomes in
interspecific hybrids. More extensive studies are required to determine the
full potential of banding techniques in *Musa* cytogenetics and its role in
Musa breeding.

Most cytological studies in *Musa* have been restricted to mitotic
metaphase chromosomes. The pachytene stage of meiosis is also ideal for
studying chromosome morphology especially in plants with very small
chromosomes. The advantages of pachytene analysis include the visibility
of centromeres, chromomeres, telomeres, knobs, nucleoli, and the possibility to distinguish eu- and hetero-chromatin. These characteristics provide
useful landmarks to identify individual chromosomes. For example, the nucleoli are associated with specific chromosomes that mark them as nucleolar organizer chromosomes. Pachytene analysis has been used in several
species with small chromosomes including tomato (Barton, 1950), rice
(Chu, 1967; Khush et al., 1984), *Brassica* (Robbelen, 1960) and soybean
(Singh and Hymowitz, 1988). Since *Musa* chromosomes are reported to be
in the range of 1 to 2 µm, pachytene analysis of *Musa* chromosomes may

overcome problems associated with analysis of mitotic metaphase chromosomes.

Many reports show that 2n gamete formation in *Musa* is common (Dodds and Simmonds, 1946: Wilson, 1946; Simmonds, 1960). Such gametes result from abnormalities during either microsporogenesis (2n pollen) or megasporogenesis (2n eggs). An array of premeiotic, meiotic, and postmeiotic mechanisms has been proposed to explain 2n gamete formation in plants. Both first- and second-division restitution mechanisms have been implicated in *Musa* (Wilson, 1946; Simmonds, 1960; Dodds and Simmonds, 1946). The formation of 2n gametes in *Musa* is influenced by genotype (Vuylsteke, Swennen, et al., 1993). Meiotic studies in *Musa* should provide valuable information on the mechanisms of 2n gamete formation and other meiotic modifications that will eventually have direct applications in *Musa* breeding.

Genome Relationships

One contribution of fundamental cytogenetic studies to breeding will be a better understanding of genome relationships. Little is known about pairing relationships and the extent of homology between the *A* and *B* genomes in bananas. The extent of variation within the *A* and *B* genomes has not been investigated. Although cultivars with various genomic combinations are available, no large-scale effort to examine meiotic configurations in these plants has been undertaken. A study in this direction would be a rewarding experience both for the cytogeneticist and the plant breeder. Agarwal (1983, 1987, 1988a, 1988b, 1988c) investigated meiotic behavior of *Musa* varieties of India. In a study of eight triploid male sterile varieties (*AAB* group) various chromosome associations ranging from univalents, bivalents, quadrivalents and higher multivalents were seen at metaphase I. Laggards were of common occurrence (Agarwal, 1987). The extent of chromosome synapsis between *A* and *B* genomes in triploids could not be distinguished. Further studies by Agarwal (1988a) showed that some diploid *AB* varieties formed 11 bivalents regularly, whereas others showed presence of univalents. Trivalents were observed in one of the *AB* varieties. These studies suggest that the level of homology between the *A* and *B* genomes is highly variable. Agarwal (1988c) found diploidlike meiosis in a pentaploid hybrid that resulted from hybridization of a double-restituted (4n) egg cell in *M. acuminata* by a haploid (n) gamete of *M. rubra* Wall. This led Ortiz (1995) to suggest that chromosome pairing may be genetically controlled in *Musa*.

Recent studies in molecular cytology involving ISH methods have made it possible to identify different genomes in plants. Multicolor genomic ISH using different fluorescent colors are able to discriminate chromosomes from different genomes. This method appears to work well even in plants such as potato and rice that have very small and similar chromosomes. Osuji and colleagues (1997) applied these techniques to identify the basic genomes in diploid, triploid, and tetraploid *Musa* genotypes. Their results showed a high degree of cross-hybridization between the *A* and *B* genomes suggesting that they may have common DNA sequences (Osuji et al., 1997). This is in agreement with our RAPD (random amplified polymorphic DNA) studies that show that *M. acuminata* (*A* genome) and *M. balbisiana* (*B* genome) have a large number of bands in common suggesting the presence of similar DNA sequences in these species (unpublished data). However, RAPD fragments unique to *M. acuminata* and *M. balbisiana* have been identified in our laboratory (Pillay et al., 1999). These fragments are useful in determining the genomic composition of *Musa* cultivars. Further studies are necessary to show if these fragments are unique to the *A* and *B* genomes or whether they are chromosome-specific bands. With the exception of the nucleolar organizer regions, there are no cytological markers that enable easy identification of all the *Musa* chromosomes. RAPD fragments that are unique to chromosomes could be used as cloned probes in ISH experiments to identify homologous chromosomes. Osuji and colleagues (1997) observed very strong ISH signals in the centromeric regions suggesting that repetitive sequences in the labeled genomic DNA probe were abundant in this area. Stronger signals were also observed on some chromosomes than on others. This type of information may be useful in identifying individual *Musa* chromosomes and needs to be explored further. Hybridization to the satellite regions was very strong. The satellites are the sites of the *Nor* locus that contain the multicopy ribosomal RNA genes and generally give strong hybridization signals. Despite the cross-hybridization observed in their study, Osuji and colleagues (1997) concluded that they were able to identify the *A* and *B* genome constituents in *AAB, ABB,* and *AAAB* genotypes of *Musa*. However, it appears that with the quality of chromosome separation and signal detection achieved in their study, it would be difficult to distinguish small genomic changes in *Musa* chromosomes. Further refinement of the ISH technique may be necessary to provide convincing evidence of its ability to distinguish the constituent genomes in *Musa*.

Aneuploid Gene Mapping

The ability of *Musa* to propagate vegetatively is advantageous for producing a complete series of aneuploids, since plants deficient for critical chromosomes could still be maintained. A complete series of aneuploids is useful in mapping genes to specific chromosomes and identifying specific linkage groups with individual chromosomes. The development of new DNA-marker systems allows large numbers of molecular markers to be placed on chromosomes. The availability of a complete set of monosomics in *Musa* should greatly facilitate gene mapping in this species. No systematic effort has been made to establish a set of aneuploids in *Musa*. However, plants from in vitro cultures suspected to be aneuploids have been established in our nurseries and await further analysis. In comparison with their siblings, these plants show abnormal morphologies such as dwarfism and unusual bunch characteristics. Osuji and colleagues (1996) identified two *Musa* hybrids (TMP2x 9722-1, TMP2x 1605-1) as having $2n = 2x + 1 = 23$ chromosomes. Caution must be exercised in counting mitotic *Musa* chromosomes that are not fully condensed as the satellites often separate from the parent chromosome and appear as distinct structures that could be mistakenly counted as extra chromosomes. This phenomenon appears to occur more often with one of the satellite chromosomes creating the appearance of 23 instead of 22 chromosomes in diploids cells. Identification of extra chromosomes must, therefore, be verified by the analysis of meiotic stages such as pachynema, diakinesis, metaphase-I, anaphase-I or telophase-I. Extra chromosomes at diakinesis or metaphase-I would create trivalents plus bivalents or a univalent plus bivalents configuration suggesting the presence of extra chromosomes (Singh, 1993).

Ploidy Determination

Since *Musa* is a multiploidy crop and ploidy manipulation is critical to breeding, ploidy determination plays an important role in *Musa* improvement. For example, crossing the triploid plantain with diploid accessions generates diploid, triploid, tetraploid, aneuploid, and hyperploid progeny (Vuylsteke, Ortiz, et al., 1993). For selection of a hybrid of a desired ploidy level, there is need for a rapid and precise method for ploidy screening at an early stage of plant development (Dolezel et al., 1997).

In the past, ploidy determination and assignment of genomic groups in *Musa* was done primarily by examining plants at the morphological level. A method was devised that uses 15 characters and a scoring system that assigns different values for characters that resemble those belonging to the two progenitor species, *M. acuminata* and *M. balbisiana* (Stover and Simmonds, 1987). This means that the plants must reach maturity (12 to 24 months) before some of these characters are fully expressed and can be measured. Morphological characters can be greatly influenced by the environment and this method can lead to inconsistent results. Alternative methods for rapid ploidy determination in *Musa* have included the use of gametophytic (pollen and stomata size) and sporophytic characters (chloroplast density) (Vandenhout et al., 1995; Tenkouano et al., 1998). It is recognized that pollen size cannot be used for early screening since plants require from nine to 12 months before anthesis (Tenkouano et al., 1998). However, Tenkouano and colleagues (1998) observed that the average number of chloroplasts in triploid and tetraploids was, respectively, 1.30 and 1.53 times higher than in diploids. These results demonstrated that the number of chloroplasts in stomatal guard cells is useful for ploidy determination in *Musa* germplasm.

Classical methods for determining ploidy are difficult due to the small size of the *Musa* chromosomes and are not practical when thousands of plants are involved. Flow cytometry has become a rapid and reliable method for ascertaining ploidy levels in *Musa* and promises to greatly increase the efficiency of *Musa* breeding. Since the DNA content of nuclei in certain phases of the cell cycle is related to the ploidy level, the flow cytometer is used to measure the size of the nuclei and produce histograms that describe the DNA content of the cells. All the cells belonging to one peak have the same quantity of DNA and would represent a ploidy level. Dolezel and colleagues (1997) listed the major advantages of this method for ploidy estimation as: (1) it is precise and rapid, making it possible to analyze several hundred samples in a day, (2) it does not require dividing cells, (3) it is not destructive since small amounts of leaf tissue is used, and (4) large populations of cells can be analyzed and mixoploidy in subpopulations can be determined. Flow cytometry would permit screening for ploidy in the early stages of development in segregating populations of banana hybrids (Dolezel et al., 1994), thus saving valuable space, time, and resources currently being used to grow out aneuploids and hyperpolyploids. The introduction of flow cytometry in *Musa* research has already questioned the ploidy status of some plants in germplasm banks. For example, Klue Tiparot that was previously classified as a natural tetraploid, *ABBB* (Simmonds and Shepherd, 1955), is now known to be a triploid with $2n = 3x = 33$ based on flow cytometry and classical cytology (Jenny et al., 1997; Horry, Dolezel, et al., 1998). Similarly, Horry, Dolezel, et al. (1998) found that 'Pisang Jambe', previously classified as a tetraploid *AAAA*, is actually a triploid ($2n = 3x = 33$), while '(Kluai) Ngoen' (*AAB*) from Malaysia is a tetraploid ($2n = 4x = 44$) instead of a triploid as originally thought. It is envisaged that careful screening of *Musa* genotypes with flow cytometric and molecular techniques will undoubtedly help resolve similar problems and provide a more consistent classification system.

DNA Content

A correlation between genome size and agronomically important traits has been observed in many plant species. Biradar and colleagues (1994) reported that growth and yield parameters were negatively correlated with nuclear DNA content in maize. These authors concluded that variation in DNA content plays an important role in determining agronomic performance in maize. Flow cytometric analysis has been used to determine DNA values in different cultivars and landraces of diploid *M. acuminata* (*AA*) and *M. balbisiana* (*BB*), and triploids of *AAA*, *AAB*, and *ABB* genome composition (Lysak et al., 1999). The 2C (C=haploid genome) DNA content ranged from 1.108 pg in *M. balbisiana* to 1.912 pg in Gran Enano (*AAA*). The nuclear DNA content of *M. balbisiana* (*BB* genome) was found to be significantly lower than that of *M. acuminata* subspecies and cultivars (*AA* genome). The haploid genome size in several *Musa* genotypes ranges from 534 to 618 Mbp (Lysak et al., 1998). This value is significantly lower than an earlier estimate of 873 Mbp (Arumuganathan and Earle, 1991). The DNA content of *Musa* is at the low end of the range of known genome sizes in plants. It is clear that a change in DNA quantity has been involved in speciation in *Musa*. Assuming that the trend has been toward an increase in DNA content, the available data suggests that *M. balbisiana (BB)* has the primitive genome size whereas the *AA* genome species are evolutionarily more advanced. Changes in DNA quantity per haploid genome are considered to involve changes in repetitive DNA. Genomic in situ hybridization showed that repetitive DNA in the *Musa* genome is concentrated in the centromeric regions (Osuji et al., 1997). C-banding experiments in *Musa* should be able to establish a correlation between repetitive DNA and genome size in *Musa* species. Presently no correlation has been established between DNA content and breeding traits in *Musa*. This relationship might be worth exploring in the future.

Technique for Analyzing Musa *Chromosomes*

There are very few reports involving methodology of chromosome techniques in *Musa*. A modified protoplast-isolation technique used in Dr. David Stelly's laboratory (Texas A&M University) is routinely used in our laboratory for mitotic analysis in *Musa*. The technique relies on the enzymatic digestion of cell walls. The released protoplasts are induced to burst on a glass slide spreading the chromosomes. The method has distinct advantages over the conventional squash method. Well-spread chromosome preparations are obtained with little or no cytoplasmic background. Little or no cytoplasmic staining occurs so that the structure of the chromosomes is not obscured. The spreading of the chromosomes without any applied physical pressure causes no distortion in the shape or size of the chromosomes. The direct exposure of the chromosomes on the slide makes them easily accessible to stain. Slides prepared in this manner are ideal for all types of cytological or other chromosome manipulations including ISH and PRINS since preparations are essentially free of cell-wall material and other debris. The molecular hybridization probes can easily reach target sites on the chromosomes.
Actively growing root tips of 2 to 3 cm in length (1 mm thick) are cut from field-grown plants and washed thoroughly in water to remove all de-

bris. Roots are pretreated on filter paper soaked with 2.5 mM 8-hydroxy-quinoline in a petri dish for 30 min to 3 h at room temperature to accumulate dividing cells at prometaphase or metaphase. Roots immersed fully in the pretreating agent did not accumulate metaphase cells and showed primarily interphase nuclei. The roots are fixed in 4:1 ethanol:acetic acid followed by a single change with the same solution after 5 to 10 min and left at room temperature overnight. Prior to slide preparation, the roots are rinsed in distilled water two to three times and then washed in water for 20 to 30 min. The root tips are then hydrolyzed in 0.1N HCl for 5 to 10 min followed by two rinses in water. Roots are subsequently washed in water for 10 min and placed in cold citrate buffer (0.01 M sodium citrate and 0.01 M citric acid, pH 4.5) for 10 min. The meristematic tips of the roots are cut and transferred to the enzyme solution cellulase-R10 (5 percent) and pectolyase (1 percent) in a microcentrifuge tube and incubated for 2 to 3 hours at 37° C. The time of incubation will vary depending on the state of the roots and the genotype and should be determined independently for each sample. Subsequently, the enzyme solution is replaced with the citrate buffer and washed with the same buffer over a period of 10 to 15 min. The root tips are washed in water for 5 to 10 min. To prepare a slide, a single root tip is transferred with a Pasteur pipet to a precleaned glass slide. The excess water is removed. Two drops of freshly prepared 3:1 ethanol:acetic acid is placed on the root tip. The root tip is macerated with a very fine tip forceps and immediately smeared over the slide. The slide is examined under a phase contrast microscope. As soon as the cells begin to adhere to the glass slide, several drops of 3:1 (alcohol:acetic acid) are added on one side of the slide and the excess allowed to flow over the slide. The slide is air-dried. Slides are stained in Leishman's stain in 0.01M phosphate buffer (KH_2PO_4 + Na_2PO_4, pH 6.8). The proportion of stain to buffer is 1:4. A 3 percent Giemsa solution in the same buffer works equally well. Staining time varies from 30 to 90 min. Slides are removed from the stain and rinsed twice in distilled water and air-dried. The slides are mounted in Canada balsam in xylene. A variation of this technique has been reported by Dolezel and colleagues (1998).

Musa *Genomics*

Much has been written recently about the application of biotechnology for the enhancement of bananas (Crouch, Vuylsteke, et al., 1998; Vuylsteke et al., 1998). Many laboratories are applying molecular genetic techniques for improving the efficiency of *Musa* breeding. DNA markers can be used for a number of applications in crop improvement including the assessment of phylogenetic relationships, germplasm characterization and fingerprinting, genetic analysis, linkage mapping, and molecular breeding. Initial DNA studies in *Musa* using restriction fragment length polymorphisms

(RFLPs) were used in taxonomic and phylogenetic relationships of *Musa* species and cultivars (Gawel et al., 1992; Bhat et al., 1994). These studies showed that RFLP analysis detected a relatively low level of polymorphism among *Musa* cultivars. Thereafter, RFLP was discounted for having limited application for routine breeding in *Musa*. Researchers then concentrated on the polymerase chain reaction (PCR) for genome analysis in *Musa*. PCR-based techniques showed a higher level of polymorphism within the *Musa* germplasm. RAPD studies (Kaemmer et al., 1997; Howell et al., 1994) were used to distinguish diverse *Musa* germplasm. Faure and colleagues (1993) developed a genetic linkage map of diploid bananas using a variety of marker systems including RAPDs. A saturated linkage map for *Musa* is essential for determining quantitative trait loci since many characters, such as bunch weight and size, are presumably determined by the combined effect of a number of loci. Although RAPD analysis is known to have several disadvantages, it appears that a well-established technique could provide useful data for marker-assisted selection in *Musa*. Jarret and colleagues (1994) developed microsatellite primers for *Musa*. The microsatellite assay for detection of polymorphisms is considered an optimal marker system due to its reliability and reproducibility. Microsatellite primers have been shown to detect a high level of polymorphism between individuals of *Musa* breeding populations (Crouch, Crouch, et al., 1998). Studies of the segregation of molecular markers have confirmed that recombination occurs during meiosis (Crouch, Crouch, et al., 1998). Molecular-marker studies have also suggested the exchange of alleles between *A* and *B* genomes in interspecific hybrids (Crouch et al., 1999). Microsatellite markers have been used to fingerprint several registered *Musa* hybrids (Ortiz et al., 1998). A comparison of microsatellite marker and pedigree-based methods of genetic analysis to estimate genetic relationships and predict hybrid performance in banana has been reported (Tenkouano et al., 1999a, 1999b). These studies of the *Musa* genomes have already proven invaluable to *Musa* improvement. These discoveries have profound implications for *Musa* breeding methods allowing for recurrent polyploid breeding and the expansion of breeding germplasm. Sequence-tagged microsatellite sites (STMS) have been proposed as anchor markers for a banana genetic core map (Kaemmer et al., 1997).

New techniques, such as amplified fragment length polymorphism (AFLP), promise to be very powerful tools in molecular breeding of banana (Crouch, Vuylsteke, et al., 1998). The technique is able to identify a larger number of polymorphic bands than could previously established DNA techniques. The AFLP technique is expected to become a common method for establishing molecular markers in *Musa*. Current work to map specific loci should result in both markers linked to loci governing important traits to be used for marker-assisted selection (MAS) but will also aid in map-based cloning of alleles. It is expected that MAS will initially be the more important of the two applications. Characters of importance include parthenocarpy, resistance to black Sigatoka, apical dominance, and resistance genes for pests and diseases.

NEW BREEDING APPROACHES

Breeding Objectives and Rationale

Rowe and Rosales (1996) and Vuylsteke and colleagues (1997) have provided rather detailed discussions on the past history of *Musa* breeding. The initial steps for genetic improvement of plantains and bananas traditionally involve crossing the heterogenomic (*AAB, ABB*) triploid accessions to the homogenomic (*AA, BB*) diploid accessions, with the major objective of producing hybrids that are high yielding per unit area and time. Hence, in addition to resistance to pests and diseases, improved hybrids should be photosynthetically efficient, early to mature in the first production cycle, and display minimum delay between consecutive harvests (Ortiz and Vuylsteke, 1994; Eckstein et al., 1995). Other desirable characteristics include short stature and strong roots for optimal nutrient uptake and greater resistance to wind damage. Quality traits are specific to the many different end uses of banana, but consumer acceptability must be maintained or improved in combination with breeding for other traits.

The agronomic and environmental aspects of *Musa* phenology and reproductive ontogeny are well understood (Turner and Lahav, 1983; Robinson and Nel, 1985, 1986; Robinson et al., 1992; Ekanayake et al., 1994), but comparatively little is known about their genetic basis. For example, the time interval between two consecutive harvests is determined by suckering behavior, a trait that is itself under control of gibberellic acid (GA_3) production (Swennen and Wilson, 1983). Ortiz and Vuylsteke (1994) suggested that gibberellic acid metabolism was regulated by a single gene that had incomplete penetrance, genetic specificity, and variable expressivity, which would allow for dosage effects to operate. In fact, complex inheritance was reported for most growth and yield characteristics of *Musa* (Ortiz and Vuylsteke, 1996) and was attributed to the irregular meiotic behavior of the species, and to unpredictable variation in genome size and structure both across and within generations. Thus, parental performance may not accurately predict progeny performance for these traits, and further genetic improvement will require control of selection over prospective male and female parents through progeny testing.

The 3x-2x breeding scheme produces hybrids with varying ploidy levels (see Figure 8.1). Among these, the most interesting selections have been (1) tetraploid hybrids that are high yielding and resistant to BLS,

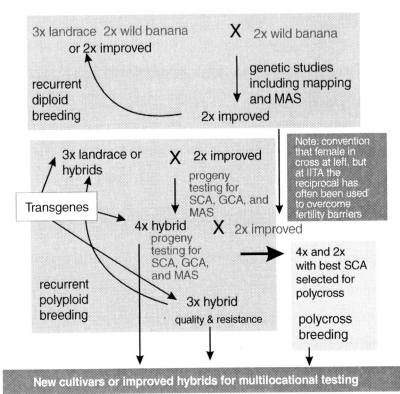

FIGURE 8.1. Flow diagram showing the breeding scheme for *Musa* improvement used at the International Institute of Tropical Agriculture (IITA). The diagram outlines three interacting phases of *Musa* improvement (1) recurrent diploid breeding, (2) recurrent polyploid breeding, and (3) polycross breeding. The diagram also shows where different types of genetic studies and interventions fit into the scheme. Marker-assisted selection (MAS) is used in both diploid and polyploid breeding and will likely be used in the identification of quantitative trait loci (QTL) associated with general and specific combining ability (GCA and SCA, respectively). Transgenes will most often be added to clones ready for cultivar release but deficient in one or a few important traits, but will also be used to introduce new variation for breeding.

and (2) diploid hybrids that have low yield potential but are resistant to BLS. The tetraploid hybrids are both female and male fertile, which often reduces fruit quality due to the presence of seeds in the pulp. Therefore, further progress in *Musa* improvement requires restoration of the seedlessness trait in a triploid background while maintaining or increasing yield and resistance to the major diseases and pests affecting this crop. This may be achieved by crossing the primary tetraploid and diploid hybrids, taking into account parental diversity and intrafamily variation to achieve genetic combinations that express maximum or progressive heterosis (Ortiz, 1997a).

The 4x-2x breeding approach has been successfully used for genetic improvement of potato and other crop species (Mok and Peloquin, 1975). Jauhar (1979) and Burton (1981) have reported the use of the 4x-2x crosses in a polycross scheme to efficiently enhance intermating and synthesize triploids in Meadow fescue and Pensacola bahiagrass, respectively. In *Musa* spp., 4x-2x crosses have been used to transfer specific attributes of the 2x germplasm (e.g., pest resistance) into 4x progenies (Rowe and Rosales, 1995). While ploidy manipulations and interspecific crosses have allowed considerable genetic gains for yield and other traits (Vuylsteke et al., 1997), the identification of 4x-2x parental combinations that would produce 3x hybrids with both high expected mean and genetic variation remains a major challenge for *Musa* breeders.

Parental Selection and Hybrid Performance

Progeny testing has been the predominant method for parental selection and designation of heterotic groups in many crop species (Hallauer and Miranda, 1988; Panter and Allen, 1995). In this strategy, potential parental genotypes are selected on the assumption that subsequent offspring will be as good as the previous ones. Thus, progeny testing is inherently postdictive. However, genome-size variation occurs within and across generations in *Musa* spp., which reduces the predictive accuracy of parental performance on progeny value for yield and other traits with complex inheritance. Furthermore, this method of parental selection cannot be routinely used for *Musa* research due to the large land requirements (6 $m^2 \cdot plant^{-1}$) and long growth cycle (12 to18 months) of this crop.

Predictive models for hybrid performance based on past performance of parents or their relatives would be most useful for breeders who seek to make the most appropriate crosses on an a priori basis. Maize (*Zea mays* L.) breeders have traditionally used a semipredictive model to select the parents of double-cross hybrids based on the mean of single crosses not involved in the double cross (Jenkins, 1934). Other methods for predicting hybrid performance have been based on the genetic relationships among prospective parents and their midparental values (Bernardo 1992, 1994; Panter and Allen, 1995). Genetic relationships among prospective parents may be estimated using Malécot's (1948) coefficient of coancestry or DNA marker polymorphisms (Staub and Serquen, 1996; Saghai Maroof et al., 1997).

The simplest approach to estimate biparental hybrid performance is to calculate midparental values, assuming (1) the traits under consideration are strictly determined by additive inheritance and (2) parents are inbred, unrelated, and contribute equally to their progeny's genotype (Panter and Allen 1995; Bernardo et al., 1996). However, current understanding of the meiotic behavior of *Musa* spp. and of inheritance of many traits suggests that these assumptions may not hold true in the species (Vuylsteke et al., 1997). Indeed, most *Musa* accessions are highly heterozygous, and there is evidence of unequal genetic contribution of parents to their progeny (Ortiz, 1997b; Tenkouano et al., 1998).

For these reasons, Tenkouano et al. (1999a) modified the midparent approach to include terms describing the relative genetic contributions of the parents to their progeny, parental heterozygosity level and genetic relatedness, giving the following formula:

$$H_{ij} = \frac{c_i(1 + f_{ii})P_I + c_j(1 + f_{jj})P_j}{c_i(1 + f_{ii}) + c_j(1 + f_{jj})} \times \left[1 - Ln \frac{2 - f_{ij}}{\sqrt{(f_{ii})(f_{jj})}} \right] \qquad (1)$$

where H_{ij} indicates the expected value of the hybrid produced from the ith and j^{th} ($i \neq j$) parents. P_i and P_j are the observed values of the ith and j^{th} parent, respectively. Equation (1) has two components: (a) an additive component, which is simply the weighted average of parental phenotypic values, and (b) a multiplicative component that reflects heterotic or inbreeding effects. The terms c_i and c_j are the relative contributions of parents i and j, respectively, to their progeny. In a disomic situation, $c_i = c_j$, that is, the genomic contributions of parents to their offspring, are assumed equal (Panter and Allen, 1995; Bernardo et al., 1996). However, it is postulated that secondary triploid *Musa* hybrids receive two chromosomes from their tetraploid maternal parent for each chromosome donated by their paternal diploid parent ($c_i \neq c_j$), provided that 2n gametes are not produced (Ortiz, 1997b). The true values of parental contributions can be estimated by using molecular markers or cytological techniques allowing for determination of the gametic composition. The terms f_{ii} and f_{jj} represent the probability of two alleles being identical at any locus in parents i and j, respectively, which is indicative of their homozygosity (inbreeding) level. The f_{ij} term is the coefficient of relationship (similarity) among parents i and j. The terms f_{ii}, f_{jj}, and f_{ij} may be calculated using pedigree or molecular marker data or a combination thereof (Tenkouano et al., 1998). The logarithmic term is analogous to Nei's (1972) formula for the calculation of genetic distance and reflects the postulate that the discrepancy between expected progeny performance and midparent value would vary as a function of the genetic distance between the parents. When parents are very similar genetically, the logarithmic term approaches zero and the performance of the progeny is expected to approach the midparent value for the trait under consideration. For example, the progeny obtained from selfing an inbred line would be expected to express the same genotypic value as the inbred line. In contrast, crossing two unrelated parents should lead to an F_1 that expresses hybrid vigor, the magnitude of which would depend also on the parental contribution and inbreeding status.

Following recent developments in DNA marker technology (Staub and Serquen, 1996; Saghai Maroof et al., 1997), genetic relationships, including parental contributions to their offspring, may be estimated as the probability of allelic identity in state using DNA marker polymorphisms. DNA marker systems based on the polymerase chain reaction (PCR) are particularly suited to applications in plant breeding (Rafalski and Tingey, 1993; Rafalski et al., 1995). Among these systems, simple sequence repeat length polymorphisms (SSRLP) have proven particularly useful in many species (Powell et al., 1996) including *Musa* (Kaemmer et al., 1997; Crouch, Vuylsteke, et al., 1998). When parents are carefully selected, tetraploid-diploid crossing may be equated to the double-cross breeding schemes that have revolutionized maize improvement by allowing the expression of heterosis while maintaining adequate levels of genetic diversity. In particular, pyramiding of resistance genes from different diploid progenitors of the $2x$ and $4x$ parents becomes possible. A most important but often overlooked feature of tetraploid-diploid breeding is that it allows bridge-crossing of triploid genotypes that are female fertile but usually male sterile.

Estimating Combining Ability in 4x–2x Crosses

Estimation of genetic parameters is often carried out using genetic models based on disomic inheritance of traits in families derived from nested or factorial mating designs (Kempthorne, 1957; Falconer and Mackay, 1996). Such models may not be suitable for $4x$-$2x$ families, since inheritance is tetrasomic with respect to the maternal parent and parental contribution to the triploid offspring may not be equal (Ortiz, 1997b; Tenkouano et al., 1998). Similarly, such models may not be appropriate for estimation of genetic covariances among triploid families, since inheritance is trisomic in those families. Recently, Tenkouano and colleagues (1998b) proposed a method for estimation of genetic effects in maternal and paternal half-sibs from tetraploid-diploid crosses.

Triploid hybrid populations cannot be maintained by random mating due to high frequency of female sterility resulting from meiotic irregularities in the formation of gametes. However, triploid *Musa* clones are often regarded as end products that can be maintained by vegetative propagation, although they may also be used as parents in future crosses following a recurrent breeding scheme (Ortiz, 1997a). Therefore, knowledge of genetic parameters is useful in defining the relationships of progeny to parents and associated breeding strategy to obtain triploid clones from tetraploid-diploid parents. In this regard, research at IITA identified traits that are primarily inherited from male ($2x$) or female ($4x$) parents, which is of practical importance for parental selection in $4x$-$2x$ cross-breeding (Tenkouano and colleagues, 1998a). For example, breeding for increased bunch weight and reduced time interval between flowering and harvest should aim at accumulating favorable alleles for these traits in a diploid male background through recurrent selection prior to crossbreeding with a tetraploid female. In contrast, plant height, number of leaves, and suckering behavior may be improved by first selecting for these traits in a tetraploid background before crossbreeding with diploid males.

New Sources of Pest and Disease Resistance

The newly developed breeding methods mentioned above are likely to have a major impact on the introgression of disease and pest resistance from wild germplasm into new cultivars. Most alleles for resistance in current cultivars have come from a very limited number of wild diploids (Hartman and Vuylsteke, 1999). Recurrent selection at both diploid and polyploid levels will allow the pyramiding of resistance alleles from multiple wild sources. This should improve both the quality and the durability of resistance. Molecular marker-assisted selection should facilitate the identification and introgression of new resistance alleles from wild accessions into more advanced breeding lines. Initially, markers will be used to screen against wild *Musa* traits such as non-parthenocarpy (Crouch, Vuylsteke, et al., 1998). The early part of the next century should see the movement of QTL between widely divergent *Musa* groups, including between *A* and *B* genomes, using MAS (Crouch et al., 1999).

In addition to introgressions from within *Musa*, transformation of banana with alien genes promises to be of great importance to banana breeding in the next century. The export industry of the large multinational corporations is based on the postharvest handling characteristics of a specific subgroup of banana clones. These companies are extremely averse to change. Transformation offers the possibility of introducing resistance with little or no impact on postharvest handling qualities.

SUMMARY AND CONCLUSIONS

More than 800 million people in the developing countries of the world were undernourished at the beginning of the 1990s. The challenge of agriculture is to feed the world's population and protect the environment simultaneously. Producers of banana can make an important contribution toward feeding the world since bananas are generally produced in regions that face the greatest food insecurity. Bananas are important components of food security in the tropical world. The large number of pests and diseases affecting banana are serious constraints to production. Great potential for increased yields in banana lies in the development of new cultivars. However, banana improvement presents several challenges to breeders and geneticists. This is evident by the fact that the past 70 years of *Musa* breeding has not produced an acceptable commercial replacement to the highly sterile 'Cavendish' cultivars. Many methods are presently available for overcoming these barriers for genetic improvement of *Musa*. For example, problems of fertility can be surmounted by genetic transformation. Although these are available, transformation targets may be limited by the scarcity of useful genes. The greatest challenge of the near future for *Musa* breeders is to breed for resistance against BSV since the virus sequences are integrated into the *Musa* genome. Whether or not the proposed strategy of breeding for lower propensity to express the integrated sequence will succeed remains to be seen. A better knowledge of *Musa* genomics including cytology, homology of *A* and *B* genomes, and a better understanding of transmission genetics will add to existing knowledge of banana breeding. This information will likely change approaches to breeding banana. In the next century, banana production systems should make the change from those dominated by landraces to those dominated by bred cultivars. These cultivars will combine superior quality attributes such as long shelf life and superior taste with multiple disease and pest resistance. The cultivars of the next century are likely to arise from a combination of classical hybridization and selection in combination with new biometrical techniques, molecular MAS, and transgenics. Although the impact of genetic transformation is currently limited by scarcity of useful genes and by social and regulatory barriers, it seems inevitable that it will have a large impact on banana improvement in the future. Much of what is considered revolutionary in banana will be the adoption of proven techniques used for improving other species. Thus, our current view of the future of *Musa* breeding borrows much from potato and alfalfa breeding methods.

REFERENCES

Agarwal, P.K. 1983. Karyotype and B-chromosomes of *Musa rubra* Wall. *Cytologia* 48(2):275-280.

Agarwal, P.K. 1987. Cytogenetical investigations in Musaceae. II. Meiotic studies in eight male sterile triploid banana varieties of India. *Cytologia* 52(3):451-454.

Agarwal, P.K. 1988a. Cytogenetical investigations in *Musaceae*. III. Meiotic studies in diploid *Musa* species and banana varieties of India. *Cytologia* 53(2):359-363.

Agarwal, P.K. 1988b. Cytogenetical investigations in *Musaceae*. IV. Cytomorphology of an interspecific triploid hybrid of *Musa acuminata* Colla. × *M. rubra* Wall. *Cytologia* 53(4):717-721.

Agarwal, P.K. 1988c. Meiotic behavior and cytology of a pentaploid *Musa acuminata* (Colla.) (2n = 22) × *M. rubra* (Wall.)(2n = 22) hybrid. *Curr. Sci.* 57(5):1079-1081.

Arumuganathan, K. and Earle, E.D. 1991. Nuclear DNA content of some important species. *Plant Mol. Biol. Rep.* 9(3):208-218.

Barton, D.W. 1950. Pachytene morphology of tomato chromosome complement. *Am. J. Bot.* 37(4):639-643.

Bernardo, R. 1992. Relationship between single-cross performance and molecular marker heterozygosity. *Theor. Appl. Genet.* 83(5):628-634.

Bernardo, R. 1994. Prediction of maize single-cross performance using RFLPs and information from related hybrids. *Crop Sci.* 34(1):20-25.

Bernardo, R., Murigneux, A., and Karaman, Z. 1996. Marker-based estimates of identity by descent and alikeness in state among maize inbreds. *Theor. Appl. Genet.* 93(1-2):262-267.

Bhat, K.V., Jarret R.L., and Liu Z-W. 1994. RFLP characterization of Indian *Musa* germplasm for clonal identification and classification. *Euphytica* 80(1/2):95-103.

Biradar, D.P, Bullock, D.G., and Rayburn, A.L. 1994. Nuclear DNA amount, growth, and yield parameters in maize. *Theor. Appl. Genet.* 88(5):557-560.

Burton, G.W. 1981. Meeting human needs through plant breeding: Past progress and prospects for the future. In K.J. Frey (Ed.), *Plant Breeding* (pp. 433-465). Iowa State University, Ames, Iowa.

Chu, Y.E. 1967. Pachytene analysis and observation of chromosome association in haploid rice. *Cytologia* 32(1):87-95.

Crouch, H.K., Crouch, J.H., Jarret, R.L., Cregan, P.B., and Ortiz, R. 1998. Segregation of microsatellite loci in haploid and diploid gametes of *Musa*. *Crop Sci.* 38(1):211-217.

Crouch J.H., Crouch, H.K., Constandt, H., Van Gysel, A., Breyne, P., Van Montagu, M., Jarret, R.L. and Ortiz, R. 1999. Comparison of PCR-based molecular marker analyses of *Musa* breeding populations. *Molecular Breeding* 5(3):233-244.

Crouch, J.H., Vuylsteke, D., and Ortiz, R. 1998. Perspectives on the application of biotechnology to assist the genetic enhancement of plantain and banana (*Musa* spp.). *Electronic Journal of Biotechnology* 1(1) 13 pages. <http:ejb.ucv.cl/content/vol1/issue1/>.

Dale, H. 1990. In G. J. Persley (Ed.), *Agricultural biotechnology opportunities for international development* (pp. 225-240). CAB International. Wallingford.

Dantas, J.L.L., Shepherd, K., S. Soares Filho W., Cordeiro, Z.J.M., de Oliveira Silva, S., and Silva Souza, A. 1993. Citogenetica e melhoramento genetico da bananeira (*Musa* spp.). Documentos EMBRAPA-CNMPF 48.

Davide, R.G. 1996. Overview of nematodes as a limiting factor in *Musa* production. In E.A. Frison, J.-P. Horry, and D. De Waele (Eds.), *New frontiers in resistance breeding for nematode, fusarium, and sigatoka. Proceedings workshop Kuala Lumpur, Malaysia October 2-5, 1995* (pp. 27-31). INIBAP, Montpellier, France.

Demeke, T. 1986. Is Ethiopia's *Ensete ventricosum* crop her greatest potential food? *Agric. Int.* 38:362-365.

Dodds, K.S. and Simmonds, N.W. 1946. Genetical and cytological studies of *Musa*. VIII. The formation of polyploid spores. *J. Genet.* 47(2):223-241.

Dolezel, J. 1998. Flow karyotyping and chromosome sorting in plants. In J. Maluszynska (Ed.), *Plant Cytogenetics* (pp. 33-50). Wydawnictwo Uniwersytetu, Slaskiego, Katowice, Czech Republic.

Dolezel, J., Dolezelova, M., and Novak, F.J. 1994. Flow cytometric estimation of nuclear DNA amount in diploid bananas (*Musa acuminata* and *M. balbisiana*). *Biol. Plant* 36(2):351-357.

Dolezel, J., Dolezelova, M., Roux, N., and Van den Houwe, I. 1998. A novel method to prepare slides for high resolution chromosome studies in *Musa* spp. *InfoMusa* 7(1):3-4.

Dolezel, J., Lysak, M.A., Van den Houwe, I., Dolezelova, M., and Roux, N. 1997. Use of flow cytometry for rapid ploidy determination in *Musa*. *InfoMusa* 6(1):6-9.

Dolezelova, M., Valarik, M., and Dolezel, J. 1998. Molecular cytogenetics of *Musa* spp. In J. Maluszynska (Ed.), *Plant Cytogenetics* (pp. 159-163). Wydawnictwo Uniwersytetu, Slaskiego, Katowice, Czech Republic.

Eckstein, K., Robinson, J.C., and Davie, S.J. 1995. Physiological responses of banana (*Musa* AAA; Cavendish sub-group) in the subtropics. III. Gas exchange, growth analysis and source-sink interaction over a complete crop cycle. *J. Hort. Sci.* 70(1):169-180.

Ekanayake, I.J., Ortiz, R., and Vuylsteke, D.R. 1994. Influence of leaf age, soil moisture, VPD, and time of day on leaf conductance of various *Musa* genotypes in a humid forest-moist savanna transition site. *Ann. Bot.* 74(2):173-178.

Falconer, D.S. and Mackay, T.F.C. 1996. *Introduction to quantitative genetics* Fourth edition. Longman, London.

Faure, S., Noyer, J.L., Horry, J.P., Bakry, F., Lanaud, C., and Gonzalez de Leon, D. 1993. A molecular marker-based linkage map of diploid bananas (*Musa acuminata*). *Theor. Appl. Genet.* 87(4):517-526.

Frison, E.A. and Sharrock, S.L. 1998. Banana streak virus: A unique virus-*Musa* interaction? *Proceedings of a workshop of the PROMUSA virology working group held in Montpellier, France, January 19-21, 1998.* IPGRI, Rome, Italy, INIBAP, Montpellier, France.

Fukui, K. 1986. Standardization of karyotyping plant chromosomes by a newly developed chromosome image analyzing system (CHIAS). *Theor. Appl. Genet.* 72(1):27-32.

Gawel, N.J., Jarret, R.L., and Whittemore, A.P. 1992. Restriction fragment length polymorphism (RFLP)-based phylogenetic analysis of *Musa*. *Theor. Appl. Genet.* 84(3/4):286-290.

Gowen, S. 1995. *Bananas and plantains.* Chapman and Hall, London.

Hallauer, A.R. and Miranda, J.B. 1988 *Quantitative genetics in maize breeding.* Iowa State University Press, Ames, Iowa.

Hartman, J.B. and Vuylsteke, D. 1999. Breeding for fungal resistance in *Musa. Euphytica.*

Horry, J-P., Dolezel, J., Dolezelova, M., and Lysak, M.A. 1998. Do natural A×B tetraploid bananas exist? *InfoMusa* 7(1):5-6.

Horry, J-P., Sharrock, S.L., Frison, E.A. 1998. Present situation of biogenetic resources for the betterment of society. In I.L.H. Arizaga (Ed.), *Memorias XIII* (pp. 679-691). Reunion ACORBAT, Ecuador.

Howell, E.C, Newbury, H.J, Swennen, R.L., Withers, L.A., and Ford-Lloyd, B.V. 1994. The use of RAPD for identifying and classifying *Musa* germplasm. *Genome* 37(2):328-332.

Iijima, K. and Fukui, K. 1991. Clarification of the conditions for the image analysis of plant chromosomes. *Bulletin of the National Institute of Agrobiological Research* (Japan) 6(1):1-58.

Jarret, R.L., Bhat, K.V., Cregan, P., Ortiz, R., and Vuylsteke, D. 1994. Isolation of microsatellite DNA markers in *Musa. InfoMusa* 3(2):3-4.

Jauhar, P.P. 1979. Synthesis and meiotic studies of triploids of Meadow fescue. In *Agronomy Abstracts* (pp. 64-65). American Society of Agronomy, Madison, Wisconsin.

Jenkins, M.T. 1934. Methods of estimating the performance of double crosses in corn. *J. Am. Soc. Agron.* 26(3):199-204.

Jenny, C., Careel, F., and Bakry, F. 1997. Revision on banana taxonomy: Klue Tiparot (*Musa* sp.) reclassified as a triploid. *Fruits* 52(1):83-91.

Jewell, D.C. and Islam-Faridi, N. 1994. A technique for somatic chromosomes preparation and C-banding of maize. In M. Freeling and V. Walbot (Eds.), *The maize handbook* (pp. 484-493). Springer-Verlag, New York.

Kaemmer, D., Fischer, D., Jarret, R.L., Baurens, F-C, Grapin, A., Dambier, D., Noyer, J.L, Lanaud, C., Kahl, G., and Lagoda, P.J.L. 1997. Molecular breeding in the genus *Musa*: A strong case for STMS marker technology. *Euphytica* 96(1):49-63.

Kempthorne, O. 1957. *An introduction to genetic statistics.* Wiley, New York.

Khush, G.S., Singh, R.J., Sur, S.C., and Librojo, A.L. 1984. Primary trisomics of rice: Origin, morphology, cytology and use in linkage mapping. *Genetics* 107(5):141-163.

Kirkby, R.A. 1990. The ecology of traditional agroecosystems in Africa. In M.A. Altieri and S.B. Hecht (Eds.), *Agroecology and small farm development* (pp. 173-180). CRC Press, Boca Raton, FL.

Kubalakova, M. and Dolezel, J. 1998. Optimization of PRINS and C-PRINS for detection of telomeric sequences in *Vicia faba. Biologia Plantarum* 41(2):177-184.

Kubalakova, M., Macas, J., and Dolezel, J. 1997. Mapping of repeated DNA sequences in plant chromosomes by PRINS and C-PRINS. *Theor. Appl. Genet.* 94(6/7):758-763.

Kubalakova, M., Nuozova, M., Dolezelova, M., Macas, J., and Dolezel, J. 1998. A combined PRINS-FISH technique for simultaneous localization of DNA sequences on plant chromosomes. *Biologia Plantarum* 41(2):293-296.

La Fleur, D.A., Lockhart, B.E.L., and Olszewski, N.E. 1996. Portions of banana streak badnavirus genome are integrated in the genome of its host, *Musa*. *Phytopathology* 86(11):S100.

Lysak, M.A, Dolezelova, M., and Dolezel, J. 1998. Flow cytometric analysis of nuclear genome size in *Musa* spp. In J. Maluszynska (Ed.), *Plant Cytogenetics* (pp. 178-183). Wydawnictwo Uniwersytetu, Slaskiego, Katowice, Czech Republic.

Lysak, M.A, Dolezelova, M., Horry, J.P., Swennen, R., and Dolezel, J. 1999. Flow cytometric analysis of nuclear DNA content in *Musa*. *Theor. Appl. Genet.* 98(8):1344-1350.

Malécot, G. 1948. *Les mathématiques de l'hérédité.* Masson et Cie, Paris.

Marriott, H. and Lancaster, P.A. 1983. Bananas and plantains. In H.T. Chan (Ed.), *Handbook of Tropical Foods* (pp. 85-143). Marcel Dekker, New York.

May, G., Afza, R., Mason, H., Wiecko, A., Novak, F., and Arntzen, C. 1995. Generation of transgenic banana (*Musa acuminata*) plants via *Agrobacterium*-mediated transformation. *Bio/Technology* 13(5):486-492.

Mitchell, G.A. 1980. *Banana entomology in the Windward Islands.* Centre for overseas pest research, London.

Mok, D.W.S. and Peloquin, S.J. 1975. Breeding value of $2n$ pollen (diplandroids) in tetraploid × diploid crosses in potato. *Theor. Appl. Genet.* 46(6):307-314.

Novak, F.J. 1992. *Musa* (Bananas and Plantains). In F.A. Hammerschlag and R.E. Litz (Eds.), *Biotechnology of Perennial Fruit Crops* (pp. 449-488). CAB International, University Press, Cambridge.

Nei, M. 1972. Genetic distance between populations. *Am. Nat.* 106(2):283-292.

Ortiz, R. 1995. *Musa* genetics. In S. Gowen (Ed.), *Bananas and Plantain* (pp. 84-109). Chapman and Hall, London.

Ortiz, R. 1997a. Secondary polyploids, heterosis and evolutionary crop breeding for further improvement of the plantain and banana (*Musa* spp. L.) genome. *Theor. Appl. Genet.* 94(8):1113-1120.

Ortiz, R. 1997b. Occurrence and inheritance of $2n$ pollen in *Musa*. *Ann. Bot.* 79(4):449-453.

Ortiz, R. and Vuylsteke, D.R. 1994. Genetics of apical dominance in plantain (*Musa* spp., AAB Group) and improvement of suckering behavior. *J. Amer. Soc. Hort. Sci.* 119(5):1050-1053.

Ortiz, R. and Vuylsteke, D.R. 1995. Factors influencing seed set in triploid *Musa* spp. L. and production of euploid hybrids. *Ann. Bot.* 75(2):151-155.

Ortiz, R. and Vuylsteke, D.R. 1996. Recent advances in *Musa* genetics, breeding and biotechnology. *Plant Breeding Abstracts* 66:1355-1363.

Ortiz, R., Vuylsteke, D., Dumpe, B., and Ferris, R.S.B. 1995. Banana weevil resistance and corm hardness in *Musa* germplasm. *Euphytica* 86(1):95-102.

Ortiz, R., Vuylsteke, D.R., Crouch, H.K., and Crouch, J.H. 1998. TM3x: Triploid black Sigatoka-resistant *Musa* hybrid germplasm. *HortScience* 33(2):362-365.

Osuji, J.O., Crouch, J., Harrison, G., and Heslop-Harrison, J.S. 1997. Identification of the genomic constitution of *Musa* L. genotypes (bananas, plantains and hybrids) using molecular cytogenetics. *Ann. Bot.* 80(6):787-793.

Osuji, J.O., Crouch, J. H., Harrison, G., and Heslop-Harrison, J.S. 1998. Molecular cytogenetics of *Musa* species, cultivars and hybrids: Location of 18S-5.8S -25S and 5S rDNA and telomere-like sequences. *Ann. Bot.* 82(2):243-248.

Osuji, J.O., Okoli, B.E., and Ortiz, R. 1996. An improved procedure for mitotic studies of the EuMusa section of the genus *Musa* L. (*Musaceae*). *InfoMusa* 5(1):12-14.

Panter, D.M. and Allen, F.L. 1995. Using best linear unbiased predictions to enhance breeding for yield in soybean: II. Selection of superior crosses from a limited number of yield trials. *Crop Sci.* 35(2):405-410.

Pich, U., Meister, A., Macas, J., Dolezel, J., Lucretti, S., and Schubert, I. 1995. Primed in situ labeling facilitates flow sorting of similar sized chromosomes. *The Plant Journal* 7(6):1039-1044.

Pillay, M., Nwakanma, D.C., and Tenkouano, A. 1999. Identification of RAPD markers linked to A and B genomes sequences in *Musa*. (In review).

Ploetz, R.C. 1994. Panama disease: Return of the first banana menace. *Int. J. Pest Management.* 40(4):326-336.

Powell, W., Mackray, G.C., and Provan, J. 1996. Polymorphism revealed by simple sequence repeats. *Trends Plant Sci.* 1(7):215-222.

Price, N.S. 1995. The origin and development of cultivation. In S. Gowen (Ed.), *Bananas and plantains* (pp. 1-13). Chapman and Hall, London.

Rafalski, J.A. and Tingey, S.V. 1993. Genetic diagnostics in plant breeding: RAPDs, microsatellites and machines. *Trends Genet.* 9(8):275-280.

Rafalski, J.A., Morgante, M., Powell, W., Vogel, J.M., and Tingey, S.V. 1995. Generating and using DNA markers in plants. In B. Birren and E. Lai (Eds.), *Nonmammalian genomic analysis: A practical guide* (pp. 75-134). Academic Press, London, New York.

Remy, S., Franscois, I., Cammue, B., Swennen, R., and Sagi, L. 1998. Cotransformation as a potential tool to create multiple and durable disease resistance in banana (*Musa* spp.) *Acta Hort.* 461(8):361-365.

Reynolds, P.K. 1951. Earliest evidence of banana culture. *J. Am. Oriental Soc.* 71: Supplement No. 12.

Robbelen, G. 1960. Beitrage zur analyse des *Brassica*-genoms. *Chromosoma* (Berlin) 11(2):205-228.

Robinson, J.C. 1996. *Bananas and plantains.* Crop Production Science in Horticulture 5. CAB International, Wallingford, UK.

Robinson, J.C., Anderson, T., and Eckstein, K. 1992. The influence of functional leaf removal at flower emergence on components of yield and photosynthetic compensation in banana. *J. Hort. Sci.* 67(3):403-410.

Robinson, J.C. and Nel, D.J. 1985. Comparative morphology, phenology and production potential of banana cultivars 'Dwarf Cavendish' and 'Williams' in the eastern Transvaal Lowveld. *Scientia Horticulturae* 25(2):149-161.

Robinson, J.C. and Nel, D. J. 1986. The influence of planting date, sucker selection, and density on yield and crop timing of bananas (cultivar 'Williams') in the eastern Transvaal. *Scientia Horticulturae* 29(4):347-358.

Rowe, P. and Rosales, F. 1996. Bananas and plantains. In J. Janick and J.N. Moore (Eds.), *Fruit Breeding, Volume I: Tree and Tropical Fruits* (pp. 167-211). John Wiley and Sons, Inc. New York.

Rowe, P. and Rosales, F.E. 1995. Current approaches and future opportunities for improving major *Musa* (ABB) types present in the Asian/Pacific region: Saba, Pisang Awak, Bluggoe. In E.A. Frison, J.-P. Horry and D. De Waele (Eds.), *New frontiers in resistance breeding for nematode, fusarium and sigatoka. Proceedings of workshop, Oct. 2-5, 1995, Kuala Lumpur, Malaysia* (pp. 129-141). INIBAP, Montpellier.

Saghai Maroof, M.A., Yang, G.P., Zhang, Q., and Gravois, K.A. 1997. Correlation between molecular marker distance and hybrid performance in U.S. southern long grain rice. *Crop Sci.* 37(1):145-150.

Sagi, L., Panis, B., Remy, S., Schoofs, H., De Smet, K., Swennen, R., and Cammue, B. 1995. Genetic transformation of banana (*Musa* spp.) via particle bombardment. *Bio/Technology* 13(5):481-485.

Simmonds, N.W. 1960. Megasporogenesis and female fertility in three edible triploid bananas. *J. Genet.* 57(243):269-278.

Simmonds, N.W. 1962. *The evolution of the bananas.* Longmans Green and Co., London.

Simmonds, N.W. 1966. *Bananas,* Second edition. Tropical Agric. Series. Longman, London.

Simmonds, N.W. and Shepherd, K. 1955. The taxonomy and origins of the cultivated bananas. *J. Linn. Soc.* (Bot) 55(3):302-312.

Singh, R.J. 1993. *Plant Cytogenetics.* CRC Press, Inc. Boca Raton, FL.

Singh, R.J. and Hymowitz, T. 1988. The genomic relationship between *Glycine max* (L.) Merr. and *G. soja* Sieb. and Zucc. as revealed by pachytene chromosome analysis. *Theor. Appl. Genet.* 76(5):705-711.

Speijer, P.R. and De Waele, D. 1997. *Screening of* Musa *germplasm for resistance and tolerance to nematodes.* INIBAP Technical Guidelines 1. IPGRI, Rome, Italy, INIBAP, Montpellier, France, IITA, Ibadan, Nigeria.

Speijer, P.R., Gold, C.S., Karamura, E.B., and Kashaija, I.N. 1994. Assessment of nematode damage in East African Highland banana systems. In R.V. Valmayor and V.N. Rao (Eds.), *Proceedings of a conference/workshop on nematodes and weevil borer affecting banana in Asia and the Pacific. Kuala Lumpur, Malaysia, April 18-22, 1994* (pp. 191-203). INIBAP, Montpellier, France.

Stace, C. 1980. *Plant Taxonomy and Biosystematics.* Edward Arnold, London.

Staub, J.E. and Serquen, F.C. 1996. Genetic markers, map construction, and their application in plant breeding. *HortScience* 31(3):729-741.

Stover, R.H. and Simmonds, N.W. 1987. *Bananas,* Third edition. Longman Scientific and Technical, Essex, England.

Swennen, R. and Wilson, G.F. 1983. La stimulation du développement du rejet baïonnette du bananier plantain (*Musa* spp. groupe AAB) par application de gibberreline (GA3). *Fruits* 38(4):261-265.

Tenkouano, A., Crouch, H.K., Crouch, J.H., Vuylsteke, D., and Ortiz, R. 1999a. Comparison of DNA marker and pedigree-based methods of genetic analysis in plantain and banana (*Musa* spp.) clones: I. Estimation of genetic relationships. *Theor. Appl. Genet.* 98(1):62-68.

Tenkouano, A., Crouch, H.K., Crouch, J.H., Vuylsteke, D., and Ortiz, R. 1999b. Comparison of DNA marker and pedigree-based methods of genetic analysis in plantain and banana (*Musa* spp.) clones: II. Predicting hybrid performance. *Theor. Appl. Genet.* 98(1):69-75.

Tenkouano, A., Crouch, J.H., Crouch, H.K., and Vuylsteke, D. 1998. Ploidy determination in *Musa* germplasm using pollen and chloroplast characteristics. *HortScience* 33 (5):889-890.

Turner, D.W. and E. Lahav. 1983. The growth of banana plants in relation to temperature. *Austr. J. Plant Physiol.* 10(1):43-53.

Vandenhout, H.R., Ortiz, R., Vuylsteke, D., Swennen, R., and Bai, K.V. 1995. Effect of ploidy on stomatal and other quantitative traits in plantain and banana hybrids. *Euphytica* 83(1):117-122.

Vuylsteke, D., Crouch, J.H., Pellegrineschi, A., and Thottappilly, G. 1998. The biotechnology case history for *Musa. Acta Hort* 461:75-86.

Vuylsteke, D., Ortiz, R., Ferris, R.S.B., and Crouch, J.H. 1997. Plantain improvement. *Plant Breed. Rev.* 14(1):267-320.

Vuylsteke, D., Ortiz, R., and Swennen, R.L. 1993. Genetic improvement of plantains at IITA. In J. Ganry (Ed.), *Breeding for resistance to diseases and pests* (pp. 267-282). CIRAD-INIBAP, Montpellier, France.

Vuylsteke, D., Swennen, R., and De Langhe, E. 1990. Tissue culture technology for the improvement of African plantains. In R.A. Fullerton and R.H. Stover (Eds.), *Sigatoka leaf spot diseases of bananas. Proceedings of an international workshop held at San José, Costa Rica, March 28–April 1, 1989* (pp. 316-337). INIBAP, Montpellier, France.

Vuylsteke, D., Swennen, R.L., and Ortiz, R. 1993. Development and performance of black sigatoka-resistant tetraploid hybrids of plantain (*Musa* spp., AAB group). *Euphytica* 65(1):33-42.

Wang, Z., Lin, Z., and Pan, K. 1993. Cytogenetical studies in *Musa* (Eumusa). In *Current banana research and development in China* (pp. 29-43). South China Agricultural University, Guangzhou.

Wilson, G.B. 1946. Cytological studies in the Musae. I. Meiosis in some triploid clones. *Genetics* 31(3):241-258.

Chapter 9

Molecular Markers, Genomics, and Cotton Improvement

M. Altaf Khan
Gerald O. Myers
J. McD. Stewart

INTRODUCTION

The Genus Gossypium

Cotton is the most important textile fiber crop in the United States and in the world as well as the second most important oilseed crop in the world (Cherry and Leffler, 1984). *Gossypium hirsutum* L. (AD genome) currently comprises 95 percent of world cotton production. But evidence shows that cotton is far behind in the areas of genomics and application of molecular markers for crop improvement.

The cotton genus, *Gossypium*, includes 44 diploid ($2n = 2x = 26$) species and five allotetraploid ($2n = 4x = 52$) species. The diploid species comprise genomic groups A, B, C, D, E, F, G, and K, and the allotetraploid species are made up of two subgenomic groups with affinity with A and D genomes (Stewart, 1995). The cultivated cottons include *G. arboreum* L. and *G. herbaceum* L., both A-genome diploid species that are native to Southeast Asia and Africa, and two allotetraploid species, *G. barbadense* L. and *G. hirsutum* L., with the AD genome, from Central America and Northern South America (Endrizzi et al., 1985).

The annual rate of genetic improvement of yield for *G. hirsutum* is estimated to be about 0.75 percent per year (Meredith and Bridge, 1984). Our genetic knowledge about cotton is inadequate relative to its impact on the U.S. and world economy. The rate of genetic progress from breeding should improve with an increased genetic knowledge of the germplasm. *Gossypium hirsutum* has nearly displaced the A-genome species from the range in which they were previously cultivated (Meredith, 1991). However, *G. arboreum* (A_2 genome) is still grown in Pakistan and India on marginal

land for use in nonwoven material and is helpful in breeding programs as a
donor of host-plant resistance genes. The A-genome cottons enhance genetic diversity for tetraploid cotton breeding programs (Stanton et al., 1994),
especially with the development of techniques for introgressing A-genome
germplasm into AD-genome cultivars (Stewart, 1992; Stewart and Hsu,
1978b). Hybrids between *G. hirsutum* and *G. arboreum* have led to the selection of genotypes with earlier maturity and an increased range of fiber
traits (Cooper, 1969; Wang et al., 1989).

At the chromosomal level, hybrids of *G. arboreum* lines form 13 bivalents at meiosis. Hybrids between *G. hirsutum* and *G. arboreum*, however,
exhibit meiotic pairing anomolies; Gerstel (1953) observed three reciprocal
translocations that distinguish the chromosomes of the latter species from
the A subgenome of the American amphidiploids, and one ring of four and
one of six chromosomes. These three naturally occurring reciprocal translocations involve chromosomes one to five (Menzel et al., 1982). The
translocations can be used to mark specific chromosomes in interspecific
hybrids of *G. hirsutum* with A-genome diploid species (SCSB, 1981). The
26 chromosomes in AD cottons have been numbered according to tests
among chromosomal translocation lines within *G. hirsutum* and between
translocations of A- and D-genome diploid species (Brown, 1980; Brown et
al., 1981; Menzel and Brown, 1954, 1978; Menzel et al., 1982). A-genome
chromosomes are numbered 1 to 13 and D-genome chromosomes, 14 to 26
(Price et al., 1990).

Genome Organization of Gossypium

The total amount of DNA in a particular organelle is described as its genome. Animals have one genome in the nucleus and a second in the mitochondria, however, plants have yet a third genome in the chloroplast. The
mitochondrial and chloroplast genomes are believed to have evolved from
genomes of independent organisms over the course of time. Eukaryotic
genomes vary considerably in the manner in which single copy and repeated
sequences are arranged and in haploid DNA content. The proportion of repeated DNA, and the extent to which it is interspersed with single copy
DNA, is generally a function of the overall DNA content of a species
(Flavell, 1980).

In cotton, the DNAs of *G. arboreum, G. herbaceum, G. thurberi, G.
raimondii, G. hirsutum, G. barbadense,* and *G. trilobum* have been analyzed
(Walbot and Dure, 1976; Wilson et al., 1976; Geever, 1980; Reinisch et al.,
1994; Baker et al., 1995; Brubaker et al., 1999). In the studies of Walbot and
Dure (1976) and of Geever (1980), the fractional components of single copy
and different classes of repetitive DNA were analyzed. Walbolt and Dure
(1976) divided the cotton genome into three major kinetic components,

namely, unique sequence DNA, middle repetitive-sequence DNA, and highly repetitive-sequence DNA. They estimated the haploid genome size of *G. hirsutum* to be 0.795 pico gram (pg), whereas Geever (1980) estimated it to be 0.950 pg. The two studies also found different amounts of single-copy and repetitive DNAs. Walbot and Dure (1976) estimated that at least 80 percent of the A and D subgenomes of *G. hirsutum* had an interspersion of single-copy and moderately repetitive DNA. Another study reported that the upland cotton genome consisted of approximately 61 percent unique sequences and low copy number DNA (Baker et al., 1995). Reinisch and colleagues (1994) estimated that the cotton genome contained about 400 kilo base (kb) of DNA per centi-Morgan (cM). In a recent study, Brubaker and colleagues (1999) observed that, in cotton, polyploidy per se was recombinogenic and that recombinational length was uncoupled from physical genome size.

The A genomes have moderately large chromosomes, and the D genomes have small chromosomes. There is roughly a twofold size difference between the smaller D chromosomes and the larger A genomes (Endrizzi and Phillips, 1960; Edwards et al., 1974). The D and A genomes do not differ in the relative amounts of single-copy (ca. 0.40 pg) or palindromic (ca. 0.05 pg) DNA, but they do differ in the amounts of moderately repetitive DNA sequences (ca. 0.2 vs. 0.56 pg) and in highly repetitive DNA sequences (0.0 vs. 0.17 pg) (Geever et al., 1989). The differences in the quantities of repetitive DNAs are thought to permit genome specificity during meiotic chromosome pairing, rather than reflect relative genetic control as is known to exist in wheat and other graminaceous species (Mursal and Endrizzi, 1976).

Genetic Diversity and Polymorphism in Gossypium

Technical improvements have made it possible to produce trispecies hybrids involving allotetraploid species and two diploid species (Stewart, 1979, 1995; Stewart and Hsu, 1977, 1978a, 1978b). As the gene pools of many domestic crop species are quite narrow, intraspecific crosses often do not exhibit sufficient DNA polymorphism to construct a high-density map. Although crosses among conspecific races or subspecies show more polymorphism, this polymorphism may not be uniformly distributed across the chromosomes because of the introgression that has occurred as a result of natural processes or designed breeding programs (e.g., Stephens et al., 1967; Rosenow and Clark, 1987). Genetic diversity created through interspecific introgression can be measured with morphological and molecular markers (Gepts, 1993). However, a large number of polymorphic markers are required to measure genetic relationships and genetic diversity in a reliable manner. This limits the use of morphological markers, which are few or lack adequate levels of polymorphism in *Gossypium* (Tatineni et al., 1996).

In *G. hirsutum*, about 80 morphological traits have been used to identify 17 linkage groups. In addition to not being useful in creating a complete linkage map of the genome, most of these morphological markers have major effects on quantitative traits, and their use in selection is limited. Also, in many cases, the heterozygous condition of most morphological traits is not identifiable. On the contrary, molecular markers are available in very large numbers, they generally have no effect in themselves on phenotypes, and the heterozygotes of some molecular markers can be easily identified (Meredith, 1995).

USE OF MOLECULAR MARKERS IN GOSSYPIUM

Today, several methods to detect polymorphic DNA markers that are suitable for genetic mapping are available. A large array of molecular-marker systems has been used to characterize the cotton genome and, to a lesser extent, in cotton improvement. These include isozymes, restriction fragment length polymorphisms (RFLPs), random amplified polymorphic DNA (RAPDs), amplified fragment length polymorphisms (AFLPs), microsatellites or simple sequence repeats (SSRs), and expressed sequence tags (ESTs). Some salient details of each of these, as applied to cotton, are presented.

Isozymes and Their Application

Despite their great potential, applications of isozyme research on the genus *Gossypium* have been few and rather limited (Cherry et al., 1972; Hancock, 1982; Centner et al., 1984; Suiter, 1988; Wendel et al., 1989). One reason for this is that the roots, stems, and leaves of cotton plants are very rich in polyphenolic products (Cherry and Leffler, 1984). Electrophoretic analysis of isozymes can be very difficult in plants that contain high levels of phenolic compounds, because these compounds interact with proteins in a variety of ways, causing inhibition of enzyme activities or otherwise obscuring isozyme banding patterns in zymograms (Wendel and Parks, 1982). As a result, genetic research on isozymes has been relatively difficult in polyphenolic-rich plants, such as cotton. Suiter and Parks (1984), studying genetic variation in *Gossypium* with isozyme markers, found allelic polymorphism in 10 of 18 isozyme systems in accessions of *G. arboreum* and *G. herbaceum*. Within the AD allotetraploids, Bourdon (1986) studied the variation of isozyme profiles among landraces and cultivars of *G. hirsutum* and *G. barbadense*.

RFLP Markers and Their Application

The development of molecular genetic markers that detect variation at the DNA sequence level has made it possible to obtain solutions to problems that were previously not amenable to genetic manipulation. Extensive genome mapping based on RFLP markers has been accomplished in many crop species (O'Brien, 1992). The making of RFLP maps may be purely of academic interest, as in model organisms; medical, as in humans; or economic, as in the crop plants. RFLP maps have been produced for a number of organisms, particularly in crop plants, such as the common bean (Vallejos et al., 1992), soybean (Keim et al., 1990), *Brassica oleracea* (Kianian and Quiros, 1992), lettuce (Kesseli et al., 1994), tomato (Tanksley and Mutschler, 1990), potato (Gebhardt et al., 1991), pea (Ellis et al., 1992), rice (McCouch et al., 1988), barley (Graner et al., 1991), maize (Coe et al., 1990), and cotton (Reinisch et al., 1994). Despite the fact that most of these maps were generated for breeding purposes, the data sets from which the maps were generated present an opportunity to study recombination in these organisms.

The availability of RFLP-based linkage maps has led to the widespread application of molecular techniques to the genetic studies of crop plants. Examples include the mapping of genes of economic importance (McCouch et al., 1990; Liu et al., 1992) to detection and genetic analysis of quantitatively inherited agronomic traits (Keim et al., 1990; Paterson et al., 1991; Stuber et al., 1992) to knowledge of the fields of population diversity and systematics (Wang et al., 1992; Zhang et al., 1993). Germplasm identification and evaluation has been greatly enhanced through RFLP studies (Dudley et al., 1992; Zhang et al., 1993).

The complexity of the cotton genome, coupled with the low levels of polymorphism, requires the use of a large number of RFLP markers to provide the basic tools for genome mapping and other genetic investigations. A wealth of knowledge has been gained in the last few years through the application of the RFLP technology. Several cotton species have been subjected to studies on evolution, population genetics, phylogenetic relationships, genome mapping, and QTL analysis (Wendel et al., 1989; Wendel and Albert, 1992; Small and Wendel 1999; Meredith, 1992; Wang et al., 1992; Stelly, 1993; Cantrell and Davis, 1993; Paterson, 1993; Wing, 1993; Reinisch et al., 1994; Brubaker and Wendel, 1994; Shappley et al., 1996, 1998a, 1998b; Yu and Kohel, 1999; Brubaker et al., 1999; Ulloa et al., 2000; Jiang et al., 2000).

The first RFLP evaluation in upland cotton, *G. hirsutum* L., was reported by Meredith (1992) in a study of heterosis and varietal origins. To detect RFLP markers, 68 individual lines from diverse genetic backgrounds were analyzed with 75 enzyme/probe combinations resulting in 179 RFLPs.

RFLP markers have subsequently been used extensively for cotton genome mapping. The first cotton genomic map was constructed utilizing a cross between G. *hirsutum* and G. *barbadense* with 705 RFLP markers (Reinisch et al., 1994). Shappley, Jenkins, et al. (1998) developed an intraspecific RFLP linkage map of upland cotton comprising 138 RFLP loci, followed by another bispecific cotton map using RFLP loci in combination with other markers (Yu and Kohel, 1999). Brubaker and colleagues (1999) used RFLP loci for comparative genetic mapping of allotetraploid cotton and its diploid progenitors to study the evolution process in upland cotton. In addition to linkage mapping, RFLP markers have been used for QTL detection in cotton (Shappley, Jenkins, et al., 1998; Wright et al., 1999; Yu and Kohel, 1999; Jiang et al., 2000).

Although, RFLPs have been used extensively, the level of RFLP variation in G. *hirsutum* is low relative to other plant taxa (Brubaker and Wendel, 1994). One additional factor limiting the use of RFLPs in *Gossypium* is the difficult task of isolating good quality DNA (that can be completely digested with endonuclease enzymes) due to the abundance of polysaccharides and phenolic compounds. Furthermore, RFLP analysis requires relatively large amounts of DNA and, in general, it requires more time and labor compared to other newly developed PCR-based techniques, such as RAPDs, AFLPs, SSRs, etc. Since cotton is an allotetraploid plant with a large genome, it is desirable to have efficient DNA assay systems to develop the large number of polymorphic markers needed to cover the complete genome in a relatively short time frame.

RAPD Markers and Their Application

The utility of DNA-based diagnostic markers is determined, to a large extent, by the technology that is used to reveal DNA-based polymorphism (Tingey and del Tufo, 1993). In the early to late 1980s, the most common DNA markers for genetic assays were RFLPs, but the detection of RFLPs by Southern blot hybridization is laborious and incompatible with the high analytical throughput required for many applications. Other polymorphism assays that require target DNA sequence information for the design of PCR amplification primers are not affordable for many large-scale genetic populations due to the time and cost of obtaining this sequence information (Williams et al., 1990).

A new genetic PCR-based assay was developed independently in two different laboratories (Welsh and McClelland, 1990; Williams et al., 1990). This procedure was called random amplified polymorphic DNA and was designed to detect nucleotide sequence polymorphism in a DNA amplification-based assay using only a single primer of an arbitrary nucleotide sequence. In this procedure, a single species of primer binds to the genomic

DNA at two different sites on opposite strands of the DNA template (Tingey and del Tufo, 1993). If these priming sites are within an amplifiable distance of each other, a discrete DNA product is produced through thermocyclic amplification. The presence of each amplification product identifies complete or partial nucleotide sequence homology between the genomic DNA and the oligonucleotide primer at each end of the amplified product. On average, each primer will direct the amplification of several discrete loci in the genome.

Sources of polymorphism may include single nucleotide changes in a primer sequence, deletions of a priming site, insertions that render priming sites too distant to support amplification, or insertions or deletions that change the size of a DNA segment without preventing its amplification (Williams et al., 1990). RAPD markers are usually noted by the presence or absence of an amplification product from a single locus, which means that the RAPD technique provides only dominant markers (Tingey and del Tufo, 1993).

RAPDs are reliable, versatile, and variable genetic markers that compare favorably with RFLPs. The main advantage of RAPDs over RFLPs and other types of markers, such as SSRs, lies in the large number of loci that can be screened quickly (Kesseli et al., 1994). RAPD markers provide a powerful tool for the automation of genome mapping, and for extending the power of genetic analysis to plant species such as cotton, which have few phenotypic markers to completely describe the whole genome. The biggest advantage of using RAPD markers for genetic mapping is that a universal set of primers can be used for genomic analysis in a wide variety of species (Williams et al., 1990). The most desirable mapping populations are those developed by backcrossing or that are recombinant inbred lines (Rafalski et al., 1994). F_2 populations, while widely used, are best to identify markers in the coupling phase to minimize error (Williams et al., 1993). Mapping with dominant markers, linked in repulsion, provides little information for the estimation of genetic distance (Allard, 1956). Despite this limitation, mapping with dominant markers, linked in coupling, is, on a per-gamete basis, as efficient for mapping as codominant markers are (Tingey and del Tufo, 1993). It is also relatively easy to turn the RAPD assay into a secondary PCR assay through DNA sequencing of the RAPD band of interest and converting it to allele-specific PCR (Wu et al., 1989). Additional sequence polymorphism may be detected in RAPD bands using restriction enzyme digestion (Williams et al., 1993).

Michelmore and colleagues (1991) described the use of RAPDs to screen efficiently for markers linked to specific regions of the genome. This method, called bulk segregant analysis (BSA), uses two bulked DNA samples collected from individuals of a segregating population. Each bulk is composed of individuals that differ for a specific phenotype and/or genotype or of individuals at either extreme of a segregating population. Chaparro and colleagues (1994) used BSA to identify RAPD markers linked to morphological traits and concluded that the combination of RAPD markers and BSA was efficient in identifying markers flanking traits of interest.

RAPD markers can be used as probes for RFLPs after reamplification of the RAPD fragment (Williams et al., 1990). RAPD markers can also be converted into SCARs (sequence characterized amplification regions) (Michelmore et al., 1992; Paran and Michelmore, 1993) that overcome some of the drawbacks of RAPDs. SCARs identify a specific locus from a defined pair of oligonucleotide primers. The sequence of these primers is derived from the termini of a fragment identified as a RAPD marker. These primers, with their increased specificity, generally amplify a single, highly repeatable band. Kesseli and colleagues (1994) reported that most SCAR primers constructed from RAPD fragments identified from one species of lettuce also amplify homologous sequences in other closely and distantly related species of *Lactuca*.

Both Mendelian segregation and stable, dominant inheritance of RAPD markers have been observed in genetically diverse crop species including *Glycine max* L. (soybean) and *Hordeum vulgare* L. (barley) (Williams et al., 1990; Tinker et al., 1993). RAPD markers have been used to verify and establish genetic relationships among species, crop cultivars, and isolates (Kresovich et al., 1992; Gonzalez and Ferrer, 1993; Lannér et al., 1996). Several research groups have used RAPD markers to study phylogenetic relationships in many plant species (Arnold et al., 1991; Hu and Quiros, 1991; Kresovich et al., 1992; Ahmad and McNeil, 1996; Millan et al., 1996; Tatineni et al., 1996).

Because of their abundance and quick generation, RAPD markers have been used extensively to construct genetic linkage maps in several plant species. Examples include *Arabidopsis thaliana* (Reiter et al., 1992), *Pinus radiata* D. Don (pine) (Chaparro et al., 1992; Devey et al., 1996), *Prunus persica* L. Batsch (peach) (Chaparro et al., 1994; Rajapakse et al., 1995), lettuce (Kesseli et al., 1994), *Theobroma cacao* L. (cocoa) (Lanaud et al., 1995), *Lycopersicon esculentum* L. (tomato) (Grandillo and Tanksley, 1996), *Saccharum officinarum* L. (sugarcane) (Mudge et al., 1996), *Cucumis melo* L. (melon) (Baudracco-Arnas and Pitrat, 1996), *Manihot esculanta* Crantz (cassava) (Fregene et al., 1997), and *Vigna ungiculata* L. Walp. (cowpea) (Mendendez et al., 1997). Many of these studies combine the use of RAPDs with other molecular and classical markers.

RAPD analysis has been reliably utilized to determine genetic relationships within a diverse array of *Gossypium* germplasm. Tatineni and colleagues (1996) used 135 RAPD markers to generate dendrograms for measuring genetic distance among genotypes of *G. hirsutum* and *G. barbadense*. They have also been used to evaluate elite cotton commercial cultivars (Iqbal et al., 1997; Multani and Lyon, 1995), to tag the *cms-D₈* restorer gene in cotton (Zhang et al., 1997), to tag genes influencing general combining ability effects for yield components in cotton (Lu and Myers, 1999), and to construct bispecific (Yu and Kohel, 1999) and trispecific (Altaf Khan et al., 1997, 1998, 1999) genomic maps of cotton. RAPDs have also served as the basis for the QTL analysis of stomatal conductance in relationship to lint yield in cotton (Ulloa et al., 2000) to investigate fiber quality properties (Yu and Kohel, 1999), and for research into various agronomic traits (Altaf Khan et al., 1998).

AFLP Markers and Their Application

Amplified fragment length polymorphism (AFLP), developed by Zabeau and Vos (1993), is an efficient PCR-based technique used to generate a large number of polymorphic DNA fragments. This technique, as described by Vos and colleagues (1995), involves three basic steps: (1) digestion of DNA and ligation of oligonucleotide adapters, (2) selective amplification of sets of restriction fragments, and (3) gel analysis of the amplified fragments. PCR amplification of restriction fragments is achieved by using the adapter and restriction site sequences as target sites for primer annealing. The selective amplification is achieved by the use of primers that extend into the restriction fragments, amplifying only those fragments in which the primer extensions match the nucleotides flanking the restriction sites. Using this method, sets of restriction fragments may be visualized using PCR without the knowledge of nucleotide sequences. The method allows specific coamplification of high numbers of restriction fragments. The number of fragments that can be analyzed simultaneously, however, is dependent on the resolution of the detection system. Typically 50 to 100 restriction fragments are amplified and detected using denaturing PAGE. The AFLP technique provides a novel and very powerful DNA fingerprinting technique for DNAs of any origin or complexity.

Becker and colleagues (1995) reported that AFLP marker technology allows efficient DNA fingerprinting and the analysis of large numbers of polymorphic restriction fragments on polyacrylamide gels. They observed that AFLP markers mapped to all parts of the barley chromosomes and filled in the gaps on barley chromosomal segments on which no RFLP loci had previously been mapped. AFLP markers seldom interrupted RFLP clusters but grouped next to them. The basic difference between RFLPs and AFLPs is that for RFLPs, only the restriction sites determine polymorphism, but in the case of AFLPs, the restriction sites plus the additional selective nucleotides determine polymorphism and, therefore, AFLPs detect more point mutations per 100 nucleotides than RFLPs (Becker et al., 1995).

AFLP is a powerful, reliable, stable, and rapid assay with potential genome-mapping applications. AFLP analysis provides a rapid and efficient technique for detecting large numbers of DNA markers and should expedite plant gene isolation by positional cloning and the construction of high-density molecular linkage maps of plant genomes (Thomas et al., 1995). The efficiency of generating AFLP markers appears to be much higher compared to RFLP mapping in the same population (Huang et al., 1994). In general, AFLP markers also cover the regions covered by RFLPs, and AFLP markers can also cover areas lacking RFLPs (Maheswaran et al., 1997). Like RAPDs, most AFLP markers are dominant and show Mendelian inheritance (Meksem et al., 1995; Maughan et al., 1996). However, the AFLP technique detects a much higher level of polymorphism than RAPD analysis, making it a more efficient marker technology than the latter for the construction of genetic linkage maps (Sharma et al., 1996). Most AFLP markers correspond to unique positions on the genome and can be utilized for genetic and physical mapping as each fragment is characterized by its size and primer combination required for amplification (Vos et al., 1995). Linkage map construction with AFLP markers is quite efficient and complementary to maps based on RFLPs (Maheswaran et al., 1997). In general, AFLP markers are randomly distributed throughout the genome, although a few clusters may be observed.

AFLP fingerprints are highly reproducible and may be used as a tool for evaluating genetic diversity (Majer et al., 1996; Tohme et al., 1996). AFLPs are found to be reproducible even against the background of different combinations of Taq DNA polymerases and buffers. However, the quantity of higher molecular weight fragments (> 400 bp) is reduced when using plant DNA of poor quality as a template (Schondelmaier et al., 1996). The capacity of AFLP analysis to detect thousands of independent genetic loci with minimal cost and time requirements make it an ideal marker system for a wide array of genetic investigations (Maughan et al., 1996).

Money and colleagues (1996) investigated the feasibility of the AFLP technique to generate mRNA fingerprints in polyploid crop plants using hexaploid wheat and one deletion mutant. The AFLP technique could be modified to allow the display of mRNAs and to isolate sequences mapping to deleted chromosome segments in hexaploid wheat. Bachem and colleagues (1996) analyzed transcriptional changes at and around the time of potato tuberization using a highly synchronous in vitro tuberization system, in combination with an AFLP-derived technique for RNA fingerprinting (cDNA-AFLP). Finally, it was shown that using cDNA-AFLP, rapid and simple verification of band identity could be achieved. The cDNA-AFLP technique is a broadly applicable technology for identifying developmentally regulated genes.

The AFLP analysis is quick, robust, requires minimal preliminary work, and has been successfully used for linkage mapping (Wang et al., 1997; Mendendez et al., 1997; Maheswaran et al., 1997; Voorrips et al., 1997), in germplasm evaluation (Tohme et al., 1996; Hongtrakul et al., 1997), and for determining genetic relationships among populations and cultivar identification (Maughan et al., 1996; Paul et al., 1997; Schut et al., 1997). Its application has been made for QTL mapping in cotton (Altaf Khan et al., 1999). AFLP primers can be easily distributed among laboratories by publishing primer sequences and common sets of primers can be established among different plant species for comparative studies. These unique characteristics make AFLP analysis an excellent new method for the detection and study of genetic polymorphism in a wide array of plant species.

In the past, the adoption of AFLP markers for cotton improvement has been slow, but their use has been increasing with the sharing of technology among various cotton research groups. So far, in cotton, the majority of AFLP markers have been used for linkage analysis and QTL mapping. Altaf Khan and colleagues (1997) used AFLP markers to study the genetics of a trispecific F_2 population and, subsequently, to construct a trispecific genomic map of the *Gossypium* (Altaf Khan et al., 1998). AFLP marker technologies have been developed by various cotton research groups for linkage and QTL analyses (Tan et al., 1999; Reddy et al., 2000; Cantrell, 2000). Vroh Bi and colleagues (1999) used AFLP markers for introgression studies of low-gossypol seed and high-gossypol plants in upland cotton. AFLP markers also have been employed to identify and map QTLs associated with fiber length and strength (Brooks et al., 2000), to study genetic diversity in diploid and tetraploid cotton species (Pillay and Myers, 1999; Abdalla et al., 2000), and to investigate somaclonal variation in cotton somatic embryogenesis (Sakhanokho et al., 2000).

SSR Markers and Their Application

Microsatellites, or simple sequence repeats (SSRs), were initially described in humans (Litt and Luty, 1989; Weber and May, 1989). Their potential as useful markers for plants was promptly recognized, resulting in their successful isolation and application in many species. Because of their hypervariability and ease of handling in comparison with RFLPs, SSRs can provide a powerful tool to construct linkage maps (Senior and Heun, 1993; Wu and Tanksley, 1993; Bell and Ecker, 1994). The principal reason for the increasing success of SSRs as a molecular tool is that they detect a higher level of polymorphism than other techniques, such as RFLPs and RAPDs (Powell et al., 1996). In addition, SSRs have an abundant and uniform distribution throughout the genome, are codominant, and segregate in a Mendelian fashion. Their screening relies on simple PCR technology and requires only small amounts of DNA. The high reproducibility of SSRs allows different laboratories to produce consistent data (Rossetto et al., 1999). Microsatellites have successfully bridged the gap between genetic mapping and genome sequencing (physical mapping) (Schuler, 1998). SSRs were converted to sequence tagged sites (STS) to facilitate the construction of the Ge´ne´thon high-resolution genetic map of human genome (Dib et al., 1996). STS markers can, thus, serve as a framework for saturation and expansion of a map with additional DNA markers to achieve genomewide coverage and facilitate integration of the genetic and physical maps (Schuler, 1998).

The upland cotton genome (AD) has a large amount of repetitive DNA fragments much like other higher eukaryotic genomes (Endrizzi et al., 1985; Baker et al., 1995; Zhao et al., 1998). Physically, the D genome is shorter than the A genome, but a larger amount of repetitive DNA is present in the A genome than in the D genome (Endrizzi et al., 1985). In a recent study (Altaf Khan et al., 1998), 60 percent of the total SSR markers investigated were assigned to the A genome as compared to 40 percent with the D genome, indicating a larger amount of repetitive DNA in the A genome compared with the D genome, as also reported by others. Geever (1980) reported that the A and D genomes have essentially the same quantity of single copy DNA, but that the A genome of *G. herbaceum* has twice the amount of moderately repetitive DNA and greater amount of highly repetitive DNA than the D genome of *G. raimondii*. Similarly, Wilson and colleagues (1976) observed greater quantities of repetitive DNA in the A genome of *G. arboreum* than in the D genome of *G. thurberi*. Endrizzi and colleagues (1985) concluded that the only quantitative difference between the chromosomal DNAs of the A and D diploid genomes was in their repetitive sequences and that all genomes of *Gossypium* had similar quantities of single copy DNA but varied in their quantities and qualities of repetitive DNAs depending on genome size. Zhao and colleagues (1998) reported that 77 percent of the dispersed repeats were restricted to the diploid taxa containing the larger A genome and accounted for about half of the difference in DNA content between Old World (A) and New World (D) diploid ancestors of the cultivated AD tetraploid cotton. Thus, SSRs could be a very powerful marker system to detect a large number of polymorphic markers in the cotton genome for future linkage and QTL mapping investigation.

In cotton, SSR markers have been used successfully for linkage and QTL analysis (Ulloa et al., 2000; Altaf Khan et al., 1999). Liu and colleagues (2000) developed anchor SSRs for cotton chromosomes to provide the basis for a framework genetic map. They used 66 primer pairs to amplify 70 marker loci and genotype 13 monosomic and 28 mono-telo-disomic cotton cytogenetic stocks. Forty-two SSR loci were assigned to cotton chromosomes or chromosome arms. Twenty-six SSRs were not located on the informative chromosomes, and 19 were clearly shown to occur on the A subgenome and 11 on the D subgenome by screening *G. herbaceum* ($2n = 2x = 26 = 2A_1$) and *G. raimondii* ($2n = 2x = 26 = 2D_5$) accessions. Some of these anchor SSR loci have been integrated in a recently developed trispecific cotton genomic map (Altaf Khan et al., 1998) to assign linkage groups to chromosomes and out of a total of 51 linkage groups, six have been assigned to different chromosomes or chromosome arms using anchor SSR loci. A large number of SSR primer pairs for cotton have been developed (Tan et al., 1999; Reddy et al., 2000; Cantrell, 2000) and are being utilized in various cotton genomic projects in the pipe line.

LINKAGE MAPS AND THEIR APPLICATIONS

Because of the cause-and-effect relationship between DNA information and the nature of organisms, complete documentation of the genes that comprise an organism affords a first step toward understanding how the organism develops. By associating phenotypic variation among different organisms with variation in genetic information, the basis of phenotype might be better understood. In view of the economic importance of cotton, much of the current research in cotton molecular genetics is focused on traits related to agricultural productivity and quality.

Detailed molecular maps provide opportunity for identification of DNA markers for rapid assay of segregating populations and selection and introgression of complex traits, even using nongerminated seed. In addition, a map offers the opportunity to undertake molecular dissection of complex measures of quality and productivity, such as fiber attributes. Detailed and discriminative genetic profiles provide extremely powerful and effective procedures to make meaningful and valid comparisons among elite lines, cultivars, and hybrids with respect to germplasm improvement.

Development of a genetic linkage map is the basic step toward the detection of factors that control the expression of economically important traits. The procedure for constructing a genetic linkage map consists of three main steps: (1) identification of polymorphic marker loci; (2) verification of Mendelian segregation; and (3) placement of loci in linear order relative to each other with the aggregation of linked markers into linkage groups (Foolad et al., 1995). These maps and their associated technologies have been used for a number of applications in plant breeding and genetics (Grandillo and Tanksley, 1996), including identification of genetic variation in germplasm collections (Figdore et al., 1988; Gawel et al., 1992), gene tagging (i.e., identification of markers tightly linked to major genes) (Schüeller et al., 1992; MacKill et al., 1993), map-based gene cloning

(Arondel et al., 1992; Martin et al., 1993), and identification of quantitative trait loci (Edwards et al., 1987; Paterson et al., 1988, 1991; Tanksley, 1993). Linkage maps can also be utilized to study evolution, genome structure and organization, inheritance, pattern of segregation, linkage pattern of different molecular marker systems in the same map, identification of introgression, identification of quantitative trait loci (QTLs) influencing agronomic traits, and to facilitate marker-assisted selection in a plant breeding program. In many plant species, molecular maps have been integrated with classical genetic maps, e.g., lettuce (Landry et al., 1987); *Arabidopsis* (Nam et al., 1989; Hauge et al., 1993); *Phaseolus vulgaris* L. (common bean) (Nodari et al., 1993); *Oryza sativa* L. (rice) (Abenes et al., 1994; Yu et al., 1995); peach (Rajapakes et al., 1995; Foolad et al., 1995); *Solanum tuberosum* (potato) (Jacobs et al., 1995); and melon (Baudracco-Arnas and Pitrat, 1996).

At present, linkage maps are being utilized for the genomic comparison of closely or distantly related species to find homologous sets of markers. If linkage relationships are conserved among species, genetic information and molecular markers produced in one species may be exploited in related species with less characterized genetic maps (Saghai Maroof et al., 1996). A high level of synteny has been identified in the genomes of related species, e.g., sugarcane; *Sorghum bicolor* (L.) Moench (sorghum); and *Zea mays* L. (maize) (Grivet et al., 1994); rice and maize (Ahn et al., 1993); maize and sorghum (Whitkus et al., 1992); potato and tomato (Gebhardt et al., 1991); barley, rice, *Triticum aestivum* L. Em. Thell. (wheat), and maize (Sherman et al., 1995); and barley and rice (Saghai Maroof et al., 1996). In cotton, genomic linkage maps of various densities have been constructed, two were RFLP-based (Reinisch et al., 1994; Shappley, Jenkins, Meredith, et al., 1998), a third one was RFLP- and isozyme-based (Brubaker et al., 1999), and a fourth map was RFLP- and RAPD-based (Yu and Kohel, 1999). Recently a trispecific genomic map also has been developed (Altaf Khan et al., 1998).

Cotton genomic map construction is complicated because of its relatively large number of chromosomes, metacentricity of chromosomes, propensity for extensive recombination per homologous chromosome arm, and polyploidy-associated redundancy. However, it is simplified by disomic patterns of inheritance, availability of crossable parents with workable rates of sequence polymorphism, and availability of relatives that allow ancestral genomic assignment (Liu et al., 2000).

The first interspecific detailed molecular map of the cotton genome was developed by Reinisch and colleagues (1994) using a cross between the two predominantly cultivated species of cotton, *G. hirsutum* and *G. barbadense*. In this map, 705 RFLP loci covered a total map distance of 4675 cM across 41 linkage groups, with an average space of about 7 cM between markers.

The overall minimum length of cotton genome was estimated to be 5000 cM, and a need for additional DNA markers was suggested to saturate the map with the ultimate objective of linking the map with 26 linkage groups corresponding to the 26 gametic chromosomes of cotton (Reinisch et al., 1994). The first intraspecific genetic map of cotton was developed by Shappley, Jenkins, Meredith, et al. (1998) by crossing two distantly related breeding lines of *G. hirsutum* (MARCABUCAG8US-1-88 and the cultivar HS46). In this map, out of 138 RFLP loci, 120 loci were linked with 31 linkage groups that covered a genetic map distance of 865 cM, which is about 8.6 percent of the cotton genome.

In another study for comparative genetic mapping of allotetraploid cotton and its diploid progenitors, Brubaker and colleagues (1999) developed genetic maps of cotton's A and D genomes. The A genome map was developed using a cross between *G. herbaceum* L. and *G. arboreum* L. with 152 RFLP and six isozyme markers covering a map distance of 856 cM across 18 linkage groups. The D genome map was developed by crossing *G. trilobum* with *G. raimondii,* and 269 RFLP loci mapped a genetic distance of 1486 cM over 17 linkage groups. These maps were used to study and characterize the evolutionary process in the diploid and allotetraploid genomes of cotton. Another bispecific *(G. hirsutum* × *G. barbadense)* map comprising 355 RAPD and RFLP markers has been completed (Yu and Kohel, 1999). This map covered a genetic distance of 4766 cM comprising over 50 linkage groups.

A first trispecific genomic map of cotton has been established (Altaf Khan et al., 1998) by using a trispecific F_2 mapping population. This population was developed by first crossing *Gossypium arboreum* L. (A_2) cv. Nanking with *G. trilobum* (D_8). The diploid hybrid was then treated with colchicine to produce a synthetic allotetraploid $2(A_2D_8)$. This hybrid was crossed with *G. hirsutum* L. (AD_1) cv. T-586, and the resulting hybrid was self-pollinated to obtain a segregating F_2 population. A total of 444 markers (332 AFLPs, 91 RAPDs, 12 SSRs, and 9 morphological markers) were scored among 90 F_2 plants. Fifty-six percent of the markers showed distorted segregation, perhaps due to divergence of the three-genome species associated with chromosomal rearrangements and areas of low recombination among the genomes. A linkage map was constructed comprising 51 linkage groups that spanned over 6961 cM of the cotton genome with an average distance of 18 cM between markers. Higher levels of polymorphism were observed in the D genome compared with the A genome. Thus, the diploid D genome appears to be more divergent from its alloploid D_h subgenome than A is from the A_h subgenome. Out of the 51 linkage groups, nine were assigned to chromosomes or chromosome arms of the cotton genome based on the position of anchor SSR marker loci and morphological markers on the chromosomes. Four linkage groups were assigned to chromosome 10, chromosome 2S, chromosome 6L, and chromosome 9L, and two linkage groups were assigned to chromosome 5L of the cotton genome based on SSR anchor loci.

Three morphological markers, pilose (T_1), red plant color (R_1), and naked seed (N_1), were found to be linked to molecular markers on three different linkage groups. These morphological markers are located on chromosomes 6, 16, and 12, respectively. The cotton molecular maps can provide opportunities for map-based gene cloning, studying chromosome evolution, assisting in conducting marker-assisted selection, identifying QTLs, and assisting in efficient introgression of traits from *Gossypium* gene pool for cotton improvement.

QTLS AND FUNCTIONAL GENOMICS

Improvement of crop species for quantitative traits is difficult because the effects of individual genes controlling the traits cannot be easily identified. The use of genetic markers in investigating the genetics of quantitative traits has been made in many studies since the pioneer work of Sax (1923), in which an association of a simply inherited genetic marker (seed coat color) with a quantitative trait (seed size) in beans was reported. These studies provided a background of theory and observation for more recent work with molecular markers. Major limitations of these studies include the limited number of markers available, undesirable effects on phenotype of many of the morphological markers, and in the case of translocation or whole chromosome effects, the extreme size of the chromosome segments being compared.

Interest in this approach was recently enhanced with the development of molecular marker techniques, such as RFLPs, RAPDs, AFLPs, and SSRs, which provide a high number of polymorphic markers among the mapping populations. The finding of a large number of RFLPs in many species has allowed the development of linkage maps with a high degree of resolution (Helentjaris et al., 1986). Since RFLP markers have no known effect on the phenotype of the plant, they are ideal for studying quantitative traits (Stuber et al., 1992). Basically, the aim of these studies has been to resolve variation of a quantitative trait at chromosomal locations (QTLs) and estimate their effects. The estimation of the contribution of each QTL to the variance of the trait of interest is then particularly interesting. For instance, it is classically reported that the QTL displaying the largest effect accounts for a given percentage of the variation of the trait of interest (Charcosset and Gallais, 1996). Associations of molecular markers with genes or gene blocks affecting quantitative traits have been reported in tomato and maize (Paterson et al., 1988, 1991; Stuber et al., 1992; Edwards et al., 1987), wheat (Quarrie et al., 1994), barley (Yin et al., 1999), rice (Ahn et al., 1992; Zhang et al., 1995; Price and Tomos, 1997), and many other species, including cotton.

In cotton, previously developed and recent linkage maps are being utilized for QTL analysis of economically important traits. Identification of QTLs for fiber traits in upland cotton (*G. hirsutum* L.) and their allelic association with molecular markers will be useful in cotton breeding. Shappley, Jenkins, Zhu, et al. (1998) used a previously developed RFLP linkage map to analyze QTLs associated with 19 agronomic and fiber traits. They mapped 100 QTLs to 60 maximum likelihood positions in 24 linkage groups. In this study, several QTLs influenced more than one trait and the most frequent association of QTLs with multiple traits was for fiber traits related to maturity and fineness. Yu and Kohel (1999) identified a total of 13 QTLs for fiber quality properties on different linkage groups or chromosomes. Out of the 13 QTLs, four were for fiber strength, three for fiber length, and six for fiber fineness.

Leaf morphology can significantly affect yield, quality, maturity, pest preference, canopy penetration of plant growth regulators, and other important production characteristics of many crops, including cotton (Jiang et al., 2000). Molecular markers have been used to detect and characterize QTLs associated with leaf morphology in cotton. Jiang and colleagues (2000) were able to detect 62 possible QTLs for 14 morphological traits, out of which 38 QTLs were mapped to the D-subgenome chromosomes compared to the A-subgenome of the upland cotton (AD) genome. In this study, morphological traits used were leaf-lobe length (19 QTLs), leaf-lobe width (19 QTLs), leaf-lobe angle (12 QTLs), leaf-hair number (1 QTL), leaf-lobe number (1 QTL), main-sublobe number (1 QTL), frego bract (5 QTLs), and nectary (4 QTLs).

Altaf Khan and colleagues (1999) reported 67 QTLs for seven agronomic traits using a trispecific cotton genomic map. Ten QTLs were detected for leaf main lobe length, nine for leaf main lobe width, 12 for length of leaf second lobe, 14 for sinus depth, 11 for bract teeth number, six for internode length, and five for petal length. Wright and colleagues (1999), using a previously developed RFLP map (Reinisch et al., 1994), reported four QTLs associated with pubescence of cotton leaves and/or stem. Ulloa and colleagues (2000) were able to identify two QTLs associated with stomatal conductance on two different linkage groups of the cotton map developed by using RAPD and SSR markers. The QTLs for stomatal conductance can be useful in cotton breeding because extended periods of high temperature can reduce lint yield, even under adequate irrigation conditions, and high stomatal conductance may confer some adaptive advantage to genotypes that experience supraoptimum temperatures.

Cotton has about 50,000 functional genes, like most other higher plants, and projects are underway to develop expressed sequence tags (ESTs) for cotton fiber development and other traits. Individual studies of these genes or their combination for cotton improvement is very tedious task, and may not even be practical. The tools, including markers, maps, and gene transfer procedures, are still very limited in cotton genomics. Integrative physical mapping and bridging with other plant models would augment current tools to advance the cotton functional genomic research.

FUTURE PROSPECTS

Although considerable interspecific introgression has been reported in cotton (Endrizzi et al., 1985; Demol et al., 1987; Meredith, 1991), modern cotton cultivars still have a relatively narrow genetic base, especially in the United States (Wendel et al., 1992). With the development of the tools of biotechnology and their application to cotton, the available gene pool is now virtually unlimited. Thus, application of the tools of molecular biology to exotic germplasm resources has the potential to greatly expand the usefulness and efficiency of manipulation of the germplasm pools (Stewart, 1995). The different applications of biotechniques for cotton improvement have been reviewed in detail by Stewart (1991). Detailed maps of DNA markers, in association with well-developed cytogenetic tools, offer the opportunity to utilize the polymorphic genus *Gossypium* and its diverse relatives, such as *Hibiscus* (Kenaf, roselle) and *Abelmoschus* (okra), to study plant chromosome evolution in detail, and to have a major impact on improvement of one of the world's oldest and most important crops (Reinisch et al., 1994). The conservation in genome structure of related plant species may provide a basis for interpreting genetic information among species. Finding conserved linkage groups among different species has important implications and practical applications. It indicates that these species share a common ancestral origin. The colinearity of markers as well as isozymes (Ahn et al., 1993; Milne and McIntosh, 1990) indicates that the location of genes of interest is most likely also conserved among species. Therefore, a species with a small genome can be exploited for isolating the corresponding gene that may not have been mapped in large genomes, such as cotton.

   Much of the effort in constructing maps is directed toward identifying useful polymorphic markers; once identified, these markers can be useful in numerous pedigrees and related taxa. Markers linked with genes controlling economically important traits may be used for marker-assisted selection in breeding programs. The linkage groups developed in these maps can be used directly for the detection and location of genes influencing quantitative trait loci.

   A basic genetic map having the maximum possible content of mapping information integrated from different sources of markers and from several parental clones will be a major tool for further genetic analyses, whether it be the identification and selection of important agronomic traits or the solutions of basic genetic questions. The integrated maps can be used to tag chromosome segments containing genes of interest and monitor their introgression into other elite genetic material in cotton and other plant breeding programs. A large number of highly variable markers also can be useful for the identification of breeding lines and detection of genetic flow in the *Gossypium* gene pool. Determination of linkage relationships of these maps in other *Gossypium* species will allow investigations of genome organization and similarity in the genus. Loci for which linkage information is available will allow appropriate selection of independent markers from across the genome for studies of diversity, population structure, and phylogeny in many *Gossypium* species. The specific needs of cotton are to develop and complete a framework genetic map with markers that are polymorphic for the general cotton germplasm. A global physical map developed and integrated with the genetic map and tools with appropriate ESTs will allow the development of functional genomics in cotton. The efforts for the development of "generic" maps have already been put forward in the *Gramineae* family (Saghai Maroof et al., 1996). Thus, overall research priority should be the development of a "generic" *Gossypium* map that could be useful in the application of marker-assisted selection and the study of chromosome evolution, speciation, and other genetic studies of the *Gossypium* genus.

REFERENCES

Abdalla, A.M., Abou-El-Zahab, A.A., Reddy, O.U.K. , Pepper, A.E., and El-Zik, K.M. 2000. Genetic diversity in diploid and tetraploid cotton species examined by amplified fragment length polymorphism (AFLP). *Proc. Beltwide Cotton Research Conferences* (pp. 499-500). National Cotton Council, Memphis, TN.

Abenes, M.L.P., Tabien, R.E., McCouch, S.R., Ikeda, R., Ronald, P., Khush, G.S., and Huang, N. 1994. Orientation and integration of the classical and molecular genetic maps of chromosome 11 in rice. *Euphytica* 76(1):81-87.

Ahmad, M. and McNeil, D. L. 1996. Comparison of crossability, RAPD, SDS-PAGE and morphological markers for revealing genetic relationship within and among *Lens* species. *Theor. Appl. Genet.* 93(5/6):788-793.

Ahn, S., Anderson, J.A., Sorrells, M.E., and Tanksley, S.D. 1993. Homeologous relationships of rice, wheat and maize chromosomes. *Mol. Gen. Genet.* 241(4):483-490.

Ahn, S.N., Bollich, C.N., and Tanksley, S.D. 1992. RFLP tagging of a gene for aroma in rice. *Theor. Appl. Genet.* 84(7/8):825-828.

Allard, R.W. 1956. Formulas and tables to facilitate the calculation of recombination of values in heredity. *Hilgardia* 24(10):235-278.

Altaf Khan, M., Myers, G.O., Stewart, J.McD., Zhang, J., and Cantrell, R.G. 1999. Addition of new markers to the trispecific cotton map. *Proc. Beltwide Cotton Research Conferences* (p. 439). National Cotton Council, Memphis, TN.

Altaf Khan, M., Stewart, J.McD., Cantrell, R.G., Wajahatullah, M.K., and Zhang, J. 1997. Molecular and morphological genetics of a trispecies F_2 population of cotton. *Proc. Beltwide Cotton Research Conferences* (pp. 448-452). National Cotton Council, Memphis, TN.

Altaf Khan, M., Zhang, J., Stewart, J.McD., and Cantrell, R.G. 1998. Integrated molecular map based on a trispecific F_2 population of cotton. *Proc. Beltwide Cotton Research Conferences* (p. 491). National Cotton Council, Memphis, TN.

Arnold, M.L., Buckner, C.M., and Robinson, J.J. 1991. Pollen-mediated introgression and hybrid speciation in Louisiana irises. *Proc. Natl. Acad. Sci. USA.* 88(2):1398-1402.

Arondel, V., Lemieux, B., Hwang, I., Gibson, S., Goodman, H.M., and Somerville, C.R. 1992. Map-based cloning of a gene controlling omega-3 fatty acid desaturation in *Arabidopsis. Science* 258(5086):1353-1355.

Bachem, C.W.B., Van der Hoeven, R.S., de Bruijn, S.M., Vreugdenhil, D., Zabeau, M., Visser, R.G.F., Van der Hoeven, R.S., and De Bruijn, S.M. 1996. Visualization of differential gene expression using a novel method of RNA fingerprinting based on AFLP: Analysis of gene expression during potato tuber development. *Plant J.* 9(5):745-753.

Baker, R.J., Longmire, J.L., and Van Den Bussche, R.A. 1995. Organization of repetitive elements in the upland cotton genome (*Gossypium hirsutum*). *J. Hered.* 86(3):178-185.

Baudracco-Arnas, S. and Pitrat, M. 1996. A genetic map of melon (*Cucumis melo* L.) with RFLP, RAPD, isozyme, disease resistance and morphological markers. *Theor. Appl. Genet.* 93(1/2):57-64.

Becker, J., Vos, P., Kuiper, M., Salamini, F., and Heun, M. 1995. Combined mapping of AFLP and RFLP markers in barley. *Mol. Gen. Genet.* 249(1):65-73.

Bell, C.J., and Ecker, J.R. 1994. Assignment of 30 microsatellite loci to the linkage map of *Arabidopsis. Genomics* 19(1):137-144.

Bourdon, C. 1986. Enzymatic polymorphism and genetic organization of two cotton tetraploid cultivated species, *G. hirsutum* and *G. barbadense. Coton et Fibres Tropicales* 41:191-210.

Brooks, T.D., Pepper, A.E., Reddy, O.U.K., Thaxton, P.M., and El-Zik, K.M. 2000. Identification and mapping of fiber length and strength QTLs in an interspecific cotton population. *Proc. Beltwide Cotton Research Conferences* (p. 484). National Cotton Council, Memphis, TN.

Brown, M.S. 1980. The identification of chromosomes of *Gossypium hirsutum* L. by means of translocations. *J. Hered.* 71(4):266-274.

Brown, M.S., Menzel, M.Y., Hasenkampf, C.A., and Naqi, S. 1981. Chromosome configurations and orientations in 58 heterozygote translocations in *Gossypium hirsutum. J. Hered.* 72(3):161-168.

Brubaker, C.L., Paterson, A.H., and Wendel, J.F. 1999. Comparative genetic mapping of allotetraploid cotton and its diploid progenitors. *Genome* 42(2):184-203.

Brubaker, C.L., and Wendel, J.F. 1994. Reevaluating the origin of domesticated cotton (*Gossypium hirsutum*; Malvaceae) using nuclear restriction fragment length polymorphisms (RFLPs). *Am. J. Bot.* 81(10):1309-1326.

Cantrell, R.G. 2000. Germplasm engineering in cotton. *Proceedings of the Beltwide Cotton Research Conferences* (p. 500). National Cotton Council, Memphis, TN.

Cantrell, R.G., and Davis, D.D. 1993. Characterization of *hirsutum* x *barbadense* breeding lines using molecular markers. *Proceedings of the Beltwide Cotton Research Conferences* (pp. 1551-1553). National Cotton Council, Memphis, TN.

Centner, M.S., Roos, E.E., and Endrizzi, J.E. 1984. Electrophoretic genome analysis of two tetraploid cultivated cottons. *Agron. Abst.* (p. 61). ASA, Madison, WI.

Chaparro, J., Wilcox, P., Grattapaglia, D., O'Malley, D., McCord, S., Sederoff, R., McIntyre, L., and Whetten, R. 1992. Genetic mapping of pine using RAPD markers: Construction of a 191 marker map and development of half-sib genetic analysis. *Advances in gene technology: Feeding the world in the twenty-first century.* Miami Winter Symp. Miami, FL.

Chaparro, J.X., Werner, D.J., O'Malley, D., and Sederoff, R.R. 1994. Targeted mapping and linkage analysis of morphological, isozyme, and RAPD markers in peach. *Theor. Appl. Genet.* 87(7):805-815.

Charcosset, A. and Gallais, A. 1996. Estimation of the contribution of quantitative trait loci (QTL) to the variance of the quantitative trait by means of genetic markers. *Theor. Appl. Genet.* 93(8):1193-1201.

Cherry, J.P., Katterman, F.R.H., and Endrizzi, J.E. 1972. Seed esterases, leucine aminopeptidases, catalases of species of the genus *Gossypium. Theor. Appl. Genet.* 42(5):218-226.

Cherry, J.P. and Leffler, H.R. 1984. Seed. In R.J. Kohel and C.F. Lewis (Eds.), *Cotton* (pp. 511-569). ASA, Madison, WI.

Coe, E.H., Hoisington, D.A., and Neuffer, M.G. 1990. Linkage map of corn (maize) (*Zea mays* L.). In O'Brien S. J. (Ed.), *Genetic maps* (pp 6.39-6.67). Cold Spring Harbor Laboratory Press, Cold Spring Harbor, NY.

Cooper, D.T. 1969. Utilization of germplasm for *Gossypium arboreum* L. × *Gossypium raimondii* Ulb. to supplement the genetic variability in upland cotton. N. Carolina State Univ Raleigh. *Diss Abst.* (69-16829).

Demol, J., Verschraege, L., and Marechal, R. 1987. Development of new *Gossypium* L. cultivars with exceptional technological characteristics introgressed from wild diploid species. Belgium Cotton Research Group, *Cotton fibers; Their Development and Properties* (2) (pp. 13-17). International Institute for Cotton, Manchester, UK.

Devey, C. E., Bell, J. C., Smith, D. N., Neale, D. B., and Moran, G.F. 1996. A genetic linkage map for *Pinus radiata* based on RFLP, RAPD, and microsatellite markers. *Theor. Appl. Genet.* 92(6):673-679.

Dib, C., Faure, S., Fizames, C., Samson, D., Drouot, N., Vignal, A., Millasseau, P., Marc, S., Hazan, J., Seboun, E., Lathrop, M., Gyapay, G., Morissette, J., and Welssenbach, J. 1996. A comprehensive genetic map of human genome based on 5,264 microsatellites. *Nature* 380(6570):152-154.

Dudley, J.W., Saghai Maroof, M.A., and Rufener, G.K. 1992. Molecular marker information and selection of parents in corn breeding programs. *Crop Sci.* 32(2):301-304.

Edwards, G.A., Endrizzi, J.E., and Stein, R. 1974. Genome DNA content and chromosome organization in *Gossypium*. *Chromosoma* 47(3):309-326.

Edwards, M.D., Stuber, C.W., and Wendel, J.F. 1987. Molecular-marker-facilitated investigations of quantitative-trait loci in maize. I. Numbers, genomic distribution and types of gene action. *Genetics* 116(1):113-125.

Ellis, T.H.N., Turner, L., Hellens, R.P., Lee, D., Harker, C.L., Enard, C., Domoney, C., and Davies, D.R. 1992. Linkage maps in pea. *Genetics* 130(3):649-663.

Endrizzi, J.E. and Phillips, L.L. 1960. A hybrid between *Gossypium arboreum* L. and *G. raimondii* Ulbr. *Can. J. Genetic Cytol.* 2(4):311-319.

Endrizzi, J.E., Turcotte, E.L., and Kohel, R.J. 1985. Genetics, cytology, and evolution of *Gossypium*. *Advances in Genet.* 23:271-375.

Figdore, S.S., Kennard, W.C., Song, K.M., Slocum, M.K., and Osborn, T.C. 1988. Assessment of the degree of restriction fragment length polymorphism in *Brassica*. *Theor. Appl. Genet.* 75(6):833-840.

Flavell, R.B. 1980. The molecular characterization and organization of plant chromosomal DNA sequences. *Annu. Rev. Plant Physiol.* 31:569-596.

Foolad, M.R., Arulsekar, S., Becerra, V., and Bliss, F.A. 1995. A genetic map of *Prunus* based on an interspecific cross between peach and almond. *Theor. Appl. Genet.* 91(2):262-269.

Fregene, M., Angel, F., Gomez, R., Rodriguez, F., Chavarriaga, P., Roca, W., Thome, J., and Bonierbale, M. 1997. A molecular genetic map of cassava (*Manihot esculenta* Crantz). *Theor. Appl. Genet.* 95(3):431-441.

Gawel, N.J., Jarret, R.L., and Whittemore, A.P. 1992. Restriction fragment length polymorphism (RFLP)-based phylogenetic analysis of *Musa*. *Theor. Appl. Genet.* 84(3/4):286-290.

Gebhardt, C., Ritter, E., Barone, A., Debener, T., Walkemeier, B., Schachtschabel, U., Kaufmann, H., Thompson, R.D., Bonierbale, M.W., Ganal, M.W., Tanksley, S.D., and Salamini, F. 1991. RFLP maps of potato and their alignment with the homeologous tomato genome. *Theor. Appl. Genet.* 83(1):49-57.

Geever, R.F. 1980. The evolution of single-copy nucleotide sequences in the genomes of *Gossypium hirsutum* L. Doctoral dissertation, University of Arizona, Tucson.

Geever, R.F., Katterman, F., and Endrizzi, J.E. 1989. DNA hybridization analyses of a *Gossypium* allotetraploid and two closely related diploid species. *Theor. Appl. Genet.* 77(4):234-244.

Gepts, P. 1993. The use of molecular and biochemical markers in crop evolution studies. In M. K. Hecht (Ed.), *Evolutionary biology*, Vol. 27 (pp. 51-94). Plenum Press, New York.

Gerstel, D.U. 1953. Chromosomal translocations in interspecific hybrids of the genus *Gossypium*. *Evolution* 7:234-240.

Gonzalez, J.M. and Ferrer, E. 1993. Random amplified polymorphic DNA analysis in *Hordeum* species. *Genome* 36(6):1029-1031.

Grandillo, S. and Tanksley, S.D. 1996. Genetic analysis of RFLPs, GATA microsatellites and RAPDs in a cross between *L. esculentum* and *L. pimpinellifolium*. *Theor. Appl. Genet.* 92(8):957-965.

Graner, A., Jahoor, A., Schondelmaier, J., Siedler, H., Pillen, K., Fischbeck, G., Wenzel, G., and Herrmann, R.G. 1991. Construction of an RFLP map of barley. *Theor. Appl. Genet.* 83(2):250-256.

Grivet, L., D'Hont, A., Dufour, P., Hamon, P., Roques, D., and Glaszmann, J.C. 1994. Comparative genome mapping of sugar cane with other species within the *Andropogoneae* tribe. *Heredity* 73(Pt. 5):500-508.

Hancock, J.F. 1982. Alcohol dehydrogenase isozymes in *Gossypium hirsutum* and its putative diploid proginators. The biochemical consequences of enzyme multiplicity. *Pl. Syst. Evol.* 140(2/3):141-149.

Hauge, B.M., Hanley, S.M., Cartinhour, S., Cherry, J.M., Goodman, H.M., Koornneef, M., Stam, P., Chang, C., Kempin, S., Medrano, L., and Meyerowitz, E.M. 1993. An integrated genetic/RFLP map of *Arabidopsis thaliana* genome. *Plant J.* 3(5):745-754.

Helentjairs, T., Slocum, T.M., Wright, S., Schaefer, A., and Nienhuis, J. 1986. Construction of genetic linkage map in maize and tomato using restriction fragment length polymorphisms. *Theor. Appl. Genet.* 72(6):761-769.

Hongtrakul, V., Huestis, G.M., and Knapp, S.J. 1997. Amplified fragment length polymorphisms as a tool for DNA fingerprinting sunflower germplasm: Genetic diversity among oilseed inbred lines. *Theor. Appl. Genet.* 95(3):400-407.

Hu, J. and Quiros, C.F. 1991. Identification of broccoli and cauliflower cultivars with RAPD markers. *Plant Cell Repts.* 10(10):505-511.

Huang, N., McCouch, S.R., Mew, T., Parco, A., and Guiderdoni, E. 1994. Development of a RFLP map from a doubled haploid population of rice. *Rice. Genet. Newsl.* 11:134-137.

Iqbal, M.J., Aziz, N., Saeed, N.A., and Zafar, Y. 1997. Genetic diversity evaluation of some elite cotton varieties by RAPD analysis. *Theor. Appl. Genet.* 94(1):139-144.

Jacobs, J.M.E., Van Eck, H.J., Arens, P., Verkerk-Bakker, B., te Lintel Hekkert, B., Bastiaanssen, H.J.M., El-Kharbotly, A., Pereira, A., Jacobsen, E., and Stiekema, W.J. 1995. A genetic map of potato (*Solanum tuberosum*) integrating molecular markers, including transposons, and classical markers. *Theor. Appl. Genet.* 91(2):289-300.

Jiang, C., Wright, R.J., Woo, S.S., DelMonte, T.A., and Paterson, A.H. 2000. QTL analysis of leaf morphology in tetraploid *Gossypium* (Cotton). *Theor. Appl. Genet.* 100(3/4):409-418.

Keim P., Diers, B.W., Olson, T.C., and Shoemaker, R.C. 1990. RFLP mapping in soybean: Association between marker loci and variation in quantitative traits. *Genetics* 126(3):735-742.

Kesseli, R., Paran, I., and Michelmore, R.W. 1994. Analysis of a detailed genetic linkage map of *Lactuca sativa* (Lettuce) constructed from RFLP and RAPD markers. *Genetics* 136(4):1435-1446.

Kianian, S.F. and Quiros, C.F. 1992. Generation of a *Brassica oleracea* composite map: linkage arrangements among various populations and evolutionary implications. *Theor. Appl. Genet.* 84(5/6):544-554.

Kresovich, S., Williams, J.G.K., McFerson, J.R., Routman, E.J., and Schaal, B.A. 1992. Characterization of genetic identities and relationships of Brassica oleracea L. via a random amplified polymorphic DNA assay. *Theor. Appl. Genet.* 85(2/3):190-196.

Lanaud, C., Risterucci, A.M., N'Goran, A.K.J., Clement, D., Flament, M.H., Laurent, V., and Flaque, M. 1995. A genetic linkage map of *Theobroma cacao* L. *Theor. Appl. Genet.* 91(6/7):987-993.

Landry, B.S., Kesseli, R.V., Farrara, B., and Michelmore, R.W. 1987. A genetic map of lettuce (*Lactuca sativa* L.) with restriction fragment length polymorphism, isozyme, disease resistance and morphological markers. *Genetics* 116(2):331-337.

Lannér, C., Bryngelsson, T., and Gustafsson, M. 1996. Genetic validity of RAPD markers at the intra- and inter-specific level in wild *Brassica* species with $n = 9$. *Theor. Appl. Genet.* 93(1/2):9-14.

Litt, M. and Luty, J.A. 1989. A hypervariable microsatellite revealed by in vitro amplification of a dinucleotide repeat within the cardiac muscle actin gene. *Am. J. Hum. Genet.* 44(3):397-401.

Liu, S., Saha, S., Stelly, D., Burr, B., and Cantrell, R.G. 2000. Chromosomal assignment of microsattelite loci in cotton. *J. Hered.* 91(4):326-332.

Liu, A., Zhang, Q., and Li, H. 1992. Location of a gene for wide compatibility in the RFLP linkage map. *Rice Genet. Newsl.* 9:134-136.

Lu, H. and Myers, G.O. 1999. DNA variation in ten influential upland cotton varieties by RAPDs. *Proceedings Beltwide Cotton Research Conferences* (p. 484). National Cotton Council, Memphis, TN.

MacKill, D.J., Salam, M.A., Wang, Z.Y., and Tanksley, S.D. 1993. A major photoperiod sensitive gene tagged with RFLP and isozymes markers in rice. *Theor. Appl. Genet.* 85(5):536-540.

Maheswaran, M., Subudhi, P.K., Nandi, S., and Xu, J.C. 1997. Polymorphism, distribution, and segregation of AFLP markers in a doubled haploid rice population. *Theor. Appl. Genet.* 94(1):39-45.

Majer, D., Mithen, R., Lewis, B.G., Vos, P., and Oliver, R.P. 1996. The use of AFLP fingerprinting for the detection of genetic variation in fungi. *Mycological Res.* 100(Pt. 9):1107-1111.

Martin, G.B., Brommonschenkel, S.H., Chunwongse, J., Frary, A., Ganal, M.W., Spivey, R., Wu, T., Earle, E.D., and Tanksley, S.D. 1993. Map-based cloning of a protein kinase gene conferring disease resistance in tomato. *Science.* 262(5138):1432-1436.

Maughan, P.J., Saghai Maroof, M.A., Buss, G.R., and Huestis, G.M. 1996. Amplified fragment length polymorphism (AFLP) in soybean: Species diversity, inheritance, and near-isogenic line analysis. *Theor. Appl. Genet.* 93(3):392-401.

McCouch, S.R., Khush, G.S., and Tanksley, S.D. 1990. Tagging genes for disease and insect resistance via linkage to RFLP markers. In IRRI (Ed.), *Rice genetics II. (Proc. 2nd Int. Rice Genet. Symp.)* (pp. 443-449). Manila, Philippines.

McCouch, S.R., Kochert, G., Yu, Z.H., Wang, Z.Y., Khush, G.S., Coffman, W.R., and Tanksley, S.D. 1988. Molecular mapping of rice chromosomes. *Theor. Appl. Genet.* 76(6):815-829.

Meksem, K., Leister, D., Peleman, J., Zabeau, M., Salamini, F., and Gebhardt, C. 1995. A high-resolution map of the vicinity of the R1 locus on chromosome V of potato based on RFLP and AFLP markers. *Mol. Gen. Genet.* 249(1):74-81.

Mendendez, C.M., Hall, A.E., and Gepts, P. 1997. A genetic linkage map of cowpea (*Vigna unguiculata*) developed from a cross between two inbred, domesticated lines. *Theor. Appl. Genet.* 95(8):1210-1217.

Menzel, M.Y. and Brown, M.S. 1954. The tolerance of *Gossypium hirsutum* for deficiences and duplications. *Am. Nat.* 88(3):407-418.

Menzel, M.Y. and Brown, M.S. 1978. Reciprocal chromosome translocations in *Gossypium hirsutum*. Arm translocation of breakpoints and recovery of duplications and deficiences. *J. Hered.* 69(6):383-390.

Menzel, M.Y., Hasenkampf, C.A., and Stewart, J.McD. 1982. Incipient genome differentiation in *Gossypium*. III. Comparison of chromosomes of *G. hirsutum* and Asiatic diploids using heterozygous translocations. *Genetics* 100(1):89-103.

Meredith Jr., W.R. 1991. Contributions of introductions to cotton improvement. In H. L. Shands and L. E. Wiesner (Eds.), *Use of plant introductions in cultivar development, part I* (pp. 127-146). CSSA Spec. Publ. 17. ASA-CSSA-SSSA, Madison, WI.

Meredith Jr., W.R. 1992. RFLP association with varietal origin and heterosis. *Proc. Beltwide Cotton Research Conferences* (p. 607). National Cotton Council, Memphis, TN.

Meredith Jr., W.R. 1995. Use of molecular markers in cotton breeding. In G.A. Constable and N.W. Forrester (Eds.), *Challenging the future: Proceedings of the World Cotton Research Conference-1* (pp. 303-308). CSIRO, Melbourne.

Meredith Jr., W.R. and Bridge, R.R. 1984. Genetic contributions to yield changes in upland cotton. In W. R. Fehr (Ed.), *Genetic contributions to yield gains of five major crop plants* (pp. 75-87). CSSA Spec. Publ. 7. Crop. Sci. Soc. Am., Madison, WI.

Michelmore, R.W., Kesseli, R.V., Francis, D.M., Paran, I., Fortin, M.G., Yang, and C.H. 1992. Strategies for cloning plant disease resistance genes. *Mol. Plant Pathology 2. A practical approach*. Oxford University Press. Oxford, UK.

Michelmore, R.W., Paran, I., and Kesseli, R.V. 1991. Identification of markers linked to disease-resistant genes by bulked segregant analysis: A rapid method to detect markers in specific genomic regions by using segregating populations. *Proc. Natl. Acad. Sci. USA* 88(21):9828-9832.

Millan, T., Osuna, F., Cobos, S., Torres, A.M., and Cubero, J.I. 1996. Using RAPDs to study phylogenetic relationships in *Rosa*. *Theor. Appl. Genet.* 92(2):273-277.

Milne, D.L. and McIntosh, R.A. 1990. *Triticum aestivum* (common wheat) In S. J. O'Brien (Ed.), *Genetic Maps*. Cold Spring Harbor Press, Cold Spring Harbor, New York.

Money, T., Reader, S., Qu, L.J., Dunford, R.P., and Moore, G. 1996. AFLP-based mRNA fingerprinting. *Nuc. Acid Res.* 24(13):2616-2617.

Mudge, J., Anderson, W.R., Kehrer, R.L., and Fairbanks, D.J. 1996. A RAPD genetic map of *Saccharum officinarum*. *Crop. Sci.* 36(5):1362-1366.

Multani, D.S., and Lyon, B.R. 1995. Genetic fingerprinting of Australian cotton cultivars with RAPD markers. *Genome* 38(5):1005-1008.

Mursal, I.E.J. and Endrizzi, J.E. 1976. A reexamination of the diploid like behaviour of polyploid cotton. *Theor. Appl. Genet.* 47(4):171-178.

Nam, H.G., Giraudat, J., den Boer, B., Moonan, F., Loos, W.D.B., Hauge, B.M., and Goodman, H.M. 1989. Restriction fragment length polymorphism linkage map of *Arabidopsis thaliana*. *Plant Cell* 1(7):699-705.

Nodari, R.O., Tsai, S.M., Gilbertson, R.L., and Gepts, P. 1993. Toward an integrated linkage map of common bean. 2. Development of an RFLP-based linkage map. *Theor. Appl. Genet.* 85(5):513-520.

O'Brien, S.J. 1992. *Genetic maps. Book 6: Plants*. Cold Spring Harbor Laboratory Press, Cold Spring Harbor, NY.

Paran, I. and Michelmore, R.W. 1993. Development of reliable PCR-based markers linked to downy mildew resistance genes in lettuce. *Theor. Appl. Genet.* 85(8):985-993.

Paterson, A.H. 1993. Molecular markers in cotton improvement. *Proc. Beltwide Cotton Research Conferences* (p. 1557). National Cotton Council, Memphis, TN.

Paterson, A.H., Damon, S., Hewitt, J.D., Zamir, D., Rabinovitch, H.D., Lincoln, S.E., Lander, E.S., and Tanksley, S.D. 1991. Mendelian factors underlying quantitative traits in tomato: Comparison across species, generations, and environments. *Genetics* 127(1):181-197.

Paterson, A.H., Lander, E.S., Hewitt, J.D., Paterson, S., Lincoln, S.E., and Tanksley, S.D. 1988. Resolution of quantitative traits into Mendelian factors by using a complete linkage map of restriction fragment length polymorphisms. *Nature* 335(6192):721-726.

Paul, S., Wachira, F.N., Powell, W., and Waugh, R. 1997. Diversity and genetic differentiation among populations of Indian and Kenyan tea [*Camellia sinensis* (L.) O. Kuntze] revealed by AFLP markers. *Theor. Appl Genet.* 94(2):1161-1168.

Pillay, M., and Myers, G.O. 1999. Genetic diversity in cotton assessed by variation in ribosomal RNA genes and AFLP markers. *Crop Sci.* 39(6):1881-1886.

Powell, W., Morgante, M., Andre, C., Hanafey, M., Vogel, J., Tingey, S., and Rafalski, A. 1996. The comparison of RFLP, RAPD, AFLP and SSR (microsatellite) markers for germplasm analysis. *Mol. Breed.* 2(3):225-238.

Price, A.H. and Tomos, A.D. 1997. Genetic dissection of root growth in rice (*Oryza sativa* L.). II: mapping quantitative trait loci using molecular markers. *Theor. Appl. Genet.* 95(1/2):143-152.

Price, H.J., Stelly, D.M., McKnight, T.D., Scheuring, C.F., Raska, D., Michaelson, M.J., and Bergey, D. 1990. Molecular cytogenetic mapping of a nucleolar organizer region in cotton. *J. Hered.* 81(5):365-370.

Quarrie S.A., Leberton, C., Gulli, M., Calestani, C., and Marmiroli, N. 1994. QTL analysis of ABA production in wheat and maize and associated physiological traits. *Russ. J. Plant Phys.* 41:565-571.

Rafalski, J.A., Hanafey, M.K., Tingey, S.V., and Williams, J.G.K. 1994. Technology for molecular breeding: RAPD markers, microstallites and machines. In P.M. Gresshoff (Ed.), *Plant genome analysis* (pp. 19-27). CRC Press Inc. Boca Raton, FL.

Rajapakse, S., Belthoff, L.E., He, G., Estager, A.E., Scorza, R., Verde, I., Ballard, R.E., Baird, W.V., Callahan, A., Monet, R., and Abbott, A.G. 1995. Genetic linkage mapping in peach using morphological, RFLP and RAPD markers. *Theor. Appl. Genet.* 90(3/4):503-510.

Reddy, O.U.K., Brooks, T.D., El-Zik, K.M., and Pepper, A.E. 2000. Development and use of PCR-based technologies for cotton mapping. *Proc. Beltwide Cotton Research Conferences* (p. 483). National Cotton Council, Memphis, TN.

Reinisch, A.J., Dong, J., Brubaker, C.L., Stelly, D.M., Wendel, J.F., and Paterson, A.H. 1994. A detailed RFLP map of cotton, *Gossypium hirsutum* × *G. barbadense*: Chromosome organization and evolution in a disomic polyploid genome. *Genetics* 138(3):829-847.

Reiter, R.S., Williams, J., Feldmann, K.A., Rafalski, J.A., Tingey, S.V., and Scolnik, P.A. 1992. Global and local genome mapping in *Arabidopsis thaliana* by using recombinant inbred lines and random amplified polymorphic DNAs. *Proc. Natl. Acad. Sci. USA.* 89(4):1477-1481.

Rosenow, D.T. and Clark, L.E. 1987. Utilization of exotic germplasm in breeding for yield stability. In *Proceedings of the 15th Biennial Grain Sorghum Research and Utilization Conference, Lubbock, TX* (pp. 49-56). American Seed Trade Association, Washington, DC.

Rossetto, M., McLauchlan, A., Harris, F.C.L., Henry, R.J., Baverstock, P.R., Lee, L.S., Maguire, T.L., and Edwards, K.J. 1999. Abundance and polymorphism of microsatellite markers in the tea tree (*Melaleuca attemifolia, Myrtaceae*). *Theor. Appl. Genet.* 98(6/7):1091-1098.

Saghai Maroof, M.A., Yang, G.P., Biyashev, R.M., Maughan, P.J., and Zhang, Q. 1996. Analysis of the barley and rice genomes by comparative RFLP linkage mapping. *Theor. Appl. Genet.* 92(5):541-551.

Sakhanokho, H.F., Zipf, A., Sharma, G.C., Karaca, M., Saha, S., and Rajasekaran, K. 2000. Induction of highly embryogenic calli and plant regeneration in diploid and tetraploid cottons. *Proceedings of the Beltwide Cotton Research Conferences* (pp. 570-574). National Cotton Council, Memphis, TN.

Sax, K. 1923. The association of maize differences with seed coat pattern and pigmentation in *Phaseolus vulgaris*. *Genetics* 8:552-560.

Schondelmaier, J., Steinrucken, G., and Jung, C. 1996. Integration of AFLP markers into a linkage map of sugar beet (*Beta vulgaris* L.). *Plant Breeding* 115(4):231-237.

Schüeller, C., Backes, G., Fischbeck, G., and Jahoor, A. 1992. RFLP markers to identify the alleles on the *Mla* locus conferring powdery mildew resistance in barley. *Theor. Appl. Genet.* 84(3/4):330-338.

Schuler, G.D. 1998. Electronic PCR: Bridging the gap between genome mapping and genome sequencing. *Trends in Biotech* 16(11):456-459.

Schut, J.W., Qi, X., and Stam, P. 1997. Association between relationship measures based on AFLP markers, pedigree data and morphological traits in barley. *Theor. Appl. Genet.* 95(7):1161-1168.

SCSB. 1981. Preservation and utilization of germplasm in cotton (1968-1980). *Southern Cooperative Series Bulletin* No. 256.

Senior, M.L. and Heun, M. 1993. Mapping maize microsatellites and polymerase chain reaction confirmation of the targeted repeats using a ct primer. *Genome* 36(5):884-889.

Shappley, Z.W., Jenkins, J.N., Meredith, W.R., and McCarty Jr., J.C. 1998. An RFLP linkage map of upland cotton, *Gossypium hirsutum* L. *Theor. Appl. Genet.* 97(5/6):756-761.

Shappley, Z.W., Jenkins, J.N., Watson Jr., C.E., Kahler, A.L., and Meredith, W.R. 1996. Establishment of molecular markers and linkage groups in two F$_2$ populations of upland cotton. *Theor. Appl. Genet.* 92(8):915-919.

Shappley, Z.W., Jenkins, J.N., Zhu, J., and McCarty Jr., J.C. 1998. Quantitative trait loci associated with agronomic and fiber traits of upland cotton. *J. Cotton Sci.* 2(4):153-163.

Sharma, S.K., Knox, M.R., and Ellis, T.H.N. 1996. AFLP analysis of diversity and phylogeny of *Lens* and its comparison with RAPD analysis. *Theor. Appl. Genet.* 93(5/6):751-758.

Sherman, J.D., Fenwick, A.L., Namuth, D.M., and Lapitan, N.L.V. 1995. A barley RFLP map: Alignment of three barley maps and comparison to *Gramineae* species. *Theor. Appl. Genet.* 91(4):681-690.

Small, R.L. and Wendel, J.F. 1999. The mitochondrial genome of allotetraploid cotton (*Gossypium* L.). *J. Hered.* 90(1):251-253.

Stanton, M.A., Stewart, J.McD., Percival, A.E., and Wendel, J.F. 1994. Morphological diversity and relationships in the A-Genome cottons, *Gossypium arboreum* and *Gossypium herbaceum*. *Crop Sci* 34(2):519-527.

Stelly, D.M. 1993. Interfacing cytogenetics with the cotton genome mapping effort. *Proceedings of the Beltwide Cotton Research Conferences* (pp. 1545-1550). National Cotton Council, Memphis, TN.

Stephens, J.C., Miller, F.R., and Rosenow, D.T. 1967. Conversion of alien sorghum to early combine genotypes. *Crop Sci.* 7(3):396.

Stewart, J.McD. 1979. Use of ovule culture to obtain interspecific hybrids of *Gossypium. Symposium of the Southern Section American Society of Plants Physiologists. New Orleans. Feb 8, 1979* (pp. 44-56).

Stewart, J.McD. 1991. *Biotechnology of cotton: Achievements and perspectives.* ICAC Review article on cotton production research No. 3. CAB International, Wallingford, UK.

Stewart, J.McD. 1992. Germplasm resources and enhancement strategies for disease resistance. *Proceedings of the Beltwide Cotton Research Conferences* (pp. 1323-1325). National Cotton Council of America, Memphis, TN.

Stewart, J.McD. 1995. Potential for crop improvement with exotic germplasm and genetic engineering. In G. A. Constable and N. W. Forrester (Eds.), *Challenging the Future: Proceedings. World Cotton Res. Conf.-1* (pp. 313-327). CSIRO, Melbourne.

Stewart, J.McD., and Hsu, C.L. 1977. *In-ovul* embryo culture and seedling development of cotton (*Gossypium hirsutum* L.). *Planta* 137(2):113-117.

Stewart, J.McD., and Hsu, C.L. 1978a. Interspecific *Gossypium* hybrids through ovule culture. *Proceedings of the Beltwide Cotton Research Conferences* (p. 68). National Cotton Council, Memphis, TN.

Stewart, J.McD, and Hsu, C.L. 1978b. Hybridization of diploid and tetraploid cottons through inovule embryo culture. *J. Hered.* 69(6):404-408.

Stuber, C.W., Lincoln, S.E., Wolff, D.W., Helentjaris, T., and Lander, E.S. 1992. Identification of genetic factors contributing to heterosis in a hybrid from elite maize inbred lines using molecular markers. *Genetics* 132(3):823-839.

Suiter, K.A. 1988. Genetics of allozyme variation in *Gossypium arboreum* L. and *Gossypium herbaceum* L. (Malvaceae). *Theor. Appl. Genet.* 75(2):259-271.

Suiter, K.A. and Parks, C.R. 1984. Genetic control and mode of inheritance of allozyme variation in Old World cotton: *G. arboreum* L. and *G. herbaceum* L. *Am. J. Bot. Abst.* 71 (No. 5, Part 2): 191.

Tan, H., Wu, J.X., Saha, S., Jenkins, J.N., McCarty, J.C., and Cantrell, R.G. 1999. Molecular marker based genetic analysis of QTLs using recombinant inbred lines in cotton. *Proceedings of the Beltwide Cotton Research Conferences* (p. 484). National Cotton Council, Memphis, TN.

Tanksley, S.D. 1993. Mapping polygenes. *Annu. Rev. Genet.* 27:205-233.

Tanksley, S.D. and Mutschler, M.A. 1990. Linkage map of the tomato (*Lycopersicon esculantum*) (2n = 24). In O'Brien S. J. (Ed.), *Genetic maps* (pp. 6.15-6.3). Cold Spring Harbor Laboratory Press, Cold Spring Harbor, NY.

Tatineni, V., Cantrell, R.G., and Davis, D.D. 1996. Genetic diversity in elite cotton germplasm determined by morphological characters and RAPDs. *Crop Sci.* 36(1):186-192.

Thomas, C.M., Vos, P., Zabeau, M., Jones, D.A., Norcott, K.A., Chadwick, B.P., Jones, J.D.G. 1995. Identification of amplified restriction fragment polymorphism (ALFP) markers tightly linked to the tomato Cf-9 gene for resistance to *Cladosporium fulvum. Plant J.* 8(5):785-794.

Tingey, S.V. and del Tufo, J.P. 1993. Genetic analysis with random amplified poly-morphic DNA markers. *Plant Physiol.* 101(2):349-352.

Tinker, N.A., Fortin, M.G., and Mather, D.E. 1993. Random amplified polymor-phic DNA and pedigree relationships in spring barley. *Theor. Appl. Genet.* 85(8):976-984.

Tohme, J., Gonzales, D.O., Beebe, S., and Duque, M.C. 1996. AFLP analysis of genepool of a wild bean core collection. *Crop Sci.* 36(5):1375-1384.

Ulloa, M., Cantrell, R.G., Percy, R.G., Zeiger, E., and Lu, Z. 2000. QTL analysis of stomatal conductance and relationship to lint yield in an interspecific cotton. *J. Cotton Sci.* 4(1):10-18.

Vallejos, C.E., Sakyiama, N.S., and Chase, C.D. 1992. Molecular marker-based linkage map of *Phaseolus vulgaris* L. *Genetics* 131(1):733-740.

Voorrips, R.E., Jongerius, M.C., and Kanne, H.J. 1997. Mapping of two genes for resistance to clubroot (*Plasmodiophora brassicae*) in a population of doubled haploid lines of *Brassica oleracea* by means of RFLP and AFLP markers. *Theor. Appl. Genet.* 94(1):75-82.

Vos, P., Hogers, R., Bleeker, M., Reijans, M., Van de Lee, T., Hornes, M., Frijters, A., Pot, J., Peleman, J., Kuiper, M., Zabeau, M., and Van de Lee, T. 1995. AFLP: a new technique for DNA fingerprinting. *Nucleic Acid Res.* 23(21):4407-4414.

Vroh Bi, I., Maquet, A., Baudoin, J.P., du Jardin, P., Jacquemin, J.M., and Mergeai, G. 1999. Breeding for "low-gossypol seed and high-gossypol plants" in upland cotton. Analysis of tri-species hybrids and backcross progenies using AFLPs and mapped RFLPs. *Theor. Appl. Genet.* 99(7/8):1233-1244.

Walbot, V. and Dure, III., L.S.1976. Developmental biochemistry of cotton seed em-bryogenesis and germination: 7. Characterization of the cotton genome. *J. Mol. Biol.* 101(4):503-536.

Wang, D.J., Li, G.P., and Li, B.L. 1989. Study on distant hybridization of cotton and character transfer through back crossing. *China Cottons* 3(1):6-7.

Wang, Y.H., Thomas, C.E., and Dean, R.A. 1997. A genetic map of melon (*Cucumis melo* L.) based on amplified fragment length polymorphism (AFLP) markers. *Theor. Appl. Genet.* 95(5/6):791-798.

Wang, Z.Y., Second, G., and Tanksley, S.D. 1992. Polymorphism and phylogenetic relationship among species in the genus Oryza as determined by analysis of nu-clear RFLPs. *Theor. Appl. Genet.* 83(5):565-581.

Weber, J.L. and May, P.E. 1989. Abundant class of human DNA polymorphisms which can be typed using the polymerase chain reaction. *Am. J. Hum. Genet.* 44(3):388-396.

Welsh, J. and McClelland, M. 1990. Fingerprinting genomes using PCR with arbi-trary primers. *Nucleic Acids Res.* 18(24):7213-7218.

Wendel, J.F. and Albert, V.A. 1992. Phylogenetics of the cotton genus (*Gossypium*): Character-state weighted parsimony analysis of chloroplast-DNA restriction site data and its systematic and biogeographic implications. *Syst. Bot.* 17(1):115-143.

Wendel, J.F., Brubaker, C.L., and Percival, A.E. 1992. Genetic diversity in *Gossypium hirsutum* and the origin of Upland cotton. *Am. J. Bot.* 79(11):1291-1310.

Wendel, J.F., Olson, P.D., and Stewart, J.McD. 1989. Genetic diversity, introgression, and independent domestication of Old World cultivated cottons. *Am. J. Bot.* 76(12):1795-1806.

Wendel, J.F. and Parks, C.R. 1982. Genetic control of isozyme variation in *Camalia japonica* L. (Theaceae). *J. Hered.* 73(3):197.

Whitkus, R., Doebley, J., and Lee, M. 1992. Comparative genome mapping of sorghum and maize. *Genetics* 132(4):1119-1130.

Wilson, J.T., Katterman, F.R.H., and Endrizzi, J.E. 1976. Analysis of repetitive DNA in three species of *Gossypium*. *Biochem. Genet.* 14(11/12):1071-1075.

Williams, J.G.K., Kubelik, A.R., Livak, K.J., Rafalski, J.A., and Tingey, S.V. 1990. DNA polymorphisms amplified by arbitrary primers are useful as genetic markers. *Nucleic Acid Res.* 18(22):6531-6535.

Williams, J.G.K., Rafalski, J.A., and Tingey, S.V. 1993. In R. Wu (Ed.), *Methods in enzymology* (pp. 704-740). Academic Press. Orlando, FL.

Wing, R.A. 1993. Prospects of physical mapping and map-based cloning of agriculturally important genes in cotton. *Proceedings of the Beltwide Cotton Research Conferences* (p. 1557). National Cotton Council, Memphis, TN.

Wright, R.J., Thaxton, P.M., El-Zik, K.M., and Paterson, A.H. 1999. Molecular mapping of genes affecting pubescence of cotton. *J. Hered.* 90(1):215-219.

Wu, K.S. and Tanksley, S.D. 1993. Abundance, polymorphism and genetic mapping of microsatellites in rice. *Mol. Gen. Genet.* 241(1/2):225-235.

Wu, D.Y., Ugozzoli, L., Pal, B.K., and Wallace, R.B. 1989. Allele-specific enzymatic amplification of β-globin genomic DNA for diagnosis of sickle cell anemia. *Proc. Natl. Acad. Sci. USA.* 86(8):2757-2760.

Yin, X., Stam, P., Dourleijn, C.J., and Kropff, M.J. 1999. AFLP mapping of quantitative trait loci for yield-determining physiological characters in spring barley. *Theor. Appl. Genet.* 99(1/2):244-253.

Yu, J., and Kohel, R.J. 1999. Update of the cotton genome mapping. *Proceedings of the Beltwide Cotton Research Conferences* (p. 439). National Cotton Council, Memphis, TN.

Yu, Z.H., McCouch, S.R., Kinoshita, T., Sato, S., and Tanksley, S.D. 1995. Association of morphological and RFLP markers in rice (*Oryza sativa* L.). *Genome* 38(3):566-574.

Zabeau, M. and Vos, P. 1993. Selective restriction fragment amplification: A general method for DNA fingerprinting. European Patent Application number: 92402629.7, publication number 0 534 858 A1.

Zhang, J.F., Wajahatullah, M.K., and Stewart, J.McD. 1997. Identification of RAPD markers linked to the D_8 CMS restorer gene in cotton. *Agron. Abst.* (pp. 152). ASA, Madison, WI.

Zhang, G.Y., Guo, Y., Chen, S.L., and Chen, S.Y. 1995. RFLP tagging of a salt tolerance gene in rice. *Plant Sci.* 110(2):227-234.

Zhang, Q., Saghai Maroof, M.A., and Kleinhofs, A. 1993. Comparative diversity analysis of RFLPs and isozymes within and among populations of *Hordeum vulgare* ssp. *Spontaneum*. *Genetics* 134(3):495-499.

Zhao, X, Si, Y., Hanson, R.E., Crane, C.F., Price, H.J., Stelly, D.M., Wendel, J.F., and Paterson, A.H. 1998. Dispersed repetitive DNA has spread to new genomes since polyploid formation in cotton. *Genome Res.* 8(6):479-492.

PART III:
CROP ISSUES

Chapter 10

Physiological and Biochemical Responses of Plants to Drought and Heat Stress

Bingru Huang
Yiwei Jiang

INTRODUCTION

Water deficit and high temperature are among the most important environmental factors that limit crop productivity in many areas of the world (Levitt, 1980). The frequency and severity of hot, dry climates may increase in the future as global warming intensifies. Understanding the mechanisms of plants' adaptations to drought and heat stress should help researchers in improving drought and heat tolerance of crop plants more effectively. Improved tolerance could sustain productivity and help extend cultivation of certain crops into areas that are currently unsuitable for crop production.

Drought stress affects many physiological processes, including carbohydrate metabolism (Massacci et al., 1996; Wang et al., 1996; Escobar-Guti et al., 1998; Hashem et al., 1998); photosynthetic efficiency (Epron, 1997; Jagtap et al., 1998); and water relations (Perdomo et al., 1996; Huang, Fry, and Wang, 1998; Paakkonen et al., 1998; Volaire, Thomas, and Leievre, 1998; Volaire et al., 1998). Drought also causes oxidative stress (Zhang and Kirkham, 1994; Gogorcena et al., 1995; Olsson et al., 1996; Sairam et al., 1997; Kronfub et al., 1998) and protein alterations (Han and Kermode, 1996; Ristic et al.,1996; Riccardi et al., 1998). Plant responses to heat stress are diverse and include many physiological and biological factors, such as protein denaturation, lipid peroxidation, and reductions in membrane stability and efficiency of photosynthesis (Jagtap and Bhargava, 1995; Gong et al., 1997; Kurganova et al., 1997; Zheng and Han, 1997; Yamane et al., 1998). High temperature also causes stomatal closure and internal water deficits (Kolb and Robberecht, 1996); reduction in carbohydrate accumulation (Howard and Watschke, 1991; Lafta and Lorenzen, 1995); and increases in respiratory rate (Al-Khatib and Paulsen, 1989; Huang, Lui, and Fry, 1998; Yamane et al., 1998).

Drought or high temperature alone can damage plants; however, these stresses usually are combined and cause more detrimental interactive effects on plants than either stress alone. Combined high temperature and drought caused reduction in grain yield of barley (*Hordeum vulgare* L.) under field conditions (Savin and Nicolas, 1996); significantly decreased the rate of CO_2 uptake and O_2 production in beans (*Phaseolus vulgaris* L.) (Yordanov et al., 1997); and reduced leaf water content, water potential, and osmotic potential in wheat *(Triticum aestivum* L.) (Shah, 1992). In addition, a reduction in leaf chlorophyll fluorescence, an indicator of effectiveness of the electron transport chain, in sunflower (*Helianthus annuus* L.) has been reported (Yordanov et al., 1997).

Various plant mechanisms have evolved to survive drought or heat. For example, some species tolerate drought stress by maintaining a positive turgor pressure at low tissue water potential through osmotic adjustment and dehydration tolerance or by induction of certain specific proteins, such as dehydrin, which protect other proteins or cell membranes from stress damage (Jacobsen and Shaw, 1989; Close et al., 1993). Heat tolerance can be conferred by many mechanisms, such as maintenance of photosynthesis and induction of heat-shock proteins that are responsible for repairing other heat-damaged proteins (Bond et al., 1988; Pelham, 1988; Heckathorn et al., 1998).

This chapter provides a review of recent literature on the responses to drought and heat stress relative to several major physiological and biochemical processes, including water relations, carbohydrate metabolism, lipid peroxidation, and protein synthesis. In addition, we discuss several widely used techniques to evaluate drought or heat-stress tolerance, which should be of interest to crop breeders.

WATER RELATIONS

Drought tolerance can be accomplished by maintenance of a positive turgor pressure at low tissue water potential through osmotic adjustment and dehydration tolerance. Drought avoidance involves maintenance of a higher tissue water potential by maintaining water uptake through development of deep, extensive root systems and by reducing water loss through modification of leaf anatomical and morphological features (Levitt, 1980; Jones et al., 1981).

Maintenance of favorable water content in leaves is related positively to drought resistance (Dedio, 1975). Huang, Fry, and Wang (1998) reported that a drought-resistant cultivar of tall fescue (*Festuca arundinacea* L.) maintained higher relative leaf water content and water potential than a susceptible cultivar as soil dried. High water content was associated with a slow rate of leaf rolling. Lower osmotic potential was observed in a drought-resistant cultivar than in a drought-sensitive cultivar of Kentucky bluegrass (*Poa pratensis* L.) (Perdomo et al., 1996). Better tolerance to dehydration was associated strongly with a low osmotic potential in orchardgrass (*Dactylis glomerata* L.) and perennial ryegrass (*Lolium perenne* L.) (Volaire et al., 1998).

Osmotic adjustment, the ability of cells to accumulate solutes and reduce osmotic potential in response to drought stress, is an important process that facilitates water retention and turgor maintenance of leaves and roots (Gunaskera and Berkowitz, 1992; White et al., 1992) and is related closely to drought tolerance (Guicherd et al., 1997). Drought-resistant cultivars maintain a higher positive turgor pressure than drought-susceptible ones as a result of osmotic adjustment (Gunaskera and Berkowitz, 1992). Accumulation of solutes, such as sugars, amino acids, organic acids, and ions, during drought stress has been observed in many plant species. Reported accumulations are of malate and mannitol in *Fraxinus excelsior* L. (Guicherd et al., 1997); soluble saccharides in wheat (Rekika et al., 1998); sorbitol in apple (*Malus domestica* L.) (Wang et al., 1995) and cherry (*Prunus* spp. L.) (Ranney et al., 1991); potassium (K^+) in cherry (Ranney et al., 1991), amino acids in strawberry (*Fragaria chiloensis* and *F. virginiana* L.) (Zhang and Archbold, 1993), glucose, fructose, and proline in black spruce (*Picea mariana* L.) (Tan et al., 1992); and fructans in cockfoot and perennial ryegrass (Volaire et al., 1998).

Heat-stress injury is associated with leaf water deficits because heat enhances evapotranspiration demand and direct tissue damage (Levitt, 1980; Gauslaa, 1984; Halgren et al., 1991). High temperature reduces leaf water potential (Graves et al., 1991; Ashraf et al., 1994) and limits water uptake (Tajima et al., 1976). Seedlings of *Pinus ponderosa* L. that survived at a relatively high temperature had significantly higher stomatal conductance and transpiration rate than seedlings that failed to survive (Kolb and Robberecht, 1996). Lehman and Engelke (1993) found that turf quality was correlated significantly with shoot water content in creeping bentgrass (*Agrostis stolonifera* L.) under elevated soil temperatures ranging from 29 to 34°C. These studies indicate the importance of maintaining favorable water status in plants for heat-stress tolerance. Maintenance of water uptake and transpiration enhance transpirational cooling and can protect plants from internal heat stress at relatively high temperatures.

Effects of high temperature on water relations also involve osmotic adjustment. Osmotic potential and turgor pressure decreased with high temperature in wheat (Ahmad et al., 1989). Turgor pressure of shaded *Piper auritum* also decreased at 37°C compared with 27°C (Schultz and Matthews, 1997). A drought-tolerant genotype of tomato *(Lycopersicon esculentum)* showed a most dramatic cellular adjustment of osmolarity in response to heat treatment (Smith et al. 1989). Ashraf and colleagues (1994) found that heat-tolerant cultivars of cotton *(Gossypium hirsutum)* significantly accumulated organic osmotica, such as soluble proteins, soluble sugars, and proline, which are important components of heat tolerance.

High temperature and drought often occur simultaneously during summer months in many areas. Transpirational water loss, induced by high temperature and dry climate in the desert, often causes the death of some desert plants (Nobel, 1988). High temperature may influence plant's ability for osmotic adjustment during drought by affecting carbohydrate accumulation. High temperature causes a decline in photosynthesis (Al-Khatib and Paulsen, 1989; Harding et al., 1990; Huang, Liu, and Fry, 1998) and an increase in dark respiration (Lawlor, 1979; Huang, Liu, and Fry, 1998), thereby causing carbohydrate depletion (Morgan, 1984; Taiz and Zeiger, 1991). Increased sugar consumption and reduced sugar accumulation during heat stress may interfere with drought tolerance by affecting osmotic adjustment (Li et al., 1993).

CARBOHYDRATE METABOLISM

Carbohydrates serve as important structural materials and energy reserves and are often associated with osmotic adjustment (Premachandra et al., 1992; Tan et al., 1992; Zhang and Archbold, 1993; Rekika et al., 1998). Therefore, carbohydrate metabolism plays an important role in plant tolerance to various environmental stresses, including drought and heat stress (Aldous and Kaufmann, 1979; Lafta and Lorenzen, 1995; Savin and Nicolas, 1996; Volaire et al., 1998).

Drought influences photosynthesis, dark respiration, carbon allocation and partitioning, and carbohydrate accumulation. Reduction in photosynthetic rate caused by drought stress has been reported in many species (Brix, 1979; Krampitz et al., 1984; Wample and Thornton, 1984; Deng et al., 1990; Nash et al., 1990; Zrenner and Stitt, 1991; Antolin and Sanchez-Diaz, 1993; Escobar-Guti et al., 1998; Faria et al., 1998; Hashem et al., 1998; Pankovic et al., 1999). A decrease in photosynthesis induced by long-term severe drought stress has been attributed to both stomatal and nonstomatal limitations. Photosynthetic rate may be inhibited mainly by stomatal closure without damage to photosynthetic apparatus under moderate water stress (Chaves, 1991). Nonstomatal control of photosynthesis is attributed partially to photoinhibition of PSII (Kaiser, 1987). Photochemical efficiency of PSII measured as chlorophyll fluorescence often decreases with drought stress (Krause and Weis, 1991; Baker, 1993; Epron, 1997). Dark respiration also decreases under drought stress (Brix, 1979; Brown and Thomas, 1980;

Wibbe and Blanke, 1997) but to a larger extent than photosynthetic rate does (Antolin and Sanchez-Diaz, 1993).

Drought stress reduces rate of assimilate export from leaves (Sung and Krieg, 1979; Deng et al., 1989) and inhibits the conversion of glucose into starch (Wang et al., 1996). Under drought stress, the allocation of carbon to leaves and expanding tissue is often reduced, whereas the proportion of carbon allocated to stem and roots is increased (McCoy et al., 1990). The carbon partitioning pattern is related to changes in shoot-to-root biomass ratios (Deng et al., 1990).

The effect of drought on the accumulation of different fractions of nonstructural carbohydrates varies with plant species and organs. Mild water deficit inhibited starch synthesis in spinach *(Spinacia oleracea)* leaves, whereas sucrose synthesis increased or remained constant (Zrenner and Stitt, 1991). Starch content also decreased significantly during water stress in peach *(Prunus persica)* seedlings (Escobar-Guti et al., 1998), but it decreased only transiently in soybean *(Glycine max)* root nodules (Muller et al., 1996). Volaire and colleagues (1998) reported that fructans accumulated in drought-sensitive cocksfoot during drought stress; as drought progressed, total water-soluble carbohydrate increased and then stabilized. Wang and colleagues (1995) found that sucrose concentration decreased in mature leaves and stems of apple, yet increased in young leaves and roots as leaf water potential declined. Sucrose concentration strongly increased in root nodules of stressed soybean plants, reaching nearly 10 percent of dry weight (Muller et al., 1996). Tan and colleagues (1992) indicated that soluble carbohydrate content (mainly glucose and fructose) increased in two progenies of *Picea mariana* with vigorous growth under drought in the field. Hays and colleagues (1991) suggested that carbohydrate accumulation in roots varied with soil depth but was not correlated significantly with turf quality when turfgrasses were exposed to drought stress.

High temperature alone reduces photosynthetic rate (Wolf et al., 1990; Wullschleger and Oosterhuis, 1990; Du and Tachibana, 1994; Lafta and Lorenzen, 1995; Lu et al., 1997; Ranney et al., 1995; Ranney and Ruter, 1997; Huang, Liu, and Fry, 1998). The effect of high temperature on net photosynthesis is the result of stomatal closure and nonstomatal metabolic inhibition (Ku et al., 1977; Dwelle et al., 1981; Markus et al., 1981). Heat stress often causes damage to the photosynthetic apparatus and electron transfer in photosystem II (PS II), as reflected by changes in chlorophyll fluorescence, and thus inhibits photosynthesis (Krause and Weis, 1991). A heat-tolerant cultivar of *Ilex cornuta* had substantially greater photosynthetic capacity than other species at 40°C (Ranney and Ruter, 1997). High temperature increases dark respiration (Duff and Beard, 1974; Deal et al., 1990; Wolf et al., 1990; Rawat and Purohit, 1991; Ranney et al., 1995; Huang, Liu, and Fry, 1998). Heat-tolerant cultivars appeared to minimize detrimental effects of high temperature through maintenance of lower respiration rate than sensitive cultivars of blue spruce *(Picea pungens)* (Deal et al., 1990) and in creeping bentgrass (Huang, Liu, and Fry, 1998).

High temperature also changes the patterns of carbon partitioning between plant organs. Accumulation of ^{14}C in heat-stressed grapevines *(Vitis vinifera)* was greater in the shoot tip and less in the trunk and roots (Sepulveda et al., 1986). High temperature increased carbon allocation to the cell wall components in the apex and the stem of potato *(Solanum tuberosum)* plants (Wolf et al., 1991) and inhibited ^{14}C export from leaves and stems and import of ^{14}C into pods of Indian mustard *(Brassica juncea)* (Subrahmanyam and Rathore, 1995). Newly photosynthesized ^{14}C is allocated more to the soluble carbohydrate pool than to insoluble carbon fractions during heat stress (Ruter and Ingram, 1990; Du and Tachibana, 1994).

Effects of heat stress on carbohydrate accumulation vary with plant species and type of carbohydrates. Sucrose accumulation significantly decreased during heat stress in winter wheat (Zemanek and Frecer, 1990), but foliar sucrose content increased in mature potato leaves (Lafta and Lorenzen, 1995). Starch content of Indian mustard and potato leaves was reduced significantly (Subrahmanyam and Rathore, 1995; Lafta and Lorenzen, 1995), but glucose content of potato leaves was unchanged under high temperature conditions (Lafta and Lorenzen, 1995). Aldous and Kaufmann (1979) indicated that carbohydrate levels at 38°C were significantly higher in roots than in the crown or verdure in a heat-sensitive cultivar of Kentucky bluegrass but root carbohydrate levels were significantly lower than those in shoots at the same temperature in a tolerant cultivar. Howard and Watschke (1991) reported that the sum of glucose, fructose, and sucrose did not change in Kentucky bluegrass cultivars during high temperature stress, but fructosan concentration was twofold lower at 30°C than at 10°C. Reduction in carbohydrate accumulation under high temperature conditions may result from the reduced photosynthetic efficiency and the imbalance between photosynthesis and respiration (Prange et al., 1990; Wolf et al., 1990; Huang, Liu, and Fry, 1998).

LIPID PEROXIDATION

Lipid peroxidation of cell membranes resulting from the production of active oxygen species occurs in plants subjected to environmental stresses such as high temperature and drought (Bowler et al., 1992; Foyer et al., 1994; Inze and Van Montagu, 1995). Oxygen species such as O_2^-, H_2O_2, and OH are toxic to proteins, membrane lipids, and other cellular components (Halliwell, 1987; Cadenas, 1989; Fridovich, 1989). Plants have evolved enzymatic and nonenzymatic mechanisms to defend themselves against oxidative damage. The enzymatic defense systems mainly include superoxide dismutase (SOD), catalase (CAT), peroxidase (POD), glutathione reductase (GR), and ascorbate peroxidase (APX). The nonenzymic systems are antioxidants, including ascorbate (ASA), glutathione (GSH), and vitamin E.

Superoxide dismutase is an essential antioxidant enzyme in the system, which protects against lipid peroxidation, because it dismutases two superoxide radicals (O_2^-) to produce hydrogen peroxide and oxygen. This process minimizes the interaction of superoxide radicals with H_2O_2 to form highly reactive and detrimental hydroxyl radicals (OH·). The reaction is called Haber-Weiss (Elstner, 1982). The other effective system against H_2O_2 is the ascorbate-glutathione cycle in chloroplasts and cytosol.

The balance between activity of scavenging antioxidants and production of oxygen species can be disturbed under environmental stress conditions, causing damage to cell membranes from lipid peroxidation. Drought-induced lipid peroxidation has been investigated in many plant species, including wheat (Zhang and Kirkham, 1994; Sairam et al., 1997, 1998); pea *(Pisum sativum)* (Moran et al., 1994; Gogorcena et al., 1995); *Vigna catjang* (Mukherjee and Choudhuri, 1985); sorghum *(Sorghum bicolor)* (Jagtap and Bhargava, 1995; Zhang and Kirkham, 1996); maize *(Zea mays)* (Pastori and Trippi, 1992; Del Longo et al., 1993; Li and Staden, 1998); sunflower (Lusia et al., 1995; Zhang and Kirkham, 1996); *Lotus corniculatus* and *Cerastium fontanum* (Olsson et al., 1996); alfalfa *(Medicago sativa)* (Irigoyen et al., 1992); soybean (Zheng and Han, 1997); and spruce *(Picea abies)* (Kronfub et al., 1998). Drought stress caused increases in the level of malondialdehyde (MDA), a product of lipid peroxidation in drought-susceptible cultivars of sorghum (Jagtap and Bhargava, 1995) and alfalfa (Irigoyen et al., 1992). Increases in MDA content were more dramatic in sunflower than in sorghum (Zhang and Kirkham, 1996).

Results on the responses of SOD and CAT to drought are inconsistent. The activities of SOD and CAT decreased in pea nodules when water potential decreased to −2.03 MPa; however, SOD activity increased by 32 to 42 percent and CAT decreased when leaf water potential decreased to −1.3 MPa (Gogorcena et al., 1995). Both SOD and CAT activities increased in a drought-tolerant sorghum cultivar when leaf water potential declined to −0.5 MPa (Moran et al., 1994; Jagtap and Bhargava, 1995). Irigoyen and colleagues (1992) reported that SOD activity was not affected by water stress (leaf water potential of −2.0 Mpa) and CAT activity significantly increased in alfalfa leaf. Zhang and Kirkham (1994) found that SOD and CAT activities in wheat cultivars increased or remained the same in the early phase of drought and then decreased with further increases in water stress. Results on the response of the ascorbate-glutathione cycle to drought stress are also contradictory. Declines in activities of GR, AP, and contents of

ASA and GSH have been observed in pea nodules subjected to water stress (Gogorcena et al., 1995). However, GR activity increased in wheat under drought stress (Sairam et al., 1997). Moran and colleagues (1994) indicated that drought had no effect on the levels of ASA and GSH in pea. The discrepancy in the responses of levels of lipid peroxidation products and enzyme activity to drought in different studies could be related to the variation in drought tolerance among plant species or the duration and severity of drought (Zhang and Kirkham, 1996).

High temperature also can cause lipid peroxidation and changes in antioxidant level and enzyme activity in various plant species, including sorghum (Jagtap and Bhargava, 1995); wheat (Mishra and Singhal, 1992); soybean (Zheng and Han, 1997); pea (Burke and Oliver, 1992; Kurganova et al., 1997); and maize (Gong et al., 1997). High temperature increased the level of MDA in both drought-tolerant and susceptible sorghum cultivars but to a larger degree in the susceptible cultivar (Jagtap and Bhargava, 1995). They also observed increases in CAT and SOD activities in two drought-tolerant cultivars and decreases in one susceptible cultivar of sorghum. Both SOD and POD activities increased in drought-tolerant and sensitive cultivars of soybean under high temperature (Zheng and Han, 1997). Heat shock induced a decrease in the activity of GR at flower bud formation in pea (Kurganova et al., 1997). A negative correlation between lipid peroxidation and electron transport activity was observed in wheat under high temperature conditions (Mishra and Singhal, 1992).

The information on oxidative stress in relation to plant tolerance to combined drought and heat stress is limited. Zheng and Han (1997) reported that a drought-sensitive soybean cultivar had lower SOD activity and higher POD activity when grown under both drought- and heat-stress conditions than when grown under drought or heat stress alone; SOD activity remained unchanged, and POD increased in a drought-tolerant cultivar. However, Jagtap and Bhargava (1995) reported that SOD and CAT activities increased in a drought-tolerant cultivar of sorghum under a combined drought and high temperature stress. Oxidative stress induced by drought and high temperature affects cell membrane integrity, enzyme activity, and antioxidant levels, which affects plant growth.

PROTEIN SYNTHESIS

Drought- or heat-induced damage to plant growth may be largely due to effects on protein and polypeptide synthesis and denaturation (Chen and Tabaeizadeh, 1992; Chandler and Robertson, 1994; Ouvrard et al., 1996). Proteins comprise one-half to two-thirds of the cell membrane dry weight and serve as important enzymes. The metabolic rate of a process can be affected by changes in protein synthesis or degradation, as well as changes in enzyme-specific activity. Therefore, proteins play important roles in a plant's adaptation to drought or heat stress.

Changes in protein synthesis during drought stress have been observed in a number of plant species, including tomato (Bray, 1988); pea (Guerrero and Mullet, 1988); castor bean *(Ricinus communis)* (Han and Kermode, 1996); geranium *(Pelargonium hortorum)* (Arora et al., 1998); and rice *(Oryza sativa)* (Perez-Molphe-Balch et al., 1996). Synthesis of many proteins generally is inhibited in leaves when plants are subjected to water deficit (Bewley et al., 1983; Dasgupta and Bewley, 1984; Bray, 1988). However, some proteins are induced during drought stress and are regarded as new gene products that could be associated with stress tolerance. Riccardi and colleagues (1998) found that 78 proteins showed significant quantitative changes (increase or decrease) in response to progressive water deficit in maize and 38 of them exhibited a different expression in two genotypes. They also found that the content of 11 proteins increased by a factor of 1.3 to 5 in stressed plants and detected eight new proteins in stressed plants. Ristic and colleagues (1996) noted that a 45 kDa protein played a role in the development of drought resistance in maize. A 66 kDa boiling-stable protein, highly expressed during water stress, was identified in cultured shoots of aspen *(Populus tremula L.)* (Pelah et al., 1995). These observations indicate that water stress-induced proteins are associated with stress tolerance of plants. However, the functional role of these stress-responsive proteins is not yet clearly established.

Dehydrin proteins accumulate in a wide range of species in response to environmental stimuli, particularly drought stress; they are hydrophilic and heat stable and are presumed to help protect other proteins and maintain physiological integrity of cells (Jacobsen and Shaw, 1989; Close et al., 1989, 1993; Bray, 1993, 1994; Chandler and Robertson, 1994; Giraudat et al., 1994; Oliver, 1996). Dehydrin gene expression in most cases is associated with abscisic acid synthesis (ABA), which is believed to act as a chemical messenger conveying water stress signals. A 15 kDa ABA-associated protein has been detected in maize leaves grown under drought stress (Gomez et al., 1988). ABA and water stress induced this type of dehydrin polypeptide accumulation in seedlings of castor bean (Han and Kermode, 1996). Arora and colleagues (1998) reported that several dehydrins accumulated in water-stressed geranium leaves that had a high stress tolerance. Therefore, dehydrin accumulation during water stress may enhance drought tolerance of plants. High temperature can reduce the rate of protein synthesis and cause denaturation of proteins in plants (Weidner and Ziemans, 1975; Wehner and Watschke, 1984). High temperature also induces accumulation of certain proteins. For example, heat shock proteins (HSPs) induced by heat shock have been observed in many plant species, such as wheat (Krishnan et al., 1989; Hendershot et al., 1992); perennial ryegrass (DiMascio et al., 1994); maize (Ristic et al.,

1991, 1996); geranium (Arora et al., 1998), tomato (Fender and O'Connell, 1990); gladiolus *(Gladiolus cormels)* (Ginzburg and Salomon, 1986); broccoli *(Brassica oleracea)* (Fabijanski et al., 1987), sorghum (Ougham and Stoddart, 1986), and cotton (Rodriguez-Garay and Barrow, 1988). Hendershot and colleagues (1992) found that the HSPs were synthesized when leaf temperature in wheat increased about 10°C above the optimum growth temperature of 18 to 23°C and before growth injury occurred.

Synthesis of HSPs during heat stress results in changes in gene expression; thus, a new set of proteins is synthesized from newly transcribed mRNA (Lindquist, 1986). In general, these proteins are classified into two groups: high molecular weight, including HSPs 110, 90, 80, 70, and 60 kDa, and low molecular weight, 30 to 15 kDa. The low molecular weight HSPs are particularly abundant and complex in plants (Mansfield and Key, 1987). The HSPs induced during drought or heat stress vary with plant species differing in heat or drought resistance. Ristic and colleagues (1991) demonstrated that HSPs of 90, 80 to 85, 70, 26 to 28 and 18 to 20 kDa were synthesized in maize leaves of both drought- and heat-resistant and sensitive lines, but a protein of approximately 45 kDa was found only in resistant lines. They further tested different hybrids and found that southern hybrids exhibited a greater ability to synthesize the 45 kDa protein than northern hybrids and also displayed much higher resistance to heat and drought than northern hybrids. The synthesis of HSPs has been found to be related closely to heat- and drought-resistance in maize hybrids (Ristic et al., 1996). The function of HSPs in stress tolerance may include interaction with other proteins to prevent their aggregation and involvement in the tagging of heat-damaged proteins (Bond et al., 1988; Pelham, 1988); stabilizing chromatin; and protecting the photosynthetic membranes (Velazquez and Lindquist, 1984; Cooper and Ho, 1987; Schuster et al., 1988).

Maintenance of photosynthesis is a critical aspect of thermotolerance, and photosynthesis is sensitive to thermal stress (Weis and Berry, 1988). The PSII in plants is particularly sensitive to heat stress, partly because heat stress changes its structure and results in the removal of oxygen-evolving enhancer proteins from the thylakoid (Takeuchi and Thornber, 1994). Several HSPs have been shown to be translocated into the chloroplast and correlated with increased thermal tolerance of photosynthesis (Vierling, 1991; Stapel et al., 1993; Clarke and Critchley, 1994). Heckathorn and colleagues (1998) demonstrated that HSP formed in small chloroplasts were involved in plant thermotolerance by protecting PSII activity during heat stress.

REGULATION OF DROUGHT AND HEAT TOLERANCE

Calcium ions have been found to be involved in the regulation of plant responses to environmental stresses, including drought and heat (Yang et al.,

1993; Gong et al., 1996; Bush, 1995; Webb et al., 1996; Gong et al., 1997). Accumulation of free Ca^{2+} in plant cells under heat shock (Biyaseheva et al., 1993; Gong et al., 1998) reduces oxidative stress (Price et al., 1994). Exogenous application of Ca^{2+} also increases drought and heat tolerance (Cooke et al., 1986; Yang et al., 1993). McAinsh and colleagues (1996) found that the increases in the internal concentration of Ca^{2+} induced by oxidative stress were related to the process of Ca^{2+}-based signal transduction in stomatal guard cells. External Ca^{2+} treatments have been observed to ameliorate the inhibitory effects of drought stress on fresh weight, chlorophyll content, and water content in *Vicia faba* (Abdel, 1998); to help maintain cell membrane stability in soybean during drought stress (Yang et al., 1993); to enhance activities of SOD, CAT, and APX in maize seedlings; and to decrease the level of lipid peroxidation induced by heat stress in maize (Gong et al., 1997). The relationship of regulation of oxidative stress Ca^{2+} to plant tolerance to combined drought and heat stress are unclear at this time.

Calcium ions may influence HSP synthesis in plant cells. Adequate Ca^{2+} concentration in cells may be necessary for thermal signal transduction and activation of individual heat-shock genes (Kuznetsov et al., 1997). Deficiency of Ca in plants reduces the synthesis of HSP 90 following heat shock (Trofimova et al., 1997). In addition to regulating HSP, Ca^{2+} has been found to facilitate attachment of a 33 kDa protein to the PSII membranes by promoting intensive aggregation of membranes during heat stress (Enami et al., 1994).

EVALUATION OF DROUGHT AND HEAT TOLERANCE

Identifying stress-resistance mechanisms is essential for genetic improvement of stress resistance in crop plants. The use of a suitable methodology for measuring stress resistance in large breeding populations is also very important. A screening method is effective if it can relate integrated plant responses to a specific response of a physiological or biochemical process. Breeding for drought and heat tolerance has been accomplished by selecting for seed yield or plant survivability under field conditions (Bouslama and Schapaugh, 1984; Brown et al., 1985), but because such procedures require full-season field data, they are not always efficient (Sammons et al., 1978). An alternative is to screen genetic materials under laboratory or greenhouse conditions using seedlings as test materials by examining physiological or biochemical parameters. Several physiological characteristics in crops have been reported to be reliable indicators for the selection of plant germplasm possessing drought and heat tolerance. These characteristics include cell membrane stability as evaluated by the degree of electrolyte leakage from drought or heat-damaged leaf cells after exposure to drought or elevated temperatures (Martineau et al., 1979; Sullivan and Ross, 1979; Blum and Ebercon, 1981); leaf chlorophyll fluorescence to provide quantitative assessment of inhibition or damage to electron transfer in the photosynthetic process or photochemical efficiency of leaves (Baker, 1993); and osmotic adjustment to evaluate turgor maintenance in leaves (Morgan 1977, 1980).

Cell Membrane Stability

The cell membrane is an initial site of stress injury. Membrane function is a physiological process drastically damaged by many environmental stresses (Levitt, 1980; McKersie and Tomes, 1980; McKersie et al., 1982). Membrane damage results in increased permeability and leakage of electrolytes, which reduces photosynthetic or mitochondrial activity and the ability of plasma lemma to retain solutes and water. Thus, evaluation of cellular membrane integrity as a measure of drought and heat-stress tolerance appears to be relevant (Sullivan, 1972).

Cell membrane stability has been evaluated using electrolyte leakage. An electrolyte leakage test for drought tolerance was developed first in sorghum by Sullivan (1972). Sullivan and Ross (1979) conducted experiments relating electrolyte leakage following a desiccation treatment to the general ability of plants to tolerate drought stress, and they found that electrolyte leakage correlates well with other plant processes used for evaluating tolerance to the stress. Electrolyte leakage has been used successfully to measure membrane integrity in plants subjected to a variety of environmental stresses (McKersie and Tomes, 1980; Blum and Ebercon, 1981; McKersie et al., 1982; Sapra and Anaele, 1991; Agarie et al., 1995). A cell membrane stability test has been found to be efficient in estimating stress tolerance of several crop plants, including wheat (Blum and Ebercon, 1981; Premachandra and Shimada, 1987); sorghum (Sullivan and Ross, 1979); and maize (Premachandra et al., 1989). Saadalla and colleagues (1990) reported that heat-tolerant genotypes of wheat, determined on the basis of electrolyte leakage, outyielded sensitive ones by 19 percent under field conditions. Kuo and colleagues (1993) showed that yields of vegetable species with low electrolyte leakage were more stable in different growing conditions.

Cell membrane stability varies with leaf age because of variations in desiccation tolerance between leaves. Younger leaf tissues were more tolerant to drought than older tissues in wheat and sorghum (Sullivan, 1972; Blum and Ebercon, 1981). For a given growth stage, leaf position also may contribute to variations. Electrolyte leakage measurement is influenced markedly by leaf age, degree of stress hardening, and plant species. Therefore, these factors should be taken into consideration during sampling.

Chlorophyll Fluorescence

Electron transfer in PSII of photosynthesis in thylakoid membranes is drought and heat sensitive. Chlorophyll fluorescence is a sensitive indicator of the state of electron transfer. Measurements of chlorophyll fluorescence can give quantitative assessment of inhibition or damage to electron transfer of the PSII (Baker, 1993). The percentage decrease in photochemical quenching of damaged leaves compared to controls, calculated during the fluorescence induction curve, indicates the severity of stress damage. Chlorophyll fluorescence has been employed widely to screen for drought and heat tolerance (Moffat et al., 1990; Smillie and Hetherington, 1990; Srinivasan et al., 1996). Hall (1992) considered fluorescence properties to be more suitable for screening than photosynthetic rates, because the former reflect both time-integrated damage to PSII and time-integrated effects on photosynthetic rates.

The chlorophyll fluorescence technique is rapid, sensitive, nondestructive, relatively cheap, and able to detect injury even before visible symptoms appear (Wilson and Greaves, 1990). Like electrolyte leakage, chlorophyll fluorescence analysis is simple and reproducible, and many samples can be analyzed within a short time. These advantages make the test ideal for large-scale screening of populations. Smillie and Hetherington (1990) and Wilson and Greaves (1990) suggested that a large number of samples is required for fluorescence analysis to reduce variability. Chlorophyll fluorescence responses to drought or heat stress vary with leaf age and growth stages, which should be considered in sampling.

Osmotic Adjustment

As described in the foregoing discussion, osmotic adjustment as an adaptational response to water stress involves an increase in the solute content of cells, which facilitates turgor maintenance in leaves and roots. Substantial differences in osmotic adjustment existed between wheat genotypes selected from widely differing backgrounds (Morgan 1977, 1980). Yield differences between two sorghum cultivars were correlated with differences in osmotic adjustment when measured in both greenhouse and field (Wright et al., 1983). Morgan (1984) reported that wheat lines selected from the high osmotic-adjustment group maintained turgor at lower water potential and produced higher grain yield than those selected from the low osmotic-adjustment group. Singh and colleagues (1972, 1973) found that barley cultivars showed substantial differences in proline accumulation at the same leaf water potential; cultivars with the highest proline accumulation rates were those with the most stable grain yields in drought-prone environments.

They suggested that capacity for proline accumulation could be used as a screening test for drought resistance in cereal breeding programs.

However, Hanson and colleagues (1979) found no evidence that accumulation of free proline in barley is of survival value during severe water stress; moreover, genotypes selected for high proline accumulation tended to be less vigorous under both favorable and stress conditions than those selected for low proline accumulation. They concluded that proline accumulation is a symptom of severe water stress in barley; in principle, the declining water status can be used to detect the genetic differences in water status that appear among barley genotypes subjected to water stress. Selecting for low proline accumulation might help identify promising material for some types of drought-prone environments. However, reliable detection of genetic differences in proline accumulation requires strict environmental control and involves a laborious chemical analysis (Hanson et al., 1977). Hanson and colleagues (1979) suggested that selecting for low leaf firing during stress might be an effective alternative.

Whether drought tolerance of plants tested under laboratory conditions reflects drought tolerance under field conditions determines the effectiveness of this technique (Singh et al., 1972). Even though greenhouse plants differ from field-grown plants in their water relations (Slatyer, 1963), the differences between cultivars or breeding lines apparently are still observable, and correlate well with field performance measured as turgor responses or grain yields (Morgan, 1977, 1980). Furthermore, the method of measuring osmoregulation in the controlled environment requires only a small number of leaf samples upon which relatively simple measurements are made. Therefore, measuring osmotic adjustment is suitable for screening drought tolerance in greenhouse and field conditions.

SUMMARY

Plant responses to drought or heat stress are diverse and complex. Mechanisms of plant tolerance to drought or heat stress deserve further investigation, although numerous studies have been conducted in various species. Drought and heat stress often occur simultaneously in many parts of the world, which can cause dramatic reductions in crop yield, particularly for C_3 plant species. However, extremely limited information is available on how plants tolerate this combined stress. It is not clear how osmotic adjustment and carbohydrate metabolism are involved in plant tolerance to combined drought and heat stress; how antioxidant defense systems are involved in oxidative stress tolerance; and how protein synthesis may be altered during the combined stress compared to a single stress. Answers to these questions are particularly important for improving plant productivity in dry and hot environments. Identifying physiological traits involved in drought and heat stress tolerance would enhance plant tolerance to these stresses by conventional breeding or biotechnology (Kang, 1998).

REFERENCES

Abdel, B.R. 1998. Calcium channels and membrane disorders induced by drought stress in *Vicia faba* plants supplemented with calcium. *Acta Physiol. Plant.* 20: 149-153.

Agarie, S., Hanaoka, N., Kubota, F., Agata, W., and Kaufman, P.B. 1995. Measurement of cell membrane stability evaluated by electrolyte leakage as a drought and heat tolerance test in rice (*Oryza sativa* L.). *J. Fac. Agric. Kyushu Univ.* 40: 233-240.

Ahmad, S., Ahmad, N., Ahmad, R., and Hamid, M. 1989. Effect of high temperature stress on wheat reproductive growth. *J. Agric. Res. Lahore.* 27: 307-313.

Aldous, D.E., and Kaufmann, J.E. 1979. Role of root temperature on shoot growth of two Kentucky bluegrass cultivars. *Agron. J.* 71: 545-547.

Al-Khatib, K., and Paulsen, G.M. 1989. Enhancement of thermal injury to photosynthesis in wheat plants and thylakoids by high light intensity. *Plant Physiol.* 90: 1041-1048.

Antolin, M.C., and Sanchez-Diaz, M. 1993. Effects of temporary droughts on photosynthesis of alfalfa plants. *J. Exp. Bot.* 44: 1341-1349.

Arora, R., Pitchay, D.S., and Bearce, B.C. 1998. Water-stress-induced heat tolerance in geranium leaf tissue: A possible linkage through stress proteins? *Physiol. Plant.* 103: 24-34.

Ashraf, M., Saeed, M.M., and Qureshi, M.J. 1994. Tolerance to high temperature in cotton at initial growth stages. *Env. Exp. Bot.* 343: 275-283.

Baker, N.R. 1993. Light-use efficiency and photoinhibition of photosynthesis in plants under environmental stress. In J.A.C. Smith and H. Griffiths (Eds.) *Water deficits: plant responses from cell to community* (pp. 221-235). BIOS Scientific, Oxford, U.K.

Bewley, J.D., Larsen, K.M., and Papp, J.E.T. 1983. Water-stressed-induced changes in the pattern of protein synthesis in maize seedling mesocotyls: A comparison with the effects of heat shock. *J. Exp. Bot.* 6: 1161-1166.

Biyaseheva, A.E., Molotkovskii, Y.G., and Mamonov, L.K. 1993. Increase of free Ca^{2+} in the cytosol of plant protoplasts in responses to heat stress as related to Ca^{2+} homeostasis. *Plant Physiol.* 40: 540-544.

Blum, A., and Ebercon, A. 1981. Cell membrane stability as a measure of drought and heat tolerance in wheat. *Crop Sci.* 21: 43-47.

Bond, U., Agell, M., Hass, A.L., Redman, K., and Schlesinger, M.J. 1988. Ubiquitin in stressed chicken embryo fibroblasts. *J. Biol. Chem.* 263: 2384-2388.

Bouslama, M., and Schapaugh, Jr., W.T. 1984. Stress tolerance in soybeans. 1. Evaluation of three screening techniques for heat and drought tolerance. *Crop Sci.* 24: 933-937.

Bowler, C., Van Montagu, M., and Inze, D. 1992. Superoxide dismutase and stress tolerance. *Ann. Rev. Plant Physiol.* 43: 83-116.

Bray, E.A. 1988. Drought- and ABA-induced changes in polypeptide and mRNA accumulation in tomato leaves. *Plant Physiol.* 88: 1210-1214.

Bray, E.A. 1993. Molecular responses to water deficit. *Plant Physiol.* 103: 1035-1041.

Bray, E.A. 1994. Alterations in gene expression in response to water deficit. In A.S. Basra (Ed.), *Stress-induced gene expression in plants* (pp. 1-23). The Netherlands: Harwood Academic Publishers.

Brix, H. 1979. Effects of plant water stress on photosynthesis and survival of four confiers. *Can. J. For. Res.* 9: 160-165.

Brown, E.A., Caviness, C.E., and Brown, D.A. 1985. Response of selected soybean cultivars to soil moisture deficit. *Agron. J.* 77: 278.

Brown, K.W., and Thomas, J.C. 1980. The influence of water stress preconditioning on dark respiration. *Physiol. Plant.* 49: 205-209.

Burke, J.J., and Oliver, M.J. 1992. Differential temperature sensitivity of pea superoxide dismutases. *Plant Physiol.* 100: 1595-1598.

Bush, D.S. 1995. Calcium regulation in plant cells and its role in signaling. *Ann. Rev. Plant Physiol. Plant Mol. Biol.* 46: 95-122.

Cadenas, E. 1989. Biochemistry of oxygen toxicity. *Ann. Rev. Biochem.* 58: 79-110.

Chandler, P.M., and Robertson, M. 1994. Gene expression regulated by abscisic acid and its relation to stress tolerance. *Ann. Rev. Plant Physiol. Plant Mol. Biol.* 45: 113-141.

Chaves, M.M. 1991. Effects of water deficits on carbon assimilation. *J. Exp. Bot.* 42: 1-16.

Chen, R.D., and Tabaeizadeh, Z. 1992. Expression and molecular cloning of drought-induced genes in the wild tomato *Lycopersicon chilense*. *Biochem. Cell Biology.* 70: 199-206.

Clarke, A.K., and Critchley, C. 1994. Characterization of chloroplast heat shock proteins in young leaves of C_4 monocotyledons. *Physiol. Plant.* 92: 118-130.

Close, T.J., Fenton, R.D., Yang, A., Asghar, R., DeMason, D.A., Crone, D.E., Meyer, N.C., and Moonan, F. 1993. Dehydrin: the protein. *Curr. Topics Plant Physiol.* 10: 104-118.

Close, T.J., Kortt, A.A., and Chandler, P.M. 1989. A cDNA-based comparison of dehydration-induced proteins (dehydrins) in barley and corn. *Plant Mol. Biol.* 13: 95-108.

Cooke, A., Cookson, A., and Earnshaw, M.J. 1986. The mechanism of action on calcium in the inhibition on high temperature-induced leakage of betacyanin from beet root discs. *New Phytol.* 102: 491-497.

Cooper, P., and Ho, T.H.D. 1987. Intracellular localization of heat shock proteins in maize. *Plant Physiol.* 84: 1197-1203.

Dasgupta, J., and Bewley, J.D. 1984. Variations in protein synthesis in different regions of greening barley seedlings and effects of imposed water stress. *J. Exp. Bot.* 42: 55-76.

Deal, D.L., Raulston, J.C., and Hinesley, L.E. 1990. High temperature effects on apical bud morphology, dark respiration, and fixed growth of blue spruce. *Can. J. For. Res.* 20: 1871-1877.

Dedio, W. 1975. Water relations in wheat leaves as screening tests for drought resistance. *Can. J. Plant Sci.* 55: 369-378.

Del Longo, O.T., Gonzalez, C.A., Pastori, G.M., and Trippi, V.S. 1993. Antioxidant defenses under hyperoxygenic and hyperosmotic conditions in leaves of two lines of maize with differential sensitivity to drought. *Plant Cell Physiol.* 347: 1023-1028.

Deng, X.M., Joly, R.J., and Hahan, D.T. 1989. Effects of plant water deficit on the daily carbon balance of leaves of cacao seedlings. *Physiol. Plant.* 77: 407-412.

Deng, X.M., Joly, R.J., and Hahan, D.T. 1990. The influence of plant water deficit on distribution of ^{14}C-labeled assimilates in cacao seedlings. *Ann. Bot.* 66: 211-217.

DiMascio, J.A., Sweeeney, P.M., Danneberger, T.K., and Kamalay, J.C. 1994. Analysis of heat shock responses in perennial ryegrass using maize heat shock protein clones. *Crop Sci.* 34: 798-804.

Du, Y.C., and Tachibana, S. 1994. Photosynthesis, photosynthate translocation and metabolism in cucumber roots held at supraoptimal temperature. *J. Jap. Soc. Hort. Sci.* 632: 401-408.

Duff, D.T., and Beard, J.B. 1974. Supraoptimal temperature effects upon *Agrostis palustris*. Part II. Influence on carbohydrate levels, photosynthetic rate, and respiration rate. *Physiol. Plant.* 32: 18-22.

Dwelle, R.B., Kleinkopf, G.E., and Pavek, J.J. 1981. Stomatal conductance and gross photosynthesis of potato (*Solanum tuberosum* L.) as influenced by irradiance, temperature, and growth stage. *Potato Res.* 24: 49-59.

Elstner, E.F. 1982. Oxygen activation and oxygen toxicity. *Ann. Rev. Plant Physiol.* 33: 73-96.

Enami, I., Kitamura, M., Tomo, T., Isokawa, Y., Ohta, H., and Katoh, S. 1994. Is the primary cause of thermal inactivation of oxygen evolution in spinach PS II membrane release of the extrinsic 33 kDa protein or of Mn *Biochim-Biophys-Acta*. 1186: 52-58.

Epron, D. 1997. Effects of drought on photosynthesis and on the thermotolerance of photosystem II in the seedlings of ceder (*Cedrus atlantica* and *C. libani*). *J. Exp. Bot.* 48: 1835-1841.

Escobar-Guti, A.J., Zipperlin, R.B., Carbonne, F., Moing, A., and Gaudill, J.P. 1998. Photosynthesis, carbon partitioning, and metabolite content during drought stress in peach seedlings. *Aust. J. Plant Physiol.* 25: 197-205.

Fabijanski, S., Altosaar, I., and Arnison, P.G. 1987. Heat shock response of *Brassica oleracea* L. (broccoli). *J. Plant Physiol.* 128: 29-38.

Faria, T., Silverio, D., Breia, E., Cabral, R., Abadia, A., Abadia, J., Pereira J.S., and Chaves, M.M. 1998. Differences in the responses of carbon assimilation to summer stress (water deficit, high light, and temperature) in four Mediterranean tree species. *Physiol. Plant.* 102: 419-428.

Fender, S.E., and O'Connell, M.A. 1990. Expression of heat shock response in a tomato interspecific hybrid is not intermediate between the two parental responses. *Plant Physiol.* 93: 1140-1146.

Foyer, C.H., Leandais, M., and Kunert, K.J. 1994. Photooxidative stress in plants. *Physiol. Plant.* 92: 696-717.

Fridovich, I. 1989. Superoxide dismutases. An adaptation to a paramagnetic gas. *J. Biol. Chem.* 264: 7761-7764.

Gauslaa, Y. 1984. Heat resistance and energy budget in different Scandinavian plants. *Hol. Ecol.* 7: 23-28.

Ginzburg, C., and Salomon, R. 1986. The effect of dormancy on the heat shock response in *Gladiolus cormels. Plant Physiol.* 81: 259-267.

Giraudat, J., Parcy, F., Bertauche, N., Gosti, F., Leung, J., Morris, P.C., Bouvier-Durand, M., and Vartanian, N. 1994. Current advances in abscisic acids action and signaling. *Plant Mol. Biol.* 26: 1557-1577.

Gogorcena, Y., Iturbe-ormaetxe, I., Escuredo, P.R., and Becana, M. 1995. Antioxidant defenses against activated oxygen in pea nodules subjected to water stress. *Plant Physiol.* 108: 753-759.

Gomez, J., Sanchez-Martinez, D., Stiefel, V., Rigau, J., Puigdomenech, P., and Pages, M. 1988. A gene induced by the plant hormone abscisic acid in response to water stress encodes a glycine-rich protein. *Nature* 334: 262-264.

Gong, M., Chen, S.N., Song, Y.Q., and Li, Z.G. 1997. Effect of calcium and calmodulin on intrinsic heat tolerance in relation to antioxidant systems in maize seedlings. *Aust. J. Plant Physiol.* 24: 371-379.

Gong, M., Du, C.K., and Xu, W.Z. 1996. Involvement of calcium and calmodulin in the regulation of drought resistance in *Zea mays* seedlings. *Acta Bot. Bor. Occid. Sin.* 16: 214-220.

Gong, M., Van der liut, A.H., Knight, M.R., and Trewavas, A.J. 1998. Heat-shock-induced changes in intracellular Ca^{2+} level in tobacco seedlings in relation to thermotolerance. *Plant Physiol.* 116: 429-437.

Graves, W.R., Joy, R.J., and Dana, M.N. 1991. Water use and growth of honey locust and tree-of-heaven at high root-zone temperature. *HortScience* 26: 1309-1312.

Guerrero, F.D., and Mullet, J.E. 1988. Reduction of turgor induces rapid changes in leaf translatable RNA. *Plant Physiol.* 88: 401-408.

Guicherd, P., Peltier, J.P., Gout, E., Bligny, R., and Marigo, G. 1997. Osmotic adjustment in *Fraxinus excelsior* (L): Malate and mannitol accumulation in leaves under drought conditions. *Trees* 11: 155-161.

Gunaskera, D., and Berkowitz, G.A. 1992. Evaluation of contrasting cellular-level acclimation responses of leaf water deficits in thee wheat genotypes. *Plant Sci.* 86: 1-12.

Halgren, J., Strand, M., and Lundmark, T. 1991. Temperature stress. In A.S. Raghavendra (Ed), *Physiology of trees* (pp. 152-201). John Wiley and Sons, New York.

Hall, A.E. 1992. Breeding for heat tolerance. *Plant Breeding Rev.* 10: 129-168.

Halliwell, B. 1987. Oxidative damage, lipid peroxidation and antioxidant protection in chloroplasts. *Chem. Phys. Lipids.* 44: 327-340.

Han, B., and Kermode, A.R. 1996. Dehydrin-like proteins in bean seeds and seedlings are differentially produced in response to ABA and water-deficit-related stresses. *J. Exp. Bot.* 47: 933-939.

Hanson, A.D., Nelson, C.E., and Everson, E.H. 1977. Evaluation of free proline accumulation as an index of drought resistance using two contrasting barley cultivars. *Crop Sci.* 17: 720-726.

Hanson, A.D., Nelsen, C.E., Pedersen, A.R., and Everson, E.H. 1979. Capacity for proline accumulation during water stress in barley and its implications for breeding for drought resistance. *Crop Sci.* 19: 489-493.

Harding, S.A., Guikema, J.A., and Paulsen, G.M. 1990. Photosynthetic decline from high temperature stress during maturation of wheat. II. Interaction with senescence processes. *Plant Physiol.* 92: 648-653.

Hashem, A., Amin Majumdar, M.N., Hamid, A., and Hossain, M.M. 1998. Drought stress effects on seed yield, yield attributes, growth, cell membrane stability and gas exchange of synthesized *Brassica napus* (L). *J. Agron. Crop Sci.* 180:129-136.

Hays, K.L., Barber, J.F., Kenna, M.P., and McCollum, T.G. 1991. Drought avoidance mechanisms of selected bermudagrass genotypes. *HortScience* 262: 180-182.

Heckathorn, S.A., Downs, C.A., Sharkey, T.D., and Coleman, J.S. 1998. The small, methionine-rich chloroplast heat-shock protein protects photosystem II electron transport during heat stress. *Plant Physiol.* 116: 439-444.

Hendershot, K.L, Weng, J., and Nguyen, H.T. 1992. Induction temperature of heat-shock proteins synthesis in wheat. *Crop Sci.* 32: 256-261.

Howard, H.F., and Wastchke, T.L. 1991. Variable high-temperature tolerance among Kentucky bluegrass cultivars. *Agron. J.* 83: 689-693.

Huang, B., Fry, J.D., and Wang, B. 1998. Water relations and canopy characteristics of tall fescue cultivars during and after drought stress. *HortScience* 335: 837-840.

Huang, B., Liu, X., and Fry, J.D. 1998. Shoot physiological responses of two bentgrass cultivars to high temperature and poor soil aeration. *Crop Sci.* 38: 1219-1224.

Inze, D., and Van Montagu, M. 1995. Oxidative stress in plants. *Curr. Opin. Biotechnol.* 6: 153-158.

Irigoyen, J.J., Emerich, D.W., and Sanchez-Diaz, M. 1992 Alfalfa leaf senescence induced by drought stress: photosynthesis hydrogen peroxide metabolism, lipid peroxidation and ethylene evolution. *Physiol. Plant.* 84: 67-72.

Jacobsen, J.V., and Shaw, D.C. 1989. Heat-stable proteins and abscisic acid action in barley aleurone cells. *Plant Physiol.* 91: 1520-1526.

Jagtap, V., and Bhargava, S. 1995. Variation in antioxidant metabolism of drought tolerant and drought susceptible varieties of *Sorghum bicolor* (L.) Moench. exposed to high light, low water and high temperature stress. *J. Plant Physiol.* 145: 195-197.

Jagtap, V., Bhargava, S., Streb, P., and Feierabend, J. 1998. Compsrstive effect of water, heat, and light stresses on photosynthetic reactions in *Sorghum bicolor* (L). Moench. *J. Exp. Bot.* 49: 1715-1721.

Jones, M.M., Turner, N.C., and Osmond, C.B. 1981. Mechanisms of drought resistance. In L.G. Paley and D. Aspinall (Eds.), *Physiology and biochemistry of drought resistance in plants* (pp. 15-38). Academic Press, Sydney, Australia.

Kaiser, W.M. 1987. Effects of water deficit on photosynthetic capacity. *Physiol. Plant.* 71: 142-149.

Kang, M.S. 1998. Using genotype-by-environment interaction for crop cultivar development. *Adv. Agron.* 62: 199-252.

Kolb, P.F., and Robberecht, R. 1996. High temperature and drought stress effects on survival of *Pinus ponderosa* seedlings. *Tree Physiol.* 16: 665-672.

Krampitz, M.J., Klug, K., and Fock, H.P. 1984. Rates of photosynthetic CO_2 uptake, photorespiratory CO_2 evolution and dark respiration in water-stresses sunflower and bean leaves. *Photosynthetica* 183: 322-328.

Krause, G.H., and Weis, E. 1991. Chlorophyll fluorescence and photosynthesis: The basics. *Ann. Rev. Plant Physiol. Plant Mol. Biol.* 42: 313-349.

Krishnan, M., Nguyen, H.T., and Burke, J.J. 1989. Heat shock protein synthesis and thermal tolerance in wheat. *Plant Physiol.* 90: 140-145.

Kronfub, G., Polle, A., Tausz, M., Havranek, W.M., and Wieser, G. 1998. Effects of ozone and mild drought stress on gas exchange, antioxidants and chloroplast pigments in current-year needles of young Norway spruce [*Picea abies* (L) Karst.]. *Trees* 12: 482-489.

Ku, S.B., Edwards, G.E., and Tanner, C.B. 1977. Effect of light, carbon dioxide, and temperature on photosynthesis, oxygen inhibition of photosynthesis, and transpiration in *Solanum tuberosum*. *Plant Physiol.* 59: 868-872.

Kuo, C.G., Chen, H.M., and Sun, H.C. 1993. Membrane thermostability and heat tolerance of vegetable leaves. In C.G. Kuo (Ed.), *Adaptation of food crops to temperature and water stress* (pp. 160-168). Asian Vegetable Research and Development Center, Shanhua, Taiwan.

Kurganova, L.N., Veselov, A.P., Goncharova, T.A, and Sinitsyna, Y.V. 1997. Lipid peroxidation and antioxidant system of protection against heat shock in pea *(Pisum sativum* L.) chloroplasts. *Russian J. Plant Physiol.* 44: 725-730.

Kuznetsov, V.V., Trofimova, M.S., and Andreev, I.M. 1997. Calcium ions as regulators of heat shock-induced protein synthesis in plant cells. *Dolady-Biological-Sci.* 354: 267-269.

Lafta, A.M., and Lorenzen, J.H. 1995. Effect of high temperature on plant growth and carbohydrate metabolism in potato. *Plant Physiol.* 109: 637-643.

Lawlor, D.W. 1979. Effect of water and heat stress on carbon metabolism of plants with C3 and C4 photosynthesis. In H. Mussel and R.C. Staples (Eds.), *Stress physiology in crop plants* (pp. 304-326). John Wiley and Sons, New York.

Lehman, V.G., and Engelke, M.C. 1993. Heritability of creeping bentgrass shoot water content under soil dehydration and elevated temperature. *Crop Sci.* 33: 1061-1066.

Levitt, J. 1980. *Responses of Plants to Environmental Stresses,* Second edition. Academic Press, New York.

Li, L., and Staden, J.V. 1998. Effects of plant growth regulators on the antioxidant system in callus of two maize cultivars subjected to water stress. *Plant Growth Reg.* 24: 55-66.

Li, X., Feng, Y., and Boersma, L. 1993. Comparison of osmotic adjustment responses to water and temperature stresses in springs wheat and sudangrass. *Ann. Bot.* 71: 303-310.

Lindquist, S. 1986. The heat-shock response. *Ann. Rev. Biochem.* 55: 1151-1191.

Lu., Z.M., Chen, J.W., Percy, R.G., and Zeiger, E. 1997. Photosynthetic rate, stomatal conductance and leaf area in two cotton species and their relation with heat resistance and yield. *Aust. J. Plant Physiol.* 24: 693.

Lusia, C., Sgherri, M., and Navari-Izzo, F. 1995. Sunflower seedlings subjected to increasing water deficit stress: oxidative stress and defense mechanisms. *Physiol. Plant.* 93: 25-30.

Mansfield, M.A., and Key, J.L. 1987. Synthesis of low molecular weight heat shock proteins in plants. *Plant Physiol.* 86: 1240-1246

Markus, V., Lurie, S., Bravdo, B., Stevens, M.A., and Rudice, J. 1981. High temperature effect on RuBP caboxylase and carbonic anhydrase activity in two tomato cultivars. *Physiol. Plant.* 53: 407-412.

Martineau, T.R., Specht, J.E., Williams, J.H., and Sullivan, C.Y. 1979. Temperature tolerance in soybeans. 1. Evaluation of a technique for assessing cellular membrane thermostability. *Crop Sci.* 19: 75-78.

Massacci, A., Battistelli, A., and Loreto, F. 1996. Effect of drought stress on photosynthetic characteristics, growth, and sugar accumulation of field-grown sweet sorghum. *Aust. J. Plant Physiol.* 23: 331-340.

McAinsh, M.R., Clayton, H., Mansfield, T.A., and Hetherington, A.M. 1996. Changes in stomatal behavior and guard cell cytosolic free calcium in response to oxidative stress. *Plant Physiol.* 111: 1031-1042.

McCoy, E.L., Boersma, L., and Ekasingh, M. 1990. Net carbon allocation in soybean seedlings as influenced by soil water stress at two soil temperatures. *Bot. Gaz.* 1514: 497-505.

McKersie, B.D., Hucl P., and Beversdorf, W.D. 1982. Solute leakage from susceptible and tolerant cultivars of Phaseolus vulgaris following ozone exposure. *Can J. Bot.* 60: 73-78.

McKersie, B.D., and Tomes, D.T. 1980. Effect of dehydration treatment on germination, vigor, and cytoplasmic leakage in wild oats and birdsfoot trefoil. *Can. J. Bot.* 58: 471-476.

Mishra, R.K., and Singhal, G.S. 1992. Function of photosynthetic apparatus of intact wheat leaves under high light and heat stress and its relationship with peroxidation of thylakoid lipids. *Plant Physiol.* 98:1-67.

Moffat, J.M., Sears, R.G., and Paulsen, G.M. 1990. Wheat high temperature tolerance during reproductive growth. II. Genetic analysis of chlorophyll fluorescence. *Crop Sci.* 30: 886-889.

Moran, J.F., Becana, M., Iturbe-ormaetxe, I., Frechilla, S., Klucas, R.V., and Aparicio-Tejo, P. 1994. Drought induces oxidative stress in pea plants. *Planta* 194: 346-352.

Morgan, J.M. 1977. Differences in osmoregulation between wheat genotypes. *Nature* 270: 234-235.

Morgan, J.M. 1980. Osmotic adjustment in the spikelets and leaves of wheat. *J. Exp. Bot.* 31: 655-665.

Morgan, J.M. 1984. Osmoregulation and water stress in higher plants. *Ann. Rev. Plant Physiol.* 35: 299-319.

Mukherjee, S.P., and Choudhuri, M.A. 1985. Implication of hydrogen peroxide-ascorbate system on membrane permeability of water stressed *Vigna* seedlings. *New Phytol.* 99: 355-360.

Muller, J., Boller, T., and Wiemken, A. 1996. Pools of non-structure carbohydrates in soybeans root nodules during water stress. *Physiol. Plant.* 98: 723-730.

Nash, T.H., Reiner, A., Demmig-Adams, B., Kilian, E., Kaiser, W.M., and Lange, O.L. 1990. The effect of atmospheric desiccation and osmotic water stress on photosynthesis and dark respiration of lichens. *New Phytol.* 116: 269-276.

Nobel, P.S. 1988. *Environmental biology of agaves and cacti.* Cambridge University Press, Cambridge

Oliver, M.J. 1996. Desiccation tolerance in vegetative plant cells. *Physiol. Plant.* 97: 779-787.

Olsson, M., Nilsson, K., Liljenberg, C., and Hendry, G.A.F. 1996. Drought stress in seedlings: Lipid metabolism and lipid peroxidation during recovery from drought in *Lotus corniculatus* and *Cerastium fontanum. Physiol. Plant.* 96: 577-584.

Oughum, H.J., and Stoddart, J.L. 1986. Synthesis of heat shock proteins and acquisition of thermotolerance in high temperature tolerant and high temperature susceptible lines of sorghum. *Plant Sci.* 44: 163-167.

Ouvrard, O., Cellier, F., Ferrare, K., Tousch, D., Lamaze, T., Dupuis, J.M., and Casse-Delbart, F. 1996. Identification and expression of water stress-and abscisic acid-regulated genes in a drought-tolerant sunflower genotype. *Plant Mol. Biol.* 31: 819-829.

Paakkonen, E., Vahala, J., Pohjolia, M., Holopainen, T., and Karenlampi, L. 1998. Physiological, stomatal and untrastructural ozone responses in birch (*Betula pendula*) are modified by water stress. *Plant Cell Env.* 21: 671-684.

Pankovic, D., Sakac, Z., Kevresan, S., and Plesnicar, M. 1999. Acclimation to long-term water deficit in the leaves of two sunflower hybrids: Photosynthesis, electron transport and carbon metabolism. *J. Exp. Bot.* 50: 127-138.

Pastori, G.M., and Trippi, V.S. 1992. Oxidative stress induces high rate glutathione reductase synthesis in a drought-resistant maize strain. *Plant Cell Physiol.* 337: 957-961.

Pelah, D., Shoseyov, O., and Altman, A. 1995. Characterization of BspA, a major boiling-stable, water-stress-responsive protein in aspen (*Populus tremula*). *Tree Physiol.* 15: 673-678.

Pelham, H.R.B. 1988. Heat shock proteins: Coming in from the cold. *Nature* 332: 776-777.

Perdomo, P., Murphy, J.A., and Berkowitz, G.A. 1996. Physiological changes associated with performance of Kentucky bluegrass cultivars during summer stress. *HortScience* 317: 1182-1186.

Perez-Molphe-Balch, E., Gidekel, M., Segura-Nieto, M., Herrera-Estrella, L., and Ochoa-Alejo, N. 1996. Effects of water stress on plant growth and root protein in three cultivars of rice (*Oryza sativa*) with different levels of drought tolerance. *Physiol. Plant.* 96: 284-290.

Prange, R.K., McRae, K.B, Midmore, D.J., and Deng, R. 1990. Reduction in potato growth at high temperature; role of photosynthesis and dark respiration. *Am. Potato J.* 67: 357-369.

Premachandara, G.S., Saneoka, H., and Ogata, S. 1989. Nutrio-physiological evaluation of polyethylene glycol test of cell membrane stability in maize. *Crop Sci.* 29: 1287-1292.

Premachandra, G.S., Saneoka, H., Fujita, H., and Ogata, S. 1992. Leaf water relations, osmotic adjustment, cell membrane stability, epicuticular wax load and growth as affected by increasing water deficits in sorghum. *J. Exp. Bot.* 43: 1569-1576.

Premachandra, G.S., and Shimada, T. 1987. The measurement of cell membrane stability using polyethylene glycol as a drought tolerance test in wheat. *Japan. J. Crop Sci.* 56: 92-98.

Price, A.H., Taylor, A., Ripley, S.J., Driffiths, A., Trewavas, A.J., and Knight, M.R. 1994. Oxidative signals in tobacco increase cytosolic calcium. *Plant Cell* 6:1301-1310.

Ranney, T.G., Bassuk, N.L., and Whitlow, T.H. 1991. Osmotic adjustment and solute constituents in leaves and roots of water-stressed cherry (*Prunus*) trees. *J. Am. Soc. Hort. Sci.* 1164: 684-688.

Ranney, T.G., Blazich, F.A., and Warren, S.T. 1995. Heat tolerance of selected species and populations of *Rhododendron*. *J. Am. Soc. Hort. Sci.* 1203: 423-428.

Ranney, T.G., and Ruter, J.M. 1997. Foliar heat tolerance of three holly species: Responses of chlorophyll fluorescence and leaf gas exchange to supraoptimal leaf temperature. *J. Am. Soc. Hort. Sci.* 1224: 499-503.

Rawat, A.S., and Purohit, A.N. 1991. CO_2 and water vapor exchange in four alpine herbs at two altitudes and under varying light and temperature conditions. *Photosyn. Res.* 28: 99-108.

Rekika, D., Nachit, M.M., Araus, J.L., and Monneveux, P. 1998. Effects of water deficit on photosynthetic rate and osmotic adjustment in tetraploid wheats. *Photosynthetica* 35: 129-138.

Riccardi, F., Gazeau, P., Vienne, D., and Zivy, M. 1998. Protein changes in response to progressive water deficit in maize. *Plant Physiol.* 117: 1253-1263.

Ristic, Z., Gifford, D.J, and Cass, D.D. 1991. Heat shock proteins in two lines of *Zea mays* L. that differ in drought and heat resistance. *Plant Physiol.* 97: 1430-1434.

Ristic, Z., Williams, G., Yang, G., Martin, B., and Fullerton, S. 1996. Dehydration, damage to cellular membranes, and heat shock proteins in maize hybrids from different climates. *J. Plant Physiol.* 149: 424-432.

Rodriguez-Garay, B., and Barrow, J.R. 1988. Pollen selection for heat tolerance in cotton. *Crop Sci.* 28: 857-859.

Ruter, J.M., and Ingram, D.L. 1990. [14]Carbon-labeled photosynthate partitioning in *Ilex crenata rotundifolia* at supraoptimal root-zone temperature. *J. Am. Soc. Hort. Sci.* 116: 1008-1013.

Saadalla, M.M., Shanahan, J.F., and Quick, J.S. 1990. Heat tolerance in winter wheat. I. Hardening and genetic effects on membrane thermostability. *Crop Sci.* 30: 1243-1247.

Sairam, P.K., Deshmukh, P.S., and Shukla, D.S. 1997. Tolerance of drought and temperature stress in relation to increases antioxidant enzymes in wheat. *J. Agron. Crop Sci.* 178: 171-178.

Sairam, P.K., Shukla, D.S., and Saxena, D.C. 1998. Stress induced injury and antioxidant enzymes in relation to drought tolerance in wheat genotypes. *Biol. Plant.* 40 3: 357-364.

Sapra, V.T., and Anaele, A.O. 1991. Screening soybean genotypes for drought and heat tolerance. *J. Agron. Crop Sci.* 167: 96-102.

Sammons, D.J., Peters, D.B., and Hymovitz, T. 1978. Screening soybeans for drought resistance. 1. Growth chamber procedure. *Crop Sci.* 18: 1050-1055.

Savin, R., and Nicolas, M.E. 1996. Effects of short periods of drought and high temperature on drain growth and starch accumulation of two malting barley cultivars. *Aust. J. Plant Physiol.* 23: 201-210.

Schultz, H.R., and Matthews, M.A. 1997. High vapor pressure deficits exacerbates xylem cavitation and photoinhibition in shade-grown *Piper auritum* H.B. and K. during prolonged sun flecks. I. Dynamics of plant water relations. *Oecologia* 110: 312-319.

Schuster, G., Even, D., Kloppstech, K., and Ohad, I. 1988. Evidence for protection by heat shock proteins against photoinhibition during heat shock. *EMBO J.* 7: 1-6.

Sepulveda, G., Kliewer, W.M., and Ryugo, K. 1986. Effect of high temperature on grapevines *Vitis vinifera* (L.). I. Translocation of ^{14}C-photosynthates. *Am. J. Ecol. Viticulture* 37: 13-19.

Shah, N.H. 1992. Responses of wheat to combined high temperature and drought or osmotic stresses during maturation. *Dissertation Abstract International -B- Science and Engineering* 52: 3984B.

Singh, T.N., Aspinall, D., and Paleg, L.G. 1972. Proline accumulation and varietal adaptability to drought in barley: A potential metabolic measure of drought resistance. *Nature New Biol.* 236: 188-190.

Singh, T.N., Aspinall, D., and Paleg, L.G. 1973. Stress metabolism. III. Variations in response to water deficit in the barley plant. *Aust. J. Biol. Sci.* 26: 65-76.

Slatyer, R.O. 1963. Climatic control of plant-water relations. In L.T. Evans (Ed.), *Environmental control of plant growth* (pp. 34-35). Academic Press, New York.

Smillie, R.M., and Hetherington, S.E. 1990. Screening for stress tolerance by chlorophyll fluorescence. In Y. Hashimoto, P.J. Kramer, H. Nonami, and B.R. Strain (Eds.), *Measurement techniques in plant science* (pp. 229-261). Academic Press, San Diego.

Smith, M.A.L, Spomer, A.L., and Skiles, E.S. 1989. Cell osmolarity adjustment in *Lycopersicon* in response to stress pretreatments. *J. Plant Nutr.* 12: 233-244.

Srinivasan, A., Takeda, H., and Senboku, T. 1996. Heat tolerance in food legumes as evaluated by cell membrane thermostability and chlorophyll fluorescence techniques. *Euphytica* 88: 35-45.

Stapel, D., Kruse, E., and Kloppstech, K. 1993. The protective effect of heat shock proteins against photoinhibition under heat shock in barley (*Hordeum vulgare*). *J. Photochem. Photobiol.* B21: 211-218.

Subrahmanyam, D., and Rathore, V.S. 1995. High temperature influences $^{14}CO_2$ assimilation and allocation of ^{14}C into different biochemical fractions in the leaves of Indian mustard. *J. Agron. Crop Sci.* 174: 319-323.

Sullivan, C.Y. 1972. Mechanisms of heat and drought resistance in grain sorghum and methods of measurement. In N.G.P. Rao and L.R. House (Eds.), *Sorghum in the seventies* (pp. 247-264). Oxford and IBH Publishing, New Delhi, India.

Sullivan, C.Y., and Ross, W.M. 1979. Selecting for drought and heat resistance in grain sorghum. In H. Mussell and R. Staples (Eds.), *Stress physiology in crop plants* (pp. 263-281). John Wiley and Sons, New York.

Sung, F.J.M., and Krieg, D.R. 1979. Relative sensitive of photosynthetic assimilation and translocation of 14C to water stress. *Plant Physiol.* 64: 852-856.

Taiz, L., and Zeiger, E. 1991. *Plant physiology.* The Benjamin/Cummings Publishing Co., Inc., Redwood City, CA.

Tajima, K., Akita, S., and Shimizu, N. 1976. Effect of high soil temperature on growth and water balance of tall fescue and perennial ryegrass. *J. Japan. Soc. Grassl. Sci.* 22: 256-260.

Takeuchi, T.S., and Thornber, J.P. 1994. Heat-induced alterations in thylakoid membrane protein composition in barley. *Aust. J. Plant Physiol.* 21: 759-770.

Tan, W.X., Blake, T.J., and Boyle, T.J.B. 1992. Drought tolerance in faster- and slower-growing black spruce (*Picea mariana*) progenies: II. Osmotic adjustment and changes of soluble carbohydrate and amino acids under osmotic stress. *Physiol. Plant.* 85: 645-651.

Trofimova, M.S., Andreev, I.M., and Kuznetsov, V.V. 1997. Calcium as an intracellular regulator of HSP96 synthesis and plant cell tolerance to high temperature. *Russian J. Plant Physiol.* 44: 511-516.

Velazquez, J.M., and Lindquist, S. 1984. Hsp70: nuclear concentration during environmental stress and cytoplasmic storage during recovery. *Cell* 36: 655-662.

Vierling, E. 1991. The roles of heat shock proteins in plants. *Ann. Rev. Plant Physiol. Plant Mol. Biol.* 42: 579-620.

Volaire, F., Thomas, H., Bertagne, N., Bourgeois, E., Gautier, M.F., and Lelievre, F. 1998. Survival and recovery of perennial forage grasses under prolonged Mediterranean drought. II. Water status, solute accumulation, abscisic acid concentration and accumulation of dehydrins transcripts in bases of immature leaves. *New Phytol.* 140: 451-460.

Volaire, F., Thomas, H., and Lelievre, F. 1998. Survival and recovery of perennial forage grasses under prolonged Mediterranean drought. I. Growth, death, water relations and solute content in herbage and stubble. *New Phytol.* 140: 439-449.

Wample, R.L., and Thornton, R.K. 1984. Differences in the response of sunflower (*Helianthus annuus*) subjected to flooding and drought stress. *Physiol. Plant.* 61: 611-616.

Wang, Z., Quebedeaux, B., and Stutte, G.W. 1995. Osmotic adjustment: Effect of water stress on carbohydrate in leaves, stems, and roots of apple. *Aust. J. Plant Physiol.* 22: 747-754.

Wang, Z., Quebedeaux, B., and Stutte, G.W. 1996. Partitioning of ^{14}C-glucose into sorbitol and other carbohydrates in apple under water stress. *Aust. J. Plant Physiol.* 23: 245-251.

Webb, A.A.R., McAinsh, M.R., Taylor, J.E., and Hetherington, A.M. 1996. Calcium ions as intercellular second messengers in higher plants. *Adv. Bot. Res.* 22: 45-96.

Wehner, D.J., and Watschke, T.L. 1984. Heat stress effects on protein synthesis and exosmosis of cell solutes in three turfgrass species. *Agron. J.* 76: 16-19.

Weidner, M., and Ziemans, C. 1975. Preadaption of protein synthesis in wheat seedlings to high temperature. *Plant Physiol.* 56: 590-594.

Weis, E., and Berry, J.A. 1988. Plants and high temperature stress. In S.P. Long and F.I. Woodward (Eds.), *Plants and temperature* (pp. 329-346). Company of Biologists Ltd, Cambridge, UK.

White, R.H., Engelke, M.C., Morton, S.J., Johnson-Cicalese, J.M., and Ruemmele, B.A. 1992. *Acremonium* endophyte effects on tall fescue drought tolerance. *Crop Sci.* 32: 1392-1396.

Wibbe, M.L., and Blanke, M.M. 1997. Effect of fruiting and drought flooding on carbon balance of apple trees. *Photosynthetica* 33: 269-275.

Wilson, J.M., and Greaves, J.A. 1990. Assessment of chilling injury by chlorophyll fluorescence analysis. In C.Y. Wang (Ed.), *Chilling injury to horticultural crops* (pp. 129-141). CRC Press, Boca Raton, FL.

Wolf, S., Marani, A., and Rudich, J. 1991. Effect of high temperature on carbohydrate metabolism in potato plants. *J. Exp. Bot.* 42: 619-625.

Wolf, S., Olesinski, A.A., Rudich, J., and Marani, A. 1990. Effect of high temperature on photosynthesis in potatoes. *Ann. Bot.* 65: 179-185.

Wright, G.C., Smith, R.C.G., and Morgan, J.M. 1983. Differences between two grain sorghum genotypes in adaptation to drought stress. III. Physiological responses. *Aust. J. Agric. Res.* 34: 637-651.

Wullschleger, S.D., and Oosterhuis, D.M. 1990. Photosynthetic and respiration activity of fruiting forms within the cotton canopy. *Plant Physiol.* 94: 463-469.

Yamane, Y., Kashino, Y., Koike, H., and Satoh, K. 1998. Effects of high temperature on the photosynthetic systems in spinach: oxygen-evolving activities, fluorescence characteristics and the denaturation process. *Photosynthesis Res.* 57: 51-59.

Yang, G.P., Gao, A.L., and Jing, J.H. 1993. The relation of calcium to cell in water stressed soybean hypocotyls. *Plant Physiol. Comm.* 29: 179-181.

Yordanov, I., Tsonev, T., Goltsev, V., Kruleva, L., and Velikova, V. 1997. Interactive effect of water deficit and high temperature on photosynthesis of sunflower and maize plants. 1. Changes in parameters of chlorophyll fluorescence induction kinetics and fluorescence quenching. *Photosynthetica* 33: 391-402.

Zemanek, M., and Frecer, R. 1990. Effect of high temperature on sucrose accumulation in the grain of winter wheat genotypes. *Rostlinna Vyroba* 36: 965-976.

Zhang, B.L., and Archbold, D.D. 1993. Solute accumulation in leaves of a *Fragaria chiloensis* and a *F. virginiana* selection responds to water deficit stress. *J. Am. Soc. Hort. Sci.* 118: 280-285.

Zhang, J.X., and Kirkham, M.B. 1994. Drought-stress-induced changes in activities of superoxide dismutase, catalase, and peroxidase in wheat species. *Plant Cell Physiol.* 35: 785-791.

Zhang, J.X., and Kirkham, M.B. 1996. Antioxidant responses to drought in sunflower and sorghum seedlings. *New Phytol.* 132: 361-373.

Zheng, Y.Z., and Han, Y.H. 1997. Effect of high temperature and/or drought stress on the activity of SOD and POD of intact leaves in two soybean (*G. max*) cultivars. *Soybean Genetics Newsletter.* 24:39.

Zrenner, R., and Stitt, M. 1991. Comparison of the effect of rapidly and gradually developing water-stress on carbohydrate metabolism in spinach leaves. *Plant Cell Env.* 14: 939-946.

Chapter 11

Stability Analysis in Crop Performance Evaluation

Hans-Peter Piepho
Fred A. van Eeuwijk

INTRODUCTION

Stability analysis is an area of research with a constantly growing number of different methods and approaches. For the practitioner, it is often difficult to choose a statistically appropriate method of analysis for his or her data set. Recently, it has been pointed out that different approaches to stability analysis can be cast into a unifying mixed modeling framework (Denis et al., 1997; Piepho, 1998a, 1999). Within such a framework, different measures of stability can be derived as functions of variance components of a specific mixed model. The choice of the class of candidate mixed models has direct implications for the choice of stability measure. We will illustrate mixed model building and subsequent stability analysis using a worked example of a real life data set. The motivation for this focus is (1) that there are already several general reviews on stability analysis (Lin et al., 1986; Becker and Léon, 1988; Piepho, 1998a,b) and (2) that most of these reviews contain no worked examples or just a small data set for demonstration. To convince the practitioner of the advantages of using statistical methods for stability analysis and to highlight the possibilities such analyses can offer, a worked realistic example seems appropriate. This example will demonstrate

We are grateful to the CIMMYT Regional Office for Latin America and INIA-Uruguay for allowing us to use their data. We are especially grateful to Sergio Ceretta (INIA-Uruguay) for compiling the data we have used and help with interpretations. The first author thanks Hugh Gauch Jr. (Department of Plant Breeding, College of Agriculture and Life Sciences, Cornell University, Ithaca, NY, USA) for very inspiring discussions on the subject of stability. Support of the first author by the Heisenberg Program of the Deutsche Forschungsgemeinschaft (DFG) is thankfully acknowledged.

that the proposed methods are relatively straightforward to apply in practice, provided software for mixed-model analysis is available.

In the analyses presented here, mixed models will be used, in which environments (years, locations) are considered as a random factor, while genotypes are a fixed factor. Such models will be shown to be useful for assessing the risk of a genotype not reaching a threshold yield and the risk for one genotype to be outperformed by another genotype. A necessary condition for an adequate evaluation of stability in terms of risk is a satisfactory model for the variance-covariance structure of the data in a mixed-model framework. Some candidate mixed models contain genotype-specific variance parameters that invite to direct but potentially misleading interpretations as stability measures (Piepho, 1998a). We, therefore, prefer to look at the implication of models for the stability of observed yield rather than inspecting individual variance components, which pertain to unobservable quantities and may be difficult to interpret as such. This perspective will involve computation of quantities from estimated model parameters, which assess stability of yield, like the risk of not attaining a threshold level of yield or the risk of being out-competed by another genotype.

The analysis comprises three consecutive steps: (1) fitting different mixed models using restricted maximum likelihood (REML) (Gilmour et al., 1995, 1999; Littell et al., 1996); (2) selecting an appropriate model from the set of fitted models; (3) drawing conclusions [interpreting variance components as stability measures, if possible (this will not be possible for many models), computing derived stability measures, e.g., the probability of outperforming a check (Eskridge, 1996)]. The number of models that can be fitted may be too large for an exhaustive search of the best fitting model. It is then useful to evaluate the initial fits for simple models to conjecture the most promising sequence of models yet to be fitted. This requires an integration of steps (1) and (2).

THE DATA

We use data from the ERCOS (Ensayos Regionales Cono Sur) regional variety trials for the south cone of South America conducted from 1975 to 1986. On average, 30 to 45 genotypes from the participating breeding programs were tested each year at several locations in seven countries: Argentina, Brazil, Bolivia, Chile, Mexico, Paraguay, and Uruguay. Fifteen genotypes with a large number of year × location combinations were chosen for stability analysis. The total number of locations for this subset was 25. Data were highly unbalanced, since newly developed cultivars entered the network each year, while those that performed badly after one or more seasons were dropped in subsequent trials. The numbers of genotypes tested in different year × location combinations are shown in Table 11.1. Only 839 of the 4500 year × location × genotype combinations in the three-way table were observed. Table 11.2 contains some descriptive information on the genotypes. The range of observed yields (measured in deci-tons per hectare: dt/ha) is

TABLE 11.1. Number of Genotypes for Different Location × Year Combinations

Location (country*)	1975	1976	1977	1978	1979	1980	1981	1982	1983	1984	1985	1986	Total
1 Bal (AR)	0	0	0	0	0	0	7	8	10	8	6	6	45
2 Bge (BR)	7	7	7	5	6	0	0	0	0	0	0	0	32
3 Bor (AR)	0	0	0	0	0	0	0	8	0	0	0	0	8
4 Bsi (BR)	0	0	0	0	0	0	0	8	10	8	0	0	26
5 Caa (PY	0	0	0	5	0	7	7	8	0	8	0	0	35
6 Chl (PY)	0	7	7	5	6	7	7	8	0	8	6	6	67
7 Chn (CH)	7	0	0	0	0	0	0	0	0	0	0	0	7
8 CMi (BO)	0	0	0	0	0	0	0	0	0	8	6	0	14
9 Dou (BR)	0	0	0	0	0	7	0	7	0	0	0	0	14
10 LEs (UY)	7	0	7	5	6	7	7	8	10	8	6	0	71
11 LGr (CH)	0	0	7	5	6	7	7	8	10	8	6	6	70
12 LPt (CH)	7	0	0	0	0	0	0	0	0	0	0	0	7
13 MJu (AR)	0	0	0	0	0	0	7	8	10	8	0	0	33
14 Obr (MX)	0	7	7	5	6	6	0	0	0	0	0	6	37
15 Pal (BR)	0	0	0	0	0	0	7	7	9	8	6	6	43
16 PFu (BR)	7	7	7	5	6	7	7	8	10	8	6	6	84
17 Pgi (PY)	0	7	0	0	0	0	0	0	0	0	0	0	7
18 Prg (AR)	7	0	7	5	6	7	7	8	10	8	6	6	77
19 SBe (BO)	7	0	7	0	6	0	0	0	0	0	0	6	26
20 SBj (BR)	7	7	7	5	6	7	7	8	10	0	0	6	70
21 SCz (BO)	7	0	0	0	0	0	0	0	0	0	0	0	7
22 Sgo (CH)	0	7	0	0	0	0	0	0	0	0	0	0	7
23 Son (MX)	7	0	0	0	0	0	0	0	0	0	0	0	7
24 Tmu (CH)	7	0	0	0	0	0	0	0	0	8	6	6	27
25 Yng (UY)	0	0	0	0	0	0	0	0	10	8	0	0	18
Total	77	49	63	45	54	62	70	102	99	104	54	60	839

Note: *AR=Argentina, BO=Bolivia, BR=Brazil, CH=Chile, MX=Mexico, PY=Paraguay, UY=Uruguay.

TABLE 11.2. Number of Observations, Minimum and Maximum for Yields (dt/ha) of Different Genotypes

Genotype	Number of observations	Minimum	Maximum	Cultivar name
1	42	4.40	90.80	Alondra 4546
2	42	7.60	74.20	Chasqui Inia
3	41	2.60	66.00	CNT 9
4	118	2.20	85.80	Diamante Inta
5	65	2.80	66.90	E. Hornero
6	44	2.00	64.50	E. Tarariras
7	100	1.50	74.20	IAS 54
8	36	1.90	70.40	Itapua 5
9	100	1.70	44.20	Jacui
10	36	4.10	63.30	Leones Inta
11	39	1.30	66.10	LE 1787
12	41	3.74	93.10	Millaleu Inia
13	42	5.41	55.06	Minuano 82
14	42	5.96	74.93	Onda Inia
15	51	1.30	72.50	281/60

very wide for all genotypes, and the number of observations differs markedly. There were 119 year × location combinations. We found that the square root transformation dramatically improved the fits of our models. Thus, all analyses are presented on this scale. Details of how the adequacy of the square-root transformation was verified are reported in Piepho and McCulloch (1999).

MODEL SELECTION

A mixed model has two important components: the expectation structure and the variance-covariance structure. Both components require an explicit model, and usually there will be several candidate models to choose from. A simple model for the expectation structure just assigns a separate expected value (mean) to each genotype. In this contribution, we will consider only this simple case. In addition, covariate information on the genotypes and the environments can be used, if available, to model the expectation structure. For details, see van Eeuwijk et al. (1996), Cullis et al. (1996), Denis et al. (1997) and Piepho et al. (1998a).

As will be demonstrated in the next section, there are a number of options for modeling the variance-covariance structure. Having fitted several different mixed models, the task then is to select an appropriate model among the ones fitted. There are several strategies for model selection. Subject matter knowledge is undoubtedly the most important guide. In addition, statistical tools are available, e.g., graphical methods and diagnostics (Christensen, Pearson, and Johnson, 1992) and likelihood-based methods (Diggle, 1988; Oman, 1991; Wolfinger, 1993). We will mainly use the latter. Provided that the expectation structure is the same and the models are estimated by REML, nested models may be compared, e.g., by a likelihood-ratio (LR) test. If models are estimated by maximum likelihood (ML), nested models can be compared by LR tests also when the expectation structure differs. In this chapter, we will only use the REML method. The LR test based on REML requires computation of the restricted log-likelihood (LL_R) for each model evaluated using the REML estimates of parameters. The LR statistic is

$$T = -2\{LL_R(\text{reduced model}) - LL_R(\text{full model})\}$$

Under the null hypothesis that the reduced model is not different from the full model, the LR statistic is distributed as χ^2 with degrees of freedom equal to the difference in the number of parameters of both models, provided the data follow a multivariate normal distribution. Because the fixed part is the same for all models, only the number of parameters in the variance-covariance structure (p) needs to be considered:

$$df = p(\text{full model}) - p(\text{reduced model})$$

For the ordinary LR test just described, the models must be nested, i.e., for the reduced model, none of the parameters of the full model may be placed at the boundary of the parameter space. If the model comparison implies a null hypothesis that a variance component is zero, the ordinary LR test is not applicable. Work by Stram and Lee (1994, case 1) suggests that comparing T to a χ^2 distribution with one df provides a conservative test for testing the null hypothesis of a variance component being zero. The appropriate asymptotic null distribution is a 50:50 mixture of

$$\chi_1^2 \text{ and } \chi_0^2.$$

Thus, one may compute the *p*-value relative to a

$$\chi_1^2$$

distribution and divide it by two.

An alternative likelihood-based mixed model selection strategy was proposed by Wolfinger (1996; also see SAS Institute, 1997), who suggested to use information criteria such as the Akaike's Information Criterion (*AIC*) and the Schwarz Bayesian Criterion (*SBC*):

$$AIC = LL_{R} - p$$
$$SBC = LL_{R} - \frac{1}{2}p \log N^*$$

where N^* is the total number of observations minus the number of fixed effects (number of genotypes in our case) and p is the number of parameters in the variance-covariance structure. *AIC* and *SBC* involve a penalty for the number of parameters, which tends to favor parsimonious models. The larger the value of the criterion, the more preferable is the model (Wolfinger, 1993, 1996). An advantage of these criteria is that they allow a straightforward comparison of nonnested models. Moreover, these criteria reduce the problem of overfitting, which can be very serious with LR testing (Gelfand and Gosh, 1998). Also, one need not consider a number of model fitting sequences as may be necessary with LR tests for complex models. We therefore believe that information criteria can be preferable to significance testing as a tool for model choice. For a good general introduction to model selection by information-theoretic criteria, see Burnham and Anderson (1998). Apart from a constant, *AIC* is a consistent estimator of the so-called Kullback-Leibler discrepancy between the distribution that generated the data and the model that approximates it. The philosophy behind *AIC* is that the true model is high dimensional, and that a working model is sought that best approximates the true model without any attempt being made to find the true model. The dimension of the model selected by *AIC* will be low if there are few data and will increase as more information becomes available. By contrast, *SBC* provides a consistent estimate of the true order of the model, assuming that a true model exists and is low dimensional (Buckland et al., 1997; Burnham and Anderson, 1998). In small samples, *AIC* tends to overfit (McQuarrie and Tsai, 1998). In this contribution, we will mainly use the *SBC*, since our assumed models are conceptually simple, with all potential causal factors influencing among-environment variation subsumed under the variance-covariance structure. The results for *AIC* will be reported for comparison. We should point out that *SBC* was developed assuming a linear regression model with only one error term (Schwarz, 1978). It is not clear that *SBC* has the same properties in the mixed model case, so the *SBC* values should not be overinterpreted. For example, in a multivariate setting, the penalty term involves the number of subjects and not the number of observations (McQuarrie and Tsai, 1998). It is not clear how to adapt *SBC* in the general mixed model case. Thus, our use of *SBC* here is of an ad hoc nature.

As will be shown, there are a large number of candidate models to consider, and it is usually not feasible to fit all of them, especially when considering three-way models. We therefore suggest a pragmatic approach based on *SBC*, which avoids fitting a large number of models. We suggest one start with the simplest model(s), gradually making the components of the model more complex. Each time a new model is fitted, the improvement is critically assessed. Based on previous fits, one may be able to guess the most promising route for further improvement. This strategy is rather informal and will require a little intuition.

TWO-WAY ANALYSIS

We first give an analysis in which year × location combinations are regarded as independent environments, allowing us to analyze a two-way table of genotypes × environment interactions. Ignoring the factorial structure of years × locations in the environments constitutes a considerable simplification. Most notably, the independence of environments is questionable. For example, observations made in the same location across different years are likely to show some covariation/correlation due to the fact that the same location is involved. This type of correlation is ignored in the two-way analysis, where, e.g., observations made at the same location but in different years are treated as independent. Usefulness and appropriateness of this simplification will be discussed after the three-way analysis.

Models

We will now introduce a few commonly used models, giving a shorthand notation for later reference. The simplest model for two-way data is

$$y_{ij} = \mu_i + u_j + f_{ij} \qquad (1)$$

where y_{ij} = yield of i-th genotype in j-th environment ($j = 1, \ldots, J$), μ_i = mean yield of i-th genotype($i=1, \ldots$ I), u_j = main effect of j-th environment and f_{ij} = residual corresponding to y_{ij}. The residual f_{ij} comprises both genotype × environment interaction and experimental error. In analysis of variance (ANOVA), it is assumed that u_j and f_{ij} are independent normal deviates with zero mean and variances

$$\sigma_u^2 \text{ and } \sigma^2$$

respectively. The resulting variance-covariance structure is sometimes labeled compound symmetry (CS). The model is restrictive in that the variance of an observation does not vary among genotypes. To relax this assumption, Shukla (1972) proposed to assume that the variance of the residual f_{ij} depends on the genotype (SH). The "stability variance,"

$$\mathrm{var}\left(f_{ij}\right) = \sigma_i^2$$

is a measure of stability of the i-th genotype, and it is closely related to Wricke's (1962) "ecovalence."

Yates and Cochran (1938) and Finlay and Wilkinson (1963) suggested regression of yields on the environmental mean. This corresponds to the model

$$y_{ij} = \mu_i + \lambda_i w_j + f_{ij} \qquad (2)$$

where λ_i is the regression coefficient of the i-th genotype and w_j is a latent environmental variable. The regression approach was first proposed in a fixed effects model context, where a regression on the environmental mean is the easiest way to estimate the model parameters. Here, we will look at the regression model from a mixed-model perspective, in which (2) induces a specific variance-covariance structure (Oman, 1991; Gogel et al., 1995). It can be assumed without loss of generality that w_j has unit variance and is independent of f_{ij} (Oman, 1991). Finlay and Wilkinson (1963) assumed that f_{ij} has constant variance (FW), while Eberhart and Russell (1966) and Shukla (1972) proposed a model with genotype-specific variances of f_{ij} (FW-H; we could denote this model as ER, but use FW-H instead to emphasize the variance heterogeneity and to be consistent with the notation for the AMMI models described below). The regression parameter λ_i and the variance are often interpreted as measures of stability. For a critical discussion, see Piepho (1998a). Model (2) has one multiplicative term. We may add more terms,

$$y_{ij} = \mu_i + \lambda_{1i} w_{1j} + \lambda_{2i} w_{2j} + \ldots + f_{ij} \qquad (3)$$

assuming the environmental scores are mutually independent. To identify all parameters, the constraint $\lambda_{ri} = 0$ for $r > i$ is imposed (SAS Institute, 1997; Piepho, 1997). Depending on the number of multiplicative terms and the assumptions about the residual variance, these models will be referred to as FWn and FWn-H, respectively, where n is the number of multiplicative terms. Finally, an environmental main effect may be added to (3):

$$y_{ij} = \mu_i + u_j + \lambda_{1i} w_{1j} + \lambda_{1i} w_{2j} + \ldots + f_{ij} \qquad (4)$$

The model given in (4) was denoted by Gauch (1988) as Additive Main Effects Multiplicative Interaction (AMMI) model within a fixed effects framework. The fixed effects AMMI model is a very useful tool for understanding and interpreting genotype-environment interaction (Gauch, 1988). Piepho (1997) first discussed these models from a mixed-model perspective. In this perspective, the AMMI model additionally offers a flexible set of models for the variance-covariance structure. To indicate the number of multiplicative terms, the models will be referred to here as AMMI1, AMMI2, etc. If the residual variance is heterogeneous, the models are denoted as AMMI1-H, AMMI2-H, etc.

It is instructive to study the implications of the different models for the variances and covariances of observations y_{ij} made in the same environment. Under the ANOVA model (CS), the variances of all genotypes are equal. Also, all pairwise covariances, and hence the correlations, are equal. By contrast, the models FWn, FWn-H, AMMIn, and AMMIn-H give rise to various forms of heterogeneity among the variances and covariances. Such heterogeneity is quite common in multienvironment trials (MET). Specifically, variances often differ, i.e., the variability (stability) of yields across environments depends on the genotype. Also, the responses of some pairs of genotypes are often more alike than those of other pairs. For example, genotypes carrying a resistance gene are more alike than those lacking this gene. Such a response pattern implies heterogeneity of covariances. It is, therefore, not usually appropriate to limit analyses of METs to the ANOVA (CS) model, which has the most restrictive variance-covariance structure. The other models considered here do allow heterogeneity, although they impose a certain structure. The most flexible variance-covariance model, which has not been mentioned yet, imposes no structure at all and allows all variances and covariances to vary freely (apart from the requirement of positive-definiteness of the variance-covariance matrix V; see following equation). This model is henceforth abbreviated as UN. The UN model is certainly the most realistic one in a majority of cases and should be the benchmark against which usefulness of simpler models may be compared. Its main disadvantage is the large number of parameters. In small data sets and/or when the number of genotypes is large relative to the number of environments, it may not even be possible to estimate this model. The parsimony principle suggests that simpler models are often preferable. We think that the main practical value of models such as SH, FW, FW-H, AMMI, and AMMI-H lies in the ability to describe actual variance-covariance structures more accurately than the restrictive CS model while avoiding the overparameterization of the UN model (Diggle, 1988).

For further discussion, it is helpful to state the variance-covariance structure in matrix form. For this purpose, yields observed in the same environment j are collected into a vector $y_j = (y_{1j}, \ldots, y_{Ij})'$. The variance-covariance matrix of observations in y_j is expressed as

$$V = \text{var}(y_j) = \{\sigma_{ii'}\} = \begin{bmatrix} \sigma_{11} & \sigma_{12} & \cdot & \cdot & \sigma_{1I} \\ \sigma_{21} & \cdot & & & \cdot \\ \cdot & & \cdot & & \cdot \\ \cdot & & & \cdot & \\ \sigma_{I1} & \cdot & \cdot & \cdot & \sigma_{II} \end{bmatrix} \quad (5)$$

where $\sigma_{ii'}$ is the covariance of genotypes i and i', while σ_{ii} is the variance of the i-th genotype. Table 11.3 contains the models for V in matrix notation. For comparison with the three-way models considered later, it is helpful to express all two-way models in a general form as

$$y_{ij} = \mu_i + e_{ij} \quad (6)$$

or in vector form as

$$y_j = \mu + e_j \quad (7)$$

where $\mu = (\mu_1, \ldots, \mu_I)'$, $e_j = (e_{1j}, \ldots, e_{Ij})'$, and $\text{var}(y_j) = \text{var}(e_j) = V$ with structures as listed in Table 11.3. Note that e_{ij} generally accounts for the correlation among observations on different genotypes in the same environment. For example, under model AMMI1, we have

$$e_{ij} = u_j + \lambda_{1j} w_{1j} + f_{ij}, \text{var}(e_{ij}) = \text{var}(y_{ij}) = \sigma_u^2 + \lambda_{1i}^2 + \sigma^2 \text{ and}$$

$$\text{cov}(e_{ij}, e_{i'j}) = \text{cov}(y_{ij}, y_{i'j}) = \sigma_u^2 + \lambda_{1i} \lambda_{1i'}.$$

The abbreviations introduced for different models will also be used to refer to specific structures for V, as in Table 11.3, and to structures used for three-way models presented in the section Three-Way Analysis.

Fitting the Models

Various two-way models were fitted to the wheat data using the MIXED procedure of the SAS System. All analyses were done on the square root scale, which for the data set at hand was found to be appropriate by likelihood-based methods (Piepho and McCulloch, 1999). The model fitting information ($-2LL_R$, AIC, SBC) is summarized in Table 11.4. Plots of AIC and SBC versus the number of model parameters are shown in Figure 11.1 and Figure 11.2.

TABLE 11.3. Variance-Covariance Structures for V

Name	Model for V
CS	$J\sigma_v^2 + I\sigma^2$
SH	$J\sigma_u^2 + diag\left(\sigma_1^2,...,\sigma_I^2\right)$
FW1	$\lambda_1\lambda_1' + I\sigma^2$
FW1-H	$\lambda_1\lambda_1' + diag\left(\sigma_1^2,...,\sigma_I^2\right)$
FW2	$\lambda_1\lambda_1' + \lambda_2\lambda_2' + I\sigma^2$
FW2-H	$\lambda_1\lambda_1' + \lambda_2\lambda_2' + diag\left(\sigma_1^2,...,\sigma_I^2\right)$
FWn	$\displaystyle\sum_{r=1}^{u}\lambda_r\lambda_r + I\sigma^2$
FWn-H	$\displaystyle\sum_{r=1}^{n}\lambda_r\lambda_r' + diag\left(\sigma_1^2,...,\sigma_I^2\right)$
AMMI1	$J\sigma_u^2 + \lambda_1\lambda_1' + I\sigma^2$
AMMI1-H	$J\sigma_u^2 + \lambda_1\lambda_1' + diag\left(\sigma_1^2,...,\sigma_I^2\right)$
AMMI2	$J\sigma_u^2 + \lambda_1\lambda_1' + \lambda_2\lambda_2' + I\sigma^2$
AMMI2-H	$J\sigma_u^2 + \lambda_1\lambda_1' + diag\left(\sigma_1^2,...,\sigma_I^2\right)$
AMMIn	$J\sigma_u^2 + \displaystyle\sum_{r=1}^{n}\lambda_r\lambda_r' + I\sigma^2$
AMMIn-H	$J\sigma_u^2 + \displaystyle\sum_{r=1}^{n}\lambda_r\lambda_r' + diag\left(\sigma_1^2,...,\sigma_I^2\right)$
UN	$\{\sigma_{ii'}\}$

Note: J is a square matrix of ones everywhere; I is an identity matrix.

The LR statistic for comparing two nested models can be obtained as the difference of $-2LL_R$. For example, the value of the LR statistic for comparing CS and AMMI1 is $2036.6 - 1876.3 = 160.3$ with $17 - 2 = 15$ df, which is clearly significant, so AMMI1 appears to be better than CS. According to *SBC*, AMMI- fitted best, while AMMI3-H was best by *AIC*. *AIC* tends to favor more complex models than *SBC* (see Figures 11.1 and 11.2). The SH model did not fit very well. The UN model could not be fitted due to convergence problems. This can indicate that there were not enough data to estimate the model. Alternatively, the model may not be appropriate for the data. It should be pointed out that the multiplicative models considered here

TABLE 11.4. Likelihood-Based Fitting Information (REMS) for Different Two-Way Models

Model	P	$-2LL_R$	AIC	SBC
CS	2	2036.6	−1020.3	−1025.0
SH	16	1958.4	−995.2	−1032.9
FW1	16	1989.2	−965.1	−1002.8
FW1-H	30	1822.0	−941.0	−1011.7
FW2	30	1838.8	−949.4	−1020.1
FW2-H	44	1772.2	−930.1	−1033.8
FW3	43	1790.1	−938.0	−1039.4
FW3-H	57	1739.8	−926.9	−1061.3
FW4	55	1737.4	−923.7	−1053.3
FW4-H	69	1706.1	−922.1	−1084.7
FW5	66	1720.7	−926.3	−1081.9
FW5-H	80	1686.8	−923.4	−1112.0
AMMI1	17	1876.3	−955.2	<u>−995.2</u>
AMMI1-H	31	1814.8	−938.4	−1011.5
AMMI2	31	1833.3	−947.7	−1020.7
AMMI2-H	45	1776.0	−933.0	−1039.1
AMMI3	44	1786.2	−928.1	−1031.8
AMMI3-H	58	1720.8	<u>−918.4</u>	−1055.1
AMMI4	56	1736.1	−924.1	−1056.1
AMMI4-H	70	1713.5	−926.7	−1091.7
AMMI5	67	1717.1	−925.6	−1083.5
AMMI5-H	81	1687.2	−924.6	−1115.5
AMMI6	77	1705.4	−929.7	−1111.2
AMMI6-H	91	1675.8	−928.9	−1143.4

Note: P = number of parameters for the variance-covariance structure. Models best according to AIC and SBC are underlined. Square-root-transformed data.

are useful approximations of the UN model. Since we prefer the *SBC* criterion (see Model Selection), we selected AMMI1 as the model on which to base stability analyses.

Stability Analysis

Stability analysis will be based on the AMMI1 model. Estimates of model parameters are shown in Table 11.5. To study stability, when a model

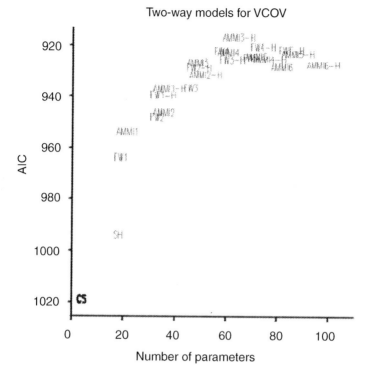

FIGURE 11.1. Plot of AIC versus number of parameters for different two-way models.

has been fitted to square-root-transformed data, it is necessary to compute the variance on the original scale. The mean and the variance of the i-th genotype on the original scale are

$$\mu_i^2 + \sigma_{ii} \text{ and } 4\mu_i^2\sigma_{ii} + 2\sigma_{ii}^2$$

respectively, where μ_i and σ_{ii} are the mean and the variance on the square-root transformed scale. This relation was used to compute means and variances on the original scale as reported in Table 11.6. Genotype 12 has the largest variance, followed by 14, while genotypes 3, 9, and 13 had the smallest variance and can thus be regarded as relatively stable (but see comments following).

Many analyses of stability are mainly concerned with the objective of minimizing the variance of yield, while failing to take into account the importance of average yield in relation to the variance. Clearly, when the aver-

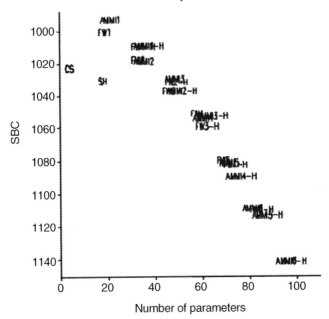

FIGURE 11.2. Plot of SBC versus number of parameters for different two-way models.

age yield of a genotype is large enough relative to the other genotypes, a large variance can usually be tolerated, whereas at low average yield levels, a small variance is more important. The argument is exemplified using a hypothetical situation (modified from Mead et al., 1986; also see Piepho, 1998a) shown below, rather than using the wheat data, just to make the point simple:

	Environment					
	A	B	C	D	E	F
Genotype 1	20	27	21	25	22	23
Genotype 2	25	37	26	35	24	32

The variances are $\hat{\sigma}_{11} = 7$ for genotype 1 and $\hat{\sigma}_{22} = 31$ for genotype 2. Thus, genotype 1 would be regarded as more stable. Note, however, that genotype 2 outyields genotype 1 at all locations, and it has the higher mean across locations, so it is clearly superior to genotype 1, despite its larger environmental variance. Obviously, looking only at the variance and ignoring mean yield may be misleading.

TABLE 11.5. Parameter Estimates (REML for AMMI1 Model in Two-Way Analysis) (Square-Root-Transformed Data)

Parameter Estimate (dt/ha)	
σ_u^2	0.885
σ^2	0.270

Parameter	Estimate	Parameter	Estimate (dt/ha)$^{0.5}$	Parameter	Estimate (dt/ha)
λ_1	1.21	μ_1	5.16	$\sigma_{1;1}$	2.62
λ_2	1.33	μ_2	4.84	$\sigma_{2;2}$	2.92
λ_3	0.53	μ_3	4.46	$\sigma_{3;3}$	1.43
λ_4	1.34	μ_4	5.00	$\sigma_{4;4}$	2.96
λ_5	1.15	μ_5	4.82	$\sigma_{5;5}$	2.48
λ_6	1.02	μ_6	4.78	$\sigma_{6;6}$	2.20
λ_7	1.07	μ_7	4.49	$\sigma_{7;7}$	2.30
λ_8	1.44	μ_8	4.81	$\sigma_{8;8}$	3.24
λ_9	0.53	μ_9	4.38	$\sigma_{9;9}$	1.44
λ_{10}	1.22	μ_{10}	4.92	$\sigma_{10;10}$	2.65
λ_{11}	1.08	μ_{11}	4.74	$\sigma_{11;11}$	2.31
λ_{12}	1.57	μ_{12}	5.39	$\sigma_{12;12}$	3.63
λ_{13}	0.54	μ_{13}	5.03	$\sigma_{13;13}$	1.45
λ_{14}	1.38	μ_{14}	5.36	$\sigma_{14;14}$	3.06
λ_{15}	1.29	μ_{15}	4.55	$\sigma_{15;15}$	2.81

TABLE 11.6. Mean and Variance Estimates on Original Scale for AMMI1 Model in Two-Way Analysis (Back-Transformation of Fit for Square-Root-Transformed Data)

Genotype	Mean (dt/ha)	Variance (dt/ha)2
1	29.23	292.5
2	26.37	290.5
3	21.31	118.2
4	27.91	313.2
5	25.76	243.0
6	25.00	210.2
7	22.44	195.7

TABLE 11.6 *(continued)*

Genotype	Mean (dt/ha)	Variance (dt/ha)2
8	26.33	320.5
9	20.63	114.5
10	26.86	270.6
11	24.74	218.2
12	32.73	449.0
13	26.74	150.4
14	31.84	371.3
15	23.55	249.0

Note: Back-transformed using the fact that mean and variance on the original scale are $\mu_i^2 + \sigma_{ii}$ and $4\mu_i^2\sigma_{ii} + 2\sigma_{ii}^2$, respectively, where μ_i and σ_{ii} are the mean and the variance on the square-root scale.

There are a number of methods that combine mean and variance into one measure. For a review see Piepho (1998a). We will focus on methods for assessing risk (Mead et al., 1986; Eskridge, 1990; Eskridge and Mumm, 1992; Piepho, 1996). Let $Y = (Y_1, \ldots, Y_I)'$ be a random vector of genotype yields in a randomly chosen environment. This vector has expectation μ and variance-covariance matrix $V = \{\sigma_{ii'}\}$. Assuming multivariate normality of Y, possibly achieved by a suitable transformation (e.g., square root transformation used in the present example), the probability (risk) for the yield of a genotype to fall below a critical (subsistence) level θ is

$$\Pr(Y_i < \theta) = \Phi\left[(\theta - \mu_i) / \sqrt{\sigma_{ii}}\right] \qquad (9)$$

where Φ is the cumulative standard normal distribution function. This depends on both mean yield (μ_i) and variance (σ_{ii}). Estimation requires inserting estimates for μ_i and σ_{ii}. In simple cases, exact confidence limits for risk estimates are easy to construct (Piepho, 2000), while in more complex cases, e.g., unbalanced data, approximate methods can be used, the usefulness of which is the subject of current research.

It should be stressed that the resulting risk estimates depend on the particular variance-covariance structures for V. For example, using the AMMI1 model we have:

$$\sigma_{ii} = \sigma_u^2 + (\lambda_1 i)^2 + \sigma^2 \qquad (10)$$

It is emphasized here that (9) can be evaluated using transformed data, while still allowing inferences about the risk on the untransformed scale. Risk statements for the original scale require back-transforming the critical level θ. For example, if θ denotes the critical level for square-root transformed data Y, the critical level on the original scale is θ^2. Clearly, $\Pr(Y_i < \theta)$ is equal to the probability of

$$Z_i = Y_i^2$$

i.e., the yield on the original scale falling below θ^2.

Computation of (9) requires specification of a relevant subsistence level. If a unique level cannot be specified, it is useful to plot the risk versus a range of critical value. For illustration, we will compare genotype 12 to genotype 13. The latter has a lower mean and also a lower variance than the former. Figure 11.3 represents a plot of $\Pr(Y_i < \theta)$ vs. θ^2 based on the AMMI1 model. Note that the critical level (θ^2) is given on the original scale (dt/ha), whereas square-root transformed data Y were used in the analysis. Up to a level of about 20 dt/ha, genotype 13 has a smaller risk, whereas above 20 dt/ha, the risk is smaller for genotype 12. Figure 11.4 displays a plot of the risk of genotype 13 versus the risk of genotype 12. This type of plot has been termed relative risk plot (Mead et al., 1986). It conveys the same information as shown in Figure 11.3. If the relative risk plot does not intersect the equal risk line, one genotype is superior for all risk levels. Here, it intersects at a level of about 20 dt/ha. The corresponding critical levels are indicated by dots.

So far, the risk to fall below a critical level was considered. Alternatively, we may study the probability that genotype i' outperforms genotype i (Eskridge and Mumm, 1992). Assuming multivariate normality of Y, this probability is given by

$$\Pr\left(Y_i - Y_{i'} < 0\right) = \Phi\left[\left(\mu_i - \mu_{i'}\right) / \sqrt{\left(\sigma_{ii} + \sigma_{i'i'} - 2\sigma_{ii'}\right)}\right] \quad (11)$$

Again, the probability depends on the model for V. For example, under the AMMI1 model, the variance is as in (10) and the covariance is

$$\sigma_{ii'} = \sigma_u^2 + \lambda_{1i}\lambda_{1i'} \quad (12)$$

Probabilities for all pairwise comparisons based on square-root-transformed data and the AMMI1 model are shown in Table 11.7. For genotype 12, which has the highest mean, the probability to outperform another genotype is always larger than 50 percent. For genotypes 1, 4, 13, and 14, the

$Pr(Y_i < \theta)$

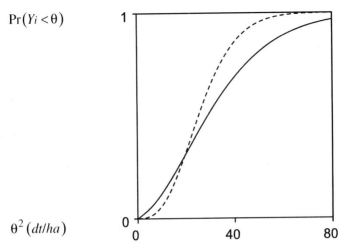

$\theta^2\,(dt/ha)$

FIGURE 11.3. Plot of risk versus critical value θ^2 (dt/ha; original scale) for two genotypes. Solid line: genotype 12; dotted line: geontype 13. Square-root-transformed data. AMMI1 model. Two-way analysis.

Risk of genotype 13

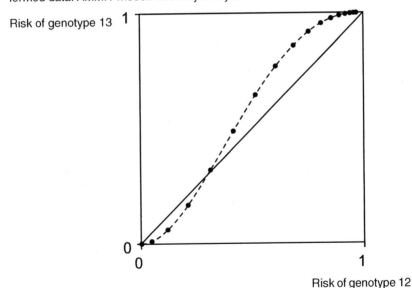

Risk of genotype 12

FIGURE 11.4. Plot of risks for genotypes 12 and 13 (dotted line). Solid line = equal risk line. Critical values are indicated by dots. First dot to the left corresponds to $\theta^2 = 0$ dt/ha original scale. The next dot corresponds to $\theta^2 = 5$ dt/ha, and so forth in increments of 5. Square-root-transformed data. AMMI1 model. Two-way analysis.

smallest probabilities are 39 percent, 30 percent, 38 percent, and 48 percent, respectively. These probabilities were relatively large compared to those for the other genotypes. Thus, these four genotypes compared favorably with genotype 12, especially genotype 14.

THREE-WAY ANALYSIS

Models

In analogy to the general form of the two-way model in equations (6) and (7), we will use the following three-way model (Piepho, 1998a; Piepho et al., 1998):

$$y_{ijk} = \mu_i + e_{ij} + a_{ik} + b_{ijk} \qquad (13)$$

where $i = 1, \ldots, I$ indexes genotypes, $j = 1, \ldots, J$ indexes locations and $k = 1, \ldots, K$ indexes years. The random terms e_{ij}, a_{ik}, and b_{ijk} are assumed to be independent. The two-way effects e_{ij} and a_{ik} as well as the three-way effect b_{ijk} play a role similar to the two-way effect e_{ij} in the two-way model (6). The effects e_{ij} (a_{ik}) in model (13) account for across-year (location) correlation among observations on different genotypes in the same location (year), while the effects b_{ijk} model correlation among observations on different genotypes in the same year × location combination. Thus, only one form of dependence is allowed: e_{ij} are dependent for observations in the same location, a_{ik} are dependent for observations in the same year and b_{ijk} are dependent for observations in the same year × location combination. This may be formalized using matrix notation. Let $e_j = (e_{1j}, \ldots, e_{Ij})'$, $a_k = (a_{1k}, \ldots, a_{Ik})'$, and $b_{jk} = (b_{1jk}, \ldots, b_{Ijk})'$. Our model assumes that e_j, a_k, and b_{jk} are mutually independent and that the variance-covariance matrices $V_e = \text{var}(e_j)$, $V_a = \text{var}(a_k)$, and $V_b = \text{var}(b_{jk})$ can have one of the structures listed in Table 11.3. For example, if all three have the CS structure, the model is equivalent to the three-way ANOVA model with

$$e_{ij} = \text{location}_j + (\text{genotype} \times \text{location})_{ij}$$
$$a_{ik} = \text{year}_k + (\text{genotype} \times \text{year})_{ik}$$
$$b_{ijk} = (\text{location} \times \text{year})_{jk} + (\text{genotype} \times \text{year} \times \text{location})_{ijk}$$

Our general matrix notation was chosen to simplify the set-up of more specific models and their comparison to the ANOVA model. For distinction of model parameters in V_e, V_a, and V_b, we will use subscripts *(e)*, *(a)*, and *(b)* on the parameters, respectively. To refer to a model, we will concatenate abbreviations of the models used for V_e, V_a, and V_b. For example, CS/-/AMMI1 refers to a model with CS for V_e, AMMI1 for V_b, while V_a is dropped.

We emphasize that under the three-way model (13), observations from the same year but different locations, and from the same location but different years, are correlated among each other. This is in contrast to the two-way model (6), where observations from all different environments (year × location combinations) are independent. A main advantage of the three-way model is that among-environment correlation, which is to be expected in METs, can be more realistically modeled. When the components of both V_e and V_a are negligible relative to those in V_b, it is reasonable to ignore such correlation and perform a two-way analysis based on model (6). Strictly speaking, the two-way model (6) as applied to three-way data is valid only when $V_e = V_a = 0$. In other words, the two-way model is a special case of the three-way model with $V_e = V_a = 0$. For example, the two-way model with AMMI1 for V is equivalent to -/-/AMMI1. This allows us to formally check the validity of the two-way model assumption by looking at the magnitude of the components in V_e and V_a relative to those in V_b.

Fitting the Models

Estimates of variance components under the CS/CS/CS model are shown in Table 11.8. The components for V_a are much smaller than those for V_e and V_b. For this reason, it seemed reasonable to assume that variance components pertaining to a_{ik} will be small in other models as well. Thus, we considered models without the a_{ik} term, i.e., models in which V_a is dropped. This made it computationally feasible to fit more complex structures for V_e and V_b, which could not be done when V_a was in the model. Since the number of genotypes exceeds that of the years, fitting any complex structure to V_a is expected to be difficult or impossible. Furthermore, it should be stressed that under the CS/CS/CS model, V_e is not negligible relative to V_b (Table 11.8), so a three-way mixed-model analysis is clearly worthwhile and preferable to a mixed-model two-way analysis (We should point out that this statement does not apply to the fixed effects case. For example, a biplot based on a fixed effects, two-way AMMI model may be quite revealing of three-way interactions.)

The likelihood information for the various models fitted to square-root-transformed data is found in Table 11.9. The very simple model CS/-/CS fitted remarkably well and was selected as the best model by the *SBC*. CS/-/AMMI1 fitted second best according to *SBC*, while FW4/-/FW4 was best by *AIC*. A comparison of models with FW*n* structures by both *SBC* and *AIC* suggests that the data support more complex models for V_b than V_e. It is instructive to compare the best fitting two-way and three-way models. According to *SBC*,

AMMI1 was the best two-way model. Its *SBC* value of –955.2 is much lower, though, than that of the two best fitting three-way models CS/-/CS (*SBC* = –947.3) and CS/-/AMMI1 (*SBC* = –948.5), again suggesting that three-way modeling is preferable. In the following analyses, we will consider the models CS/-/CS and CS/-/AMMI1.

TABLE 11.8. Variance Components for Three-Way ANOVA Model (CS Structure for V_e, V_a, and V_b)

Source	Estimate (dt/ha)	s.e. (dt/ha)
Year *(Va)*	0.102	0.091
Year × genotype *(Va)*	0.00643	0.00599
Location *(Ve)*	1.30	0.50
Location × genotype *(Ve)*	0.147	0.023
Year × location *(Vb)*	0.842	0.140
Year × location × genotype *(Vb)* (Residual)	0.232	0.016

Note: Square-root-transformed data.

TABLE 11.9. Likelihood-Based Fitting Information (REML) for Different Three-Way Models

Models for					
e_j	a_k	b_{jk}	$-2LL_R$	AIC	SBC
CS	CS	CS	1864.1	–938.1	–952.3
CS	CS	SH	1809.6	–924.8	–972.1
SH	CS	CS	1823.1	–931.5	–978.9
CS	SH	CS	1852.5	–946.2	–993.6
SH	SH	CS	1779.4	–923.7	–1004.1
CS	SH	SH	1811.8	–939.9	–1020.3
SH	CS	SH	1802.0	–935.0	–1015.5
SH	SH	SH	1773.0	–934.5	–1048.1
FW1-H	CS	FW1-H	1657.1	–890.5	–1037.2
FW1-H	CS	SH	1726.4	–911.2	–1024.8
FW1-H	CS	CS	1773.6	–920.8	–1001.3
CS	CS	FW1-H	1729.0	–898.5	–979.0
CS	-	CS	1867.6	–937.8	–947.3
CS	-	SH	1812.5	–923.3	–963.5
SH	-	CS	1828.0	–931.0	–971.2
SH	-	SH	1783.8	–923.9	–999.6

TABLE 11.9 *(continued)*

Models for e_j	a_k	b_{jk}	$-2LL_R$	AIC	SBC
FW1	-	CS	1809.8	−922.9	−965.3
FW1-H	-	CS	1777.5	−920.8	−996.2
CS	-	FW1	1793.8	−914.9	−957.3
CS	-	FW1-H	1732.9	−898.5	−974.2
AMMI1	-	CS	1807.4	−922.7	−967.5
AMMI1-H	-	CS	1777.4	−921.7	−999.5
CS	-	AMMI1	1769.5	−903.7	<u>−948.5</u>
CS	-	AMMI1-H	1722.8	−894.4	−972.5
FW2	-	CS	1746.3	−905.1	−980.6
FW2-H	-	CS	1720.7	−906.3	−1014.8
CS	-	FW2	1722.1	−893.1	−968.5
CS	-	FW2-H	1689.1	−890.5	−999.0
AMMI2	-	CS	1744.3	-905.1	−982.9
AMMI2-H	-	CS	1718.3	−906.1	−1016.9
CS	-	AMMI2	1711.7	−888.9	−966.6
CS	-	AMMI2-H	1683.9	−889.0	−999.7
FW3	-	CS	1713.2	−901.6	−1007.7
FW3-H	-	CS	1708.9	−913.5	−1052.5
CS	-	FW3	1700.9	−895.4	−1001.5
CS	-	FW3-H	1645.5	−881.8	−1020.8
AMMI3	-	CS	1710.4	−901.2	−1009.6
AMMI3-H	-	CS	1707.4	−913.7	−1055.1
CS	-	AMMI3	1686.6	−889.3	−997.7
CS	-	AMMI3-H	1652.4	−886.2	−1027.7
FW4	-	CS	1696.6	−905.3	−1039.7
FW4-H	-	CS	1694.9	−918.5	−1085.8
CS	-	FW4	1664.9	−889.5	−1023.8
CS	-	FW4-H	1634.4	−888.2	−1055.5
AMMI4	-	CS	1693.7	−904.9	−1041.6
AMMI4-H	-	CS	1692.9	−918.4	−1088.1
CS	-	AMMI4	1660.8	−888.4	−1025.1
CS	-	AMMI4-H	1631.3	−887.7	−1057.4
FW1	-	FW1	1747.1	−905.5	−981.0
FW1	-	FW2	1689.1	−890.6	−999.0
FW2	-	FW1	1690.8	−891.4	−999.8

	Models for		$-2LL_R$	AIC	SBC
e_j	a_k	b_{jk}			
FW2	-	FW2	1630.6	−875.3	−1016.8
FW2	-	FW3	1601.8	−873.9	−1045.9
FW3	-	FW2	1607.4	−876.7	−1048.8
FW3		FW3	1566.9	−869.5	−1072.2
FW3		FW4	1542.4	−869.4	−1100.4
FW4		FW3	1544.1	−870.1	−1101.1
FW4		FW4	1514.0	−867.0	−1126.3

Note: Models best according to *AIC* and *SBC* are underlined. Square-root-transformed data.

Stability Analysis

Distinguishing years and locations allows tailoring stability analysis to the relevant perspective. Consider, for example, a research station or breeder extending a new cultivar to a target region. In this situation, both years and locations can be taken as random, since interest is in predicting yield for a new location in a new year. Thus, in assessing stability, all three components, V_e, V_a, and V_b, are relevant. By contrast, for the farmer, the location is usually fixed, and the variance of yield at a location is only governed by V_a and V_b. For a detailed discussion of this issue, see Piepho (1998a). Here, we will restrict attention to the breeders perspective, which seems reasonable, considering the purpose of the MET used as an example.

Under the model CS/-/CS for the square-root-transformed data, the mean and the variance of observations on the original scale are

$$\mu_i^2 + \sigma_{ii} \text{ and } 4\mu_i^2\sigma_{ii} + 2\sigma_{ii}^2$$

respectively, where μ_i and σ_{ii} are the mean and the variance on the transformed scale, and σ_{ii} does not depend on the genotype. Thus, the genotype with the smallest mean on the transformed scale also has the smallest mean and the smallest variance on the untransformed scale. In the example, genotype 9 can be considered to be the most stable one (Table 11.10). Genotype 12 is very unstable, whereas genotype 13 is rather stable.

The smallest mean on the untransformed scale is not necessarily associated with the smallest variance on that scale in models that are heteroscedastic on the transformed scale. The fit of the CS/-/AMMI1 model, which is heteroscedastic, is shown in Tables 11.11 and 11.12. Genotype 12 has the largest variance on the back-transformed scale, whereas genotype 13 has a relatively small variance and is, therefore, quite stable. Thus,

TABLE 11.10. Parameter Estimates (REML) for Model CS/-/CS in Three-Way Analysis

Parameter	Estimate (dt/ha)
$\sigma^2_{u(e)}$	0.144
$\sigma^2_{(e)}$	1.236
$\sigma^2_{u(b)}$	0.950
$\sigma^2_{(b)}$	0.239

Parameter	Estimate (dt/ha)$^{0.5}$	Genotype	Mean on original scale (dt/ha)	Variance on original scale (dt/ha)^2
μ_1	5.07	1	28.29	276.5
μ_2	4.85	2	26.10	254.0
μ_3	4.44	3	22.26	214.8
μ_4	4.94	4	26.97	263.0
μ_5	4.81	5	25.68	249.7
μ_6	4.71	6	24.77	240.4
μ_7	4.45	7	22.36	215.8
μ_8	4.72	8	24.85	241.3
μ_9	4.28	9	20.85	200.3
μ_{10}	4.86	10	26.19	255.0
μ_{11}	4.66	11	24.25	235.2
μ_{12}	5.40	12	31.71	311.5
μ_{13}	4.89	13	26.42	257.4
μ_{14}	5.35	14	31.21	306.3
μ_{15}	4.49	15	22.73	219.6

Notes: Square-root-transformed data. *Back-transformed using the fact that mean and variance on the original scale are $\mu_i^2 + \sigma_{ii}$ and $4\mu_i^2\sigma_{ii} + 2\sigma_{ii}^2$, respectively, where μ_i and σ_{ii} are the mean and the variance on the square-root scale.

the conclusions drawn from the CS/-/CS model and the CS/-/AMMI1 model are quite similar.

We now turn to probability methods. Let $Y = (Y_1, \ldots, Y_l)'$ be a random vector of genotype yields in a randomly chosen year and location. This vector has expectation μ and variance-covariance matrix

$$V = \text{var}(Y) = \{\sigma ii'\} = Ve + Va + Vb \qquad (14)$$

TABLE 11.11. Parameter Estimates (REML) for Model CS/-/AMMI1 in Three-Way Analysis

Parameter	Estimate (dt/ha)
$\sigma^2_{u(e)}$	0.992
$\sigma^2_{(e)}$	0.119
$\sigma^2_{u(b)}$	0.709
$\sigma^2_{(b)}$	0.172

Parameter	Estimate	Parameter	Estimate (dt/ha)$^{0.5}$	Parameter	Estimate (dt/ha)
$\lambda_{1(b)}$	0.501	μ_1	5.043	$\sigma_{1;1}$	2.243
$\lambda_{2(b)}$	0.677	μ_2	4.795	$\sigma_{2;2}$	2.450
$\lambda_{3(b)}$	−0.327	μ_3	4.406	$\sigma_{3;3}$	2.099
$\lambda_{4(b)}$	0.685	μ_4	4.910	$\sigma_{4;4}$	2.462
$\lambda_{5(b)}$	0.515	μ_5	4.776	$\sigma_{5;5}$	2.258
$\lambda_{6(b)}$	0.419	μ_6	4.680	$\sigma_{6;6}$	2.167
$\lambda_{7(b)}$	0.571	μ_7	4.425	$\sigma_{7;7}$	2.318
$\lambda_{8(b)}$	0.869	μ_8	4.707	$\sigma_{8;8}$	2.747
$\lambda_{9(b)}$	0.230	μ_9	4.248	$\sigma_{9;9}$	2.045
$\lambda_{10(b)}$	0.724	μ_{10}	4.862	$\sigma_{10;10}$	2.517
$\lambda_{11(b)}$	0.104	μ_{11}	4.636	$\sigma_{11;11}$	2.003
$\lambda_{12(b)}$	0.842	μ_{12}	5.318	$\sigma_{12;12}$	2.702
$\lambda_{13(b)}$	0.049	μ_{13}	4.902	$\sigma_{13;13}$	1.994
$\lambda_{14(b)}$	0.664	μ_{14}	5.295	$\sigma_{14;14}$	2.432
$\lambda_{15(b)}$	0.812	μ_{15}	4.475	$\sigma_{15;15}$	2.652

Note: Square-root-transformed data.

The probability of falling below a critical level θ can be computed from eq. (9) and the probability that genotype i outperforms genotype i' is as given in eq. (11). The probabilities are estimated by inserting estimates for μ_i, σ_{ii}, and $\sigma_{ii'}$. Again, the probabilities must be evaluated for transformed data. For (9), the transformation has to be applied to the critical level θ. The resulting

TABLE 11.12. Mean and Variance Estimates on Original Scale for CS/-/AMMI1 in Three-Way Analysis

Genotype	Mean* (dt/ha)	Variance*(dt/ta)
1	27.67	238.23
2	25.44	237.26
3	21.51	171.78
4	26.57	249.47
5	25.07	216.17
6	24.07	199.29
7	21.90	192.31
8	24.90	258.53
9	20.09	155.97
10	26.16	250.63
11	23.50	180.20
12	30.99	320.25
13	26.03	199.64
14	30.47	284.58
15	22.67	226.43

Note: Back-transformation of fit for square-root-transformed data. *Back-transformed using the fact that mean and variance on the original scale are

$$\mu_i^2 + \sigma_{ii} \text{ and } 4\mu_i^2\sigma + \sigma_{ii} + 2\sigma_{ii}^2$$

respectively, where μ_i and σ_{ii} are the mean and the variance on the square-root scale.

risk estimates depend on the particular variance-covariance structures used. For example, under the CS/-/AMMI1 model,

$$\sigma_{ii} = \sigma_{ii(e)} + \sigma_{ii(b)} = \sigma_{u(e)}^2 + \sigma_{(e)}^2 + \sigma_{u(b)}^2 + \left(\lambda_{1i(b)}\right)^2 + \sigma_{(b)}^2 \quad (15)$$

$$\sigma_{ii'} = \sigma_{ii'(e)} + \sigma_{ii'(b)} = \sigma_{u(e)}^2 + \sigma_{u(b)}^2 + \lambda_{1i(b)}\lambda_{1i'(b)} \quad (16)$$

The estimated risk $\Pr(Y_i < \theta)$ of genotypes 12 and 13 based on the CS/-/AMMI1 model is plotted against critical levels in Figure 11.5. A relative risk plot is shown in Figure 11.6. These plots are discussed later. Note that, again, the critical level (θ^2) is given on the original scale, whereas square-root-transformed data Y were used in the analysis.

Under the CS/-/CS model, the risk is a monotone function of the mean, and the genotype with the largest mean has the smallest risk for all critical levels θ^2. As a result, the diagonal relative risk curve does not intersect the equal risk line (see Figure 11.7). Under the CS/-/AMMI1 model, the lines

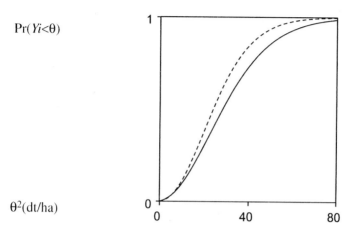

FIGURE 11.5. Plot of risk versus critical value θ^2(dt/ha; original scale) for two genotypes. Solid line: genotype 12; dotted line: genotype 13. Square-root-transformed data. CS/-/AMMI1 model. Three-way analysis.

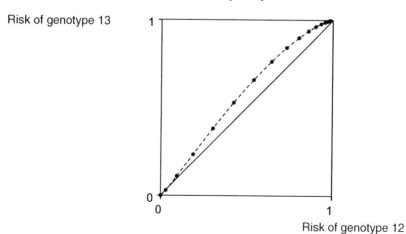

FIGURE 11.6. Plot of risks for genotypes 12 and 13 (dotted line). Solid line = equal risk line. Critical values are indicated by dots. First dot to the left corresponds to $\theta^2 = 0$ dt/ha (original scale). The next dot corresponds to $\theta^2 = 5$ dt/ha, and so forth by increments of 5. Square-root-transformed data. CS/–/AMMI1 model. Three-way analysis.

Risk of genotype 13

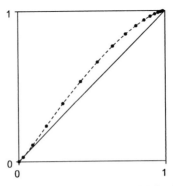

Risk of genotype 12

FIGURE 11.7. Plot of risks for genotypes 12 and 13 (dotted line). Solid line = equal risk line. Critical values are indicated by dots. First dot to the left corresponds to θ^2=0dt/ha (original scale). The next dot corresponds to $\theta^2 = 5$ dt/ha, and so forth by increments of 5. Square-root-transformed data. CS/-/CS model. Three-way analysis.

do intersect, between $\theta^2 = 5$ dt/ha and $\theta^2 = 10$ dt/ha, though this is barely visible on the plot (Figures 11.5 and 11.6). Nevertheless, the relative risk curves are quite similar under both models: For larger critical levels θ^2, genotype 13 is more risky under both models. Thus, in this example the conclusion regarding the relative risk of both genotypes is robust to the choice of model for the variance-covariance structure. The plots differ notably, however, from that obtained in the two-way analysis (Figures 11.3 and 11.4), where the point of intersection is close to

$$\theta^2 = 20 \text{ dt/ha}$$

and there is a rather pronounced inflection. Thus, the three-way analysis yields a somewhat different result.

Risks of yield falling below 10 dt/ha and 30 dt/ha are given in Table 11.13. Interestingly, genotype 14 has the lowest risk to fall below 10 dt/ha, whereas genotype 12 has the smallest risk to fail 30 dt/ha. Note that both genotypes have a very similar mean. Figure 11.8 depicts the relative risk plot for these two genotypes. Up to a critical level of about 25 dt/ha, genotype 14 is better, whereas genotype 12 is less risky above that value. The differences in risk are seldom large, however, and both genotypes show broad adaptation.

The probability $\Pr(Y_i - Y_{i'} < 0)$ under the CS/-/AMMI1 model is tabulated for all pairs of genotypes in Table 11.14. For genotype 12, the probability to outperform another genotype is greater than 50 percent in all comparisons. For genotypes 1, 4, 13, and 14, the smallest probabilities are 37 percent, 30 percent, 34 percent, and 49 percent, respectively, so these are not much worse than genotype 12. This conclusion is comparable to that

Risk of genotype 14

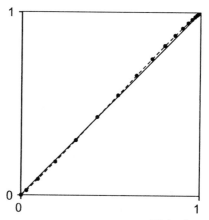

Risk of genotype 12

FIGURE 11.8. Plot of risks for genotypes 12 and 14 (dotted line). Solid line = equal risk line. Critical values are indicated by dots. First dot to the left corresponds to Θ^2=0 dt/ha (original scale). The next dot corresponds to $\Theta^2 = 5$ dt/ha, and so forth by increments of 5. Square-root-transformed data. CS/-/CS model. Three-way analysis.

TABLE 11.13. Risk of Genotypes to Fall Below 10 dt/ha and 30dt/ha

Genotype	Cultivar name	Risk at 10 dt/ha	Risk at 30 t/ha
1	Alondara 4546	0.105	0.614
2	Chasqui Inia	0.148	0.669
3	CNT 9	0.195	0.770
4	Diamante Inta	0.133	0.641
5	E. Hornero	0.141	0.680
6	E. Tarariras	0.151	0.706
7	IAS 54	0.204	0.755
8	Itapua 5	0.176	0.679
9	Jacui	0.224	0.805
10	Leones Inta	0.142	0.651
11	LE 1787	0.149	0.724
12	Millaleu Inia	0.095	0.539
13	Minuano 82	0.109	0.658
14	Onda Inia	0.086	0.547
15	281/60	0.210	0.731

Note: Square-root-transformed data. CS/–/AMMI1 model.

from the two-way analysis. So in this case, conclusions with respect to risk are rather insensitive to slight misspecifications of the model.

Finally, we computed the probability that a genotype outperforms all others in the set, assuming that on the square-root scale yields follow a multivariate normal with variance-covariance structure given by (14), using the CS/-/AMMI1 model with parameter values given in Table 11.11. Since this probability cannot generally be computed analytically, we employed Monte Carlo simulation based on a Cholesky decomposition of the estimated variance-covariance matrix. The result for the simulated probabilities based on 1,000,000 runs is shown in Table 11.15. Genotypes 12 and 14 emerge as the best genotypes. The probability that the best genotype (12) wins is remarkably low (23.18 percent), though. This example shows that even though we may have selected the best genotype, the probability that this is outperformed by some other genotype can be rather high.

TABLE 11.14. Probability That Genotype i Outperforms Genotype i'

i/i'	1	2	3	4	5	6	7	8	9	10	11	12	13	14	15
1	*	0.62	0.71	0.57	0.64	0.68	0.79	0.65	0.84	0.59	0.68	0.37	0.56	0.37	0.75
2	0.38	*	0.62	0.44	0.51	0.56	0.68	0.54	0.73	0.46	0.57	0.25	0.46	0.26	0.60
3	0.29	0.38	*	0.35	0.37	0.40	0.49	0.42	0.57	0.36	0.40	0.26	0.28	0.24	0.48
4	0.43	0.56	0.65	*	0.57	0.61	0.74	0.60	0.77	0.52	0.61	0.30	0.50	0.31	0.71
5	0.36	0.49	0.63	0.43	*	0.55	0.68	0.53	0.74	0.46	0.56	0.26	0.44	0.25	0.64
6	0.32	0.44	0.60	0.39	0.45	*	0.63	0.49	0.71	0.41	0.52	0.23	0.40	0.22	0.59
7	0.21	0.32	0.51	0.26	0.32	0.37	*	0.37	0.58	0.29	0.41	0.13	0.30	0.13	0.48
8	0.35	0.46	0.58	0.40	0.47	0.51	0.63	*	0.68	0.42	0.53	0.21	0.43	0.23	0.62
9	0.16	0.27	0.43	0.23	0.26	0.29	0.42	0.32	*	0.25	0.31	0.14	0.20	0.12	0.41
10	0.41	0.54	0.64	0.48	0.54	0.59	0.71	0.58	0.75	*	0.59	0.28	0.48	0.29	0.69
11	0.32	0.43	0.60	0.39	0.44	0.48	0.59	0.47	0.69	0.41	*	0.26	0.36	0.24	0.56
12	0.63	0.75	0.74	0.70	0.74	0.77	0.87	0.79	0.86	0.72	0.74	*	0.65	0.51	0.87
13	0.44	0.54	0.72	0.50	0.56	0.60	0.70	0.57	0.80	0.52	0.64	0.35	*	0.34	0.65
14	0.63	0.74	0.76	0.69	0.75	0.78	0.87	0.77	0.88	0.71	0.76	0.49	0.66	*	0.85
15	0.25	0.34	0.52	0.29	0.36	0.41	0.52	0.38	0.59	0.31	0.44	0.13	0.35	0.15	*

Note: CS/-/AMMI1 model. Square-root-transformed data. Smallest probability in a row is underlined.

TABLE 11.15. Simulated Probability of Outperforming All Genotypes

Genotype	Probability to win
1	0.0858
2	0.0385
3	0.0769
4	0.0585
5	0.0341
6	0.0253
7	0.0081
8	0.0389
9	0.0060
10	0.0517
11	0.0401
12	0.2318
13	0.1038
14	0.1864
15	0.0141

Note: CS/-/AMMI1 model. Square-root-transformed data. 1,000,000 simulation runs.

DISCUSSION

We used the MIXED procedure (SAS Institute, 1997) and ASREML (Gilmour et al., 1999) to fit a variety of three-way mixed models. A great practical problem in three-way analyses is the enormous number of models that can be considered. Suppose we are modeling only V_e and V_b, while dropping V_a, and are willing to contemplate 20 different models for each V_e and V_b, then there is a total of $20^2 = 400$ variance-covariance structures to be fitted. Adding V_a to the model, we would have $20^3 = 8,000$ models to be fitted. Therefore, some selective strategy as in multiple regression is needed to avoid the need to fit all possible models. We are not aware of a well-established selection strategy for variance-covariance structures. One option seems to be to fix two of the three components at a certain structure, e.g., CS or UN, and find the best fitting model for the third. Alternatively, one may start with the simplest case (CS for all three components) and gradually make the components more complicated. Based on previous fits, one may be able to guess the most promising component among V_e, V_a, and V_b for further improvement and also the most promising types of model for that component. This is the approach we have taken in this chapter.

Another problem is the great computational burden for fitting models with all three variance-covariance structures, especially when the structures are more complex than CS. Models involving SH for one or more of the components were fitted by ASREML, which was much faster than MIXED. With MIXED, we were not able to fit (within reasonable time) models for FW, FW-H, AMMI, AMMI-H or UN, when all three variance-covariance structures were in the model. With ASREML, only a few models could be fitted. Choice of starting values was the primary problem with ASREML, especially with FW-H and AMMI-H models that contained more than one multiplicative term, while computational resources were the main limiting factor with MIXED. ASREML has no option for fitting FW and AMMI models, but this will be added in the near future (Arthur Gilmour, e-mail, spring 1999). When dropping V_a, a number of models could be fitted with MIXED, excluding those which had variance heterogeneity (SH, FW-H, AMMI-H) in more than one of the components V_e, V_a, and V_b: here, MIXED (Release 6.12) aborted with an error message. It appears that this problem is solved in SAS Version 8 (Russ Wolfinger, e-mail, spring 1999). ASREML was able to fit some of these models.

The three-way mixed model analysis of MET data is usually difficult because the number of years and locations is small relative to the number of genotypes. As a result, it is often difficult to fit models to V_e and V_a that are more complicated than CS. There will always be more information on V_b than on V_e and V_a. In a complete table of J locations $\times K$ years, there are J observations on e_j, K observations on a_k and JK on b_{jk}. Thus, it is expected that the scope for modeling is usually largest for V_b. This was the case for the wheat data. A two-way analysis of three-way data has the virtue of simplicity, but before considering such a simplified analysis, it is prudent to study the magnitude of the components in V_e and V_a. Only if these are small is it justifiable to set $V_e = V_a = 0$, which is the assumption implied by a two-way analysis. For the wheat data, V_a was small and could be ignored, but V_e was rather large and clearly not negligible. Therefore, the three-way analysis is preferable to the two-way analysis, which is underscored by the differences in results obtained by both forms of analysis. For example, the relative risk plot for genotypes 12 and 13 differs markedly for the two-way model AMMI1 (Figure 11.4) and the three-way models (Figures 11.6 and 11.7). Similar differences were obtained for the other relative risk plots (results not shown).

When fitting variance-covariance structures, which involve the term

$$J\sigma_{ii}^2, \text{ where } J \text{ is a square matrix}$$

of ones everywhere, e.g., models SH, AMMI, AMMI-H, one can consider an alternative fitting strategy. Take, for example, the SH two-way model, for which

$$V = J\sigma_{ii}^2 + diag\left(\sigma_1^2, ..., \sigma_1^2\right).$$

The corresponding linear model is $y_{ij} = \mu_i + u_j + f_{ij}$. The variance of the i-th genotype is

$$\sigma_{ii} = \sigma_u^2 + \sigma_i^2.$$

The difference of the variance (stability) of two genotypes i and i' is

$$\sigma_i^2 - \sigma_{i'}^2.$$

The variance of the difference of y_{ij} and $y_{i'j}$ is

$$\sigma_i^2 + \sigma_{i'}^2.$$

This variance is needed to compute the probability that genotype i outperforms genotype i'. Also, the probability of the i-th genotype outperforming all other genotypes can be computed based on the multivariate normal distribution of differences $Y_i - Y_{i'}$ ($i \neq i'$), although this requires a separate simulation for each genotype. Thus, if these probabilities are of interest or if we just want to compare variances, the variance of u_j is not needed, and we may fit the model

$$y_{ij} = \mu_i + u_j + f_{ij},$$

formally treating u_j as a fixed (nuisance) effect. This analysis, which was proposed by Shukla (1972), will yield estimates of the stability variances

$$\sigma_i^2 \text{ and of } \mu_i,$$

which is all that is needed for further analysis. The advantage of treating u_j as fixed is the simplified estimation of the remaining variance components. For example, if the SH model with random u_j does not converge, it may be easier to obtain estimates of

$$\sigma_i^2, \text{ when } u_j \text{ is treated as fixed.}$$

If, however, we are interested in computing the probability of yield falling below a level θ, we need an estimate of

$$\sigma_u^2 \text{ and cannot treat } u_j \text{ as a fixed effect.}$$

Also, when u_j is fixed, the comparison with other models via the log-likelihood is no longer appropriate. It is possible that due to a large standard error for the estimate of σ_u^2 under a model with random u_j, modeling u_j as fixed provides better estimates of

$$\sigma_i^2 \text{ and of } \mu_i$$

especially when the number of locations is limited, but this remains to be investigated.

In our analysis, a normalizing transformation of the data has been very important. With untransformed data, we obtained large probabilities (around 5 to 10 percent) of yield falling below zero for many cultivars (results not shown; also see Piepho and McCulloch, 1999), which is nonsensical, indicating that the normality assumption is not tenable. We suggest that a transformation be routinely considered. The normality assumption is particularly important when computing probabilities/risks based on this assumption. When the variance of the data is large relative to the mean, a transformation will certainly be needed. In our example, the square-root-transformation worked remarkably well. The Box-Cox family of transformations (Box and Cox, 1964), of which the logarithmic and the square-root-transformation are special cases, is quite flexible and allows to formally test the need for transformation by maximum likelihood. An option, which we have not considered, but which is an interesting alternative for modeling skewed data, is to transform random terms in the mixed model, giving rise to nonlinear mixed models. It appears, however, that the Box-Cox-transformation compares favorably with these models regarding the ability to handle non-normality (Piepho and McCulloch, 1999).

Our analysis has not explored the possibility of grouping locations. CIMMYT advocates the classification of locations into mega-environments (Braun et al., 1996). Alternatively, locations may be grouped by geographical regions. There is a host of multivariate techniques for grouping locations into mega-environments (see, e.g., Lin et al., 1986; Cooper and Hammer, 1996; Gauch and Zobel, 1997). In order to derive more specific recommendations, the methods advocated in this chapter may be applied separately to each mega-environment or region. Also, quantitative covariate information, such as fertility scores, may be exploited for risk assessment (Piepho, 2000). We have not considered these options but have tried to convey just the basic principles of our ideas. A full analysis, which we do not claim to have presented here, certainly should exploit covariate information, which will require a good knowledge of the locations sampled.

We have placed emphasis on risk methods (probability to fall below a level θ and probability to outperform other genotypes). These methods are highly relevant, especially in developing countries. The mixed modeling framework naturally links these methods to a variety of stability measures, which, we believe, are not of so much interest in themselves. Rather, the wide variety of stability measures and models opens up a host of options for modeling variance-covariance structures. Clearly, mixed models involving stability parameters are more flexible and often more realistic than the usual ANOVA mixed model, which is so commonly used.

REFERENCES

Becker, H.C. and Léon, J. 1988. Stability analysis in plant breeding. *Plant Breeding* 101:1-23.

Box, G. E. P. and Cox, D. R. 1964. An analysis of transformations. *Journal of the Royal Statistical Society* B 26:211-246.

Braun, H.-J., Rajaram, S., and van Ginkel, M. 1996. CIMMYT's approach to breeding for wide adaptation. *Euphytica* 54:175-183.

Buckland, S. T., Burnham, K. P., and Augustin, N. H. 1997. Model selection: An integral part of inference. *Biometrics* 53:603-618.

Burnham, K. P. and Anderson, D. R. 1998. *Model selection and inference.* Springer, New York.

Christensen, R., Pearson, L. M., and Johnson, W. 1992. Case-deletion diagnostics for mixed models. *Technometrics* 34:38-45.

Cooper, M. and Hammer, G. L. (Eds.) 1996. *Plant adaptation and crop improvement.* CAB International, Wallingford, U.K.

Cullis, B. R., Thomson, F. M., Fisher, J. A., Gilmour, A. R., and Thompson, R. 1996. The analysis of the NSW wheat variety data base. II. Variance component estimation. *Theoretical and Applied Genetics* 92:28-39.

Denis, J.B., Piepho, H.P., and van Eeuwijk, F.A. 1997. Modeling expectation and variance for genotype by environment data. *Heredity* 79:162-171.

Diggle, P. 1988. An approach to the analysis of repeated measurements. *Biometrics* 44:959-971.

Eberhart, S. A. and Russell, W. A. 1966. Stability parameters for comparing varieties. *Crop Sci.* 6:36-40.

Eskridge, K. M. 1990. Selection of stable cultivars using a safety first rule. *Crop Science* 30:369-374.

Eskridge, K. M. 1996. Analysis of multiple environment trials using the probability of outperforming a check. In M. S. Kang and H. G. Gauch (Eds.), *Genotype-by-environment interaction* (pp. 273-307). CRC Press, Boca Raton, FL.

Eskridge, K. M. and R. F. Mumm. 1992. Choosing plant cultivars based on the probability of outperforming a check. *Theoretical and Applied Genetics* 84:894-900.

Finlay, K. W. and Wilkinson, G. N. 1963. The analysis of adaptation in a plant breeding program. *Australian Journal of Agricultural Research* 14:742-754.

Gauch, H. G. 1988. Model selection and validation for yield trials with interaction. *Biometrics* 44:705-715.

Gauch, H. G. Jr. and Zobel, R. W. 1997. Identifying mega-environments and targeting genotypes. *Crop Science* 37:311-326.

Gelfand, A. E. and Ghosh, S. K. 1998. Model choice: A minimum posterior predictive loss approach. *Biometrika* 85:1-11.

Gilmour, A. R., Thompson, R., and Cullis, B. R. 1995. Average information REML, an efficient algorithm for variance parameter estimation in linear mixed models. *Biometrics* 51:1440-1450.

Gilmour, A. R., Cullis, B. R., Welham, S. J., and Thompson, R. 1999. *ASREML. User manual.* <ftp://ftp.res.bbsrc.ac.uk/pub/aar/>

Gogel, B. J., Cullis, B., and Verbyla, A. 1995. REML estimation of multiplicative effects in multi-environment variety trials. *Biometrics* 51:744-749.

Lin, C. S., Binns, M. R., and Lefkovitch, L. P. 1986. Stability analysis: Where do we stand? *Crop Science* 26:894-900.

Littell, R. C., Milliken, G. A., Stroup, W. W., and Wolfinger, R. D. 1996. *SAS system for mixed models.* SAS Institute, Cary, NC.

McQuarrie, A. D. R. and Tsai, C.-L. 1998. *Regression and time series model selection.* World Scientific Publishers, Singapore.

Mead, R., Riley, J., Dear, K., and Singh, S. P. 1986. Stability comparison of intercropping and monocropping systems. *Biometrics* 42:253–266.

Oman, S. D. 1991. Multiplicative effects in mixed model analysis of variance. *Biometrika* 78:729-739.

Piepho, H. P. 1996. A simplified procedure for comparing the stability of cropping systems. *Biometrics* 52:378-383.

Piepho, H. P. 1997. Analyzing genotype-environment data by mixed models with multiplicative effects. *Biometrics* 53:761-766.

Piepho, H. P. 1998a. Methods for comparing the yield stability of cropping systems: A review. *Journal of Agronomy and Crop Science* 180:193-213.

Piepho, H. P. 1998b. Empirical best linear unbiased prediction in cultivar trials using factor analytic variance-covariance structures. *Theoretical and Applied Genetics* 97:195-201.

Piepho, H. P. 1999. Stability analysis using the SAS system. *Agronomy Journal* 91:154-160.

Piepho H. P. 2000. Exact confidence limits for covariate-dependent risk in cultivar trials. (To appear in Journals of Agricultural, Biological and Environmental Statistics).

Piepho, H. P., Denis, J. B., and van Eeuwijk, F. A. 1998. Predicting cultivar differences using covariates. *Journal of Agricultural, Biological, and Environmental Statistics* 3:151-162.

Piepho, H. P. and McCulloch, C. E. 1999. Transformations in mixed models: Application to risk analysis for a multi-environment trial. (In review).

SAS Institute. 1997. *SAS/STAT software: Changes and enhancements through release 6.12.* SAS Institute, Cary, NC.

Schwarz, G. 1978. Estimating the dimension of a model. *Annals of Statistics* 6:461-464.

Shukla, G. K. 1972. Some statistical aspects of partitioning genotype-environmental components of variability. *Heredity* 29:237-245.

Stram, D. O. and Lee, J. W. 1994. Variance components testing in the longitudinal mixed effects setting. *Biometrics* 50:1171-1177.

van Eeuwijk, F. A., Denis, J. B., and Kang, M. S. 1996. Incorporating additional information on genotypes and environments in models for two-way genotype by environment tables. In M. S. Kang and H. G. Gauch (Eds.), *Genotype-by-environment interaction* (pp. 15-50). CRC Press, Boca Raton, FL.

Wolfinger, R. D. 1993. Covariance structure selection in general mixed models. *Communications in Statistics A* 22:1079-1106.

Wolfinger, R. D. 1996. Heterogeneous variance-covariance structures for repeated measures. *Journal of Agricultural, Biological, and Environmental Statistics* 1:205-230.

Wricke, G. 1962. Über eine Methode zur Erfassung der ökologischen Streubreite. *Zeitschrift für Pflanzenzüchtung* 47:92-96.

Yates, F., and Cochran, W. G. 1938. The analysis of groups of experiments. *Journal of Agricultural Science* 28:556-580.

Chapter 12

Best Linear Unbiased Prediction: A Mixed-Model Approach in Multienvironment Trials

Monica Balzarini
Scott B. Milligan
Manjit S. Kang

INTRODUCTION

The best linear unbiased predictor (BLUP) approach has been extensively used for evaluating and predicting the genetic merit in animals (Henderson, 1975). The BLUP has relatively recently been introduced to and used in plant breeding for predicting cross performance and choosing parental lines and elite progeny with a relatively high probability of success (Bernardo, 1994,1995,1996a, 1996b; Panter and Allen, 1995; Balzarini, 2000). The BLUP is a traditional estimation procedure under the mixed linear-model approach (Searle et al., 1992). The BLUP procedure is not only being routinely used by maize breeders for selecting single crosses but also for choosing F_2 populations for inbred development (Bernardo, 1999).

As the BLUP procedures become more common and are better understood, additional applications are expected to evolve. For example, this chapter illustrates a new application of BLUP, viz., that it can be used to improve prediction of genotype performance in multienvironment trials (METs).

Multi-environment, replicated yield trials are often conducted in advanced stages of breeding programs to select genotypes based on yield and other economically important traits. The METs are also common in agricultural research to evaluate cropping systems and other selected treatments. A usual feature of all METs is the representation of a relatively large number of representative elements (Littell et al., 1996). In METs, environments might be reasonably assumed to be random effects. However, the genotype effects might be treated as fixed since only a few highly selected genotypes are usually involved in the late breeding stages. Therefore, the mixed-model approach, with environmental effects and genotype-by-environment effects as random, and genotype effects as fixed, is most appropriate.

The main aim of METs in plant breeding is to compare genotype performance of new cultivars. In addition to comparing mean genotypic performance, there is an interest in analyzing genotype-by-environment interaction (GEI) (Kang, 1990; Kang and Gauch, 1996) and quantifying stability of genotype performance across diverse environments (Lin et al., 1986; Becker and Leon, 1988; Crossa, 1990; Lin and Binns, 1994; Kang and Gauch, 1996). The mixed-model approach allows an analysis of METs relative to mean performance, GEI, and genotype stability in a unique framework.

Two types of inference about mean genotype performance are of interest in METs: (1) broad inference, i.e., general performance of a genotype, and (2) environment-specific or narrow inference, i.e., performance of a genotype in a specific environment. The traditional analytical approach relates to multiple pairwise comparisons of genotype means. The narrow inference from METs relies on comparisons of genotypic means in specific environments. Unfortunately, this procedure does not use all the available information. It is only possible to make inferences about performance of genotypes that have been tested in a specific environment. Mixed-model prediction uses information from an entire data set to obtain environment-specific inferences, allowing prediction of genotype performance even in environments where the genotype was not tested.

Most of the analytical procedures to quantify a genotype's contribution to the overall GEI are based on the fixed-effects model approach. Such fixed models are applicable only to balanced data. However, a common feature of most yield trials is that test entries vary from year to year because new entries are included as they become available and those with poor performance are deleted from further consideration (Hill and Rosenberg, 1985). The deletion and substitution results in unbalanced data. Even within a year, a balanced set of data may not be possible because certain replications and/or locations may not include all genotypes. Mixed model and restricted maximum likelihood-based estimation procedures relative to parameters in the models provide a more flexible analytical approach for the analysis of METs because balanced data are not required (Hill and Rosenberg, 1985; Stroup and Mulitze, 1991; Piepho, 1994, 1997, 1998a).

Magari and Kang (1997) used the restricted maximum likelihood (REML) method under a mixed model to estimate stability variances in unbalanced data sets when analyzing GEI for ear moisture loss rate in corn (*Zea mays* L.). The REML variance components, assignable to each genotype, estimate the same statistics as Shukla's stability variance (Shukla, 1972). The mixed model with heterogeneous GEI terms is a priori more tenable than the traditional mixed analysis of variance because it allows different stability statistics for each genotype while still assuming independence among the GEI effects.

By further modeling the variance-covariance structure of environment and interaction random effects, well known stability measures can be expressed as parameters of closely related mixed models (Piepho, 1998a). The common regression approaches for studying genotype sensitivities to environmental changes with multiplicative models for GEI (Yates and Cochran, 1938; Finlay and Wilkinson, 1963; Eberhart and Russell, 1966), including AMMI models (Gauch, 1988; Zobel et al., 1988), can be handled by integrating a factor-analytic variance-covariance structure into a mixed model (Oman, 1991; Piepho, 1997; Piepho, 1998b).

MIXED MODELS FOR MULTIENVIRONMENT TRIALS

The models employed to analyze yield trials use a variation of the following model:

$$y_{ijk} = \mu + E_i + R_{k(i)} + G_j + GE_{ji} + \varepsilon_{ijk}$$

where y_{ijk} is the k-th observation for the j-th genotype in the i-th environment, μ, E_i, G_j, $R_{k(i)}$, and GE_{ji} denote, respectively, the overall mean, the environmental effect $[i = 1, \ldots, s]$, the genotype effect $[j = 1, \ldots, g]$, the replication-within-environment effect $[k = 1, \ldots, r]$, and the genotype-by-environment interaction effect, and ε_{ijk} is the error term associated with y_{ijk}. Generally, statistical models treat genotypes as fixed effects. In addition to the regular fixed model that considers all effects as fixed, except the error term, the mixed models for METs treat environments, blocks-within-environment effects, interaction, and the error terms as random. The assumptions for the random effects are: environmental effects,

$$E_i, \text{ are } iid \ N\left(0, \sigma_E^2\right)$$

and replication effects,

$$R_{k(i)}, \text{ are } iid \ N\left(0, \sigma_R^2\right) iid \ N(0, \sigma).$$

Error terms are usually assumed to be

$$iid \ N\left(0, \sigma_e^2\right),$$

but a heterogeneous variance model for the error terms is permissible. Environment, replication, interaction, and error effects are independent of one another.

The GEI terms are also regarded as normal random effects with zero means but with a variance-covariance matrix not necessarily implying independence and homogeneity of variances. In modeling variances and covariances of the random GEI terms, one might compare the mean performance in a more realistic manner and obtain stability statistics and GEI analysis as a by-product of the mixed-model approach.

The means of any two genotypes in a specific environment, y_{ij} and $y_{ij'}$, have the covariance

$$Cov\left(y_{ij}, y_{ij'}\right) = \sigma_E^2 + Cov\left(GE_{ij}, GE_{ij'}\right), \text{ for } j \neq j'$$

A set of potential mixed models for METs is made by modifying variance-covariance structures that can be imposed on the interaction term, $Cov(GE_{ij}, GE_{ij'})$ (see Table 12.1). For example, Model [1] (Mixed ANOVA) assumes that the GEI terms have the same variance and are independent. Model [2] (Mixed Shukla) provides flexibility to GEI terms to have different variances but assumes that they are independent. Model [2] assumes that all GEI terms involving a particular genotype have the same GEI variance; thus, there will be as many different GEI variance components as the number of genotypes, which is analogous to Shukla's stability variances. Therefore, even though the model was designated as "Shukla," it is important to note that the original Shukla's proposal (Shukla, 1972) did not express stability variances as parameters of a mixed model. Model [3] (Mixed AMMI) considers multiplicative GEI effects,

$$GE_{ij} = \sum_{m=1}^{M} \lambda_{mj} x_{mi} + d_{ji}$$

where the first part,

$$\left(\sum_{m=1}^{M} \lambda_{mj} x_{mi}\right)$$

is the sum of multiplicative terms used to explain interaction signals and d_{ji} is the residual interaction term. Each multiplicative term represents a linear regression model of the residuals from the main effects model for the j-th genotype on a latent unobservable variable related to the i-th environment. A sum of multiplicative terms is used to model *GEI* variability pattern in more than one dimension. The subscript m indexes the axis of variability from which the fixed genotype and random environment scores are obtained. Thus, for each axis of variation, the genotypic score λ_j can be interpreted as the response of the j-th genotype to changes in some latent environmental variable with value x_i in the i-th environment. The model for the GEI terms resembles the nonadditive part of the traditional AMMI models

(Gauch, 1988; Zobel et al., 1988), but in the fixed AMMI models, environment scores are fixed. The sum of multiplicative terms is part of the expected value of y_{ijk} in the fixed-effects approach, whereas under the mixed model, this relates to its covariance structure.

Model [4] (Mixed E&R) does not contain the main effect for environment and also considers multiplicative *GEI* effects,

$$GE_{ij} = \lambda_j x_i + d_{ji}$$

where λ_j is the sensitivity of the j-th genotype to a nonobserved environmental variable x_i and d_{ij} is the unexplained part of the *GEI*. The deviations d_{ij} are allowed to have a separate variance for each genotype.

$$\sigma_d^2(j)$$

Despite the fact that environmental variable is assumed to be random, the model resembles the Eberhart-Russell (1966) regression model. A genotype with large λ_j absolute value shows a large sensitivity to changes in the underlying random environmental variable x_i.

The models imposed on the GEI terms generate specific variance-covariance matrix types for the vector $y_{(i)}$ containing the genotypic means in the i-th environment (Table 12.1). Model parameters can be estimated by REML (Searle et al., 1992). Traditional stability parameters are obtained from the covariance parameters. All calculations, including fixed and random effects estimates and traditional biplots to represent multiplicative models, can be easily obtained using "PROC Mixed/SAS" (SAS Institute, 1997) that solves the mixed-model equations for REML estimates (Balzarini, 2000).

Because the parameters of interest are integrated in a model, the decision about the appropriate variance-covariance model to use, and, consequently, the appropriate stability parameter to choose, can be based on the difference between -2 Residual Log Likelihood (-2 Res LL) of nested mixed models. Differences between -2ResLL can be compared with a χ^2 variable (degrees of freedom equal to the difference in the number of covariance parameters between the two models being compared to detect the significance from a more complicated model—higher number of covariance parameters). Other criteria for model selection, such as Akaike's criterion, may be used for nonnested models. In this way, the selection of the stability and *GE* measure will not depend on the researcher's preference for one or other methodology. The likelihood ratio tests will help obtain these measures in the framework of the best model for a given data set.

Generalized least-squares means for each genotype can be used for broad inference. Pairwise comparisons among these genotype means should use a

TABLE 12.1. Some Mixed Models for the Analysis of Mulitenvironment Yield Trials

Model	Model equation*	Interaction effects assumptions	Covariance structure for $y_{(i)}$ †
[1] Mixed ANOVA	$y_{ijk} = \mu + E_i + R_{k(i)} + G_j + GE_{ij} + \varepsilon_{ijk}$	$GE_{ij} \sim iid\ N\left(0, \sigma^2_{GE}\right)$	$\Sigma/\mathbf{env} = \mathbf{J}\sigma^2_E + \mathbf{I}\sigma^2_{GE}$
[2] Mixed Shukla	$y_{ijk} = \mu + E_i + R_{k(i)} + G_j + GE_{ij} + \varepsilon_{ijk}$	$GE_{ij} \sim iid\ N\left(0, \sigma^2_{GE(j)}\right)$	$\Sigma/\mathbf{env} = \mathbf{J}\sigma^2_E + \mathbf{I}\sigma^2_{GE(j)}$
[3] Mixed AMMI	$y_{ijk} = \mu + E_i + R_{k(i)} + G_j + GE_{ij} + \varepsilon_{ijk}$ $$GE_{ij} = \sum_{m=1}^{M} \lambda_{mj} x_{mi} + d_{ij}$$	$\dot{GE}_j \sim N\left(0, \sum_{m=1}^{M}\lambda^2_{mj} + \sigma^2_d\right)$ for all i. $Cov(GE_{ij}, GE_{ij'}) = \sum_{m=1}^{M}\lambda_{mj}\lambda_{mj'}$ for $j \neq j'$	$\Sigma/\mathbf{env} = \mathbf{J}\sigma^2_E + \Lambda\Lambda' + \mathbf{I}\sigma^2_d$
[4] Mixed E&R	$y_{ijk} = \mu + R_{k(i)} + G_j + GE_{ij} + \varepsilon_{ijk}$ $GE_{ij} = \lambda_j x_i + d_{ij}$	$GE_j \sim N\left(0, \lambda^2_j + \sigma^2_{d(j)}\right)$ for all i. $Cov(GE_{ij}, GE_{ij'}) = \lambda_j \lambda_{j'}$ for $j \neq j'$	$\Sigma/\mathbf{env} = \Lambda\Lambda' + \mathbf{diag}\left(\sigma^2_{d(j)}\right)$

Notes: *μ: overall mean; E_i: random environment i effect; $R_{k(i)}$: random replication-within-environment effect; G_j: fixed genotype j effect, GE_{ij} random genotype-by-environment interaction; λ_{mj} $(j = 1, \ldots, g)$ genotype factor loading on the m-th multiplicative interaction term, x_{mi}: m-th predicted score for a latent environmental variable in environment i; d_{ji}: residual interaction term; ε_{ijk}: error terms associated with the response y_{ijk}; † $y_{(i)}$: vector of genotype means in environment i; Σ/env: variance-covariance matrix of $y_{(i)}$, $\mathbf{J}$: $g{\times}g$ matrix of 1's; $\mathbf{I}$: $g{\times}g$ identity matrix; Λ: $g{\times}M$ matrix of genotype factor loadings for each multiplicative term $m = 1, \ldots, M$.

sampling error variance for the mean difference that incorporates all covariance parameters, thus stability and interaction measures participate, in this way, in the comparison of mean genotype performance. This chapter illustrates the use of BLUP in MET analysis relative to narrow inferences; broad inferences are not treated here.

BLUPs are used for narrow or environment-specific inferences under a mixed-model approach. We are interested in the BLUP of the conditional expectation μ_{ij}, BLUP(μ_{ij}). The BLUP of genotype j in environment i is a linear combination of the estimated mean for genotype j (estimated fixed effect), the estimated random effects for environment i, and the estimated GEI term ji. Each random term effect is weighted by a factor w that depends on the repeatability of the random effect represented by a ratio of variance-covariance components. In the simplest mixed ANOVA model, w represents broad-sense heritability of the random effect.

$$BLUP\left(\mu_{ij}\right) = \overline{y}_{\cdot j} + w_1\left(\overline{y}_{i\cdot} - \overline{y}_{\cdot\cdot}\right) + w_2\left(\overline{y}_{ij} - \overline{y}_{i\cdot} - \overline{y}_{\cdot j} + \overline{y}_{\cdot\cdot}\right)$$

As mentioned before, in the fixed-model approach, the narrow inference is based only on the genotype mean within the environment, which complicates interpretations when not all genotypes are grown in all environments. The BLUPs of specific genotype performance use all available information, allowing predictions in the case of unbalanced data.

PREDICTIVE ACCURACY OF BLUPS

Plant-cane data from outfield trials (1996 to 1998) of the Louisiana Sugarcane Variety Development Program (Quebedeaux et al., 1996, 1997; Guillot et al., 1998) are used to compare prediction accuracy of the four mixed models (Table 12.1) against a fixed-model approach. Regular outfield tests include 10 to 12 genotypes per trial. Each year, trials are conducted at several (7 to 10) commercial farms distributed throughout the 158,000-ha crop region. Each trial is laid out in a randomized complete-block design with three replications. Prediction accuracy measures (Mean Square Prediction Errors or MSPEs) of cane yield (Mg·ha[1]) were obtained by a "leave-one-block-out" cross-validation procedure. Independent cross-validation was conducted for each test year (MET). For each MET, the data set was split into two subsets, one with two replications per environment (calibration data) and the other with one replication per environment (validation data set). The calibration data set was used to predict variety performance in each environment. Predicted performance was compared with the observed yield for each variety in the validation data set. The process was repeated 30 times for different randomizations, i.e., sets of two blocks per environment. The mean of squared differences between predicted and observed values of each genotype were used to approximate the MSPE of narrow inference under each model.

The fixed-model approach consistently produced larger prediction errors in cane yield than the other models (Table 12.2). On average, the fixed-

TABLE 12.2. Prediction Error for Environment-Specific Inferences for Four Models to Analyze Cane Yield in Multienvironment Sugarcane Yield Trials (Mg·ha^{-1})*

Farm†	Test Year	Fixed Model	Mixed ANOVA [1]	Mixed Shukla [2]	Mixed AMMI [3]
1:ALL.	1998	4.079	3.74	3.91	3.66
	1996	4.707	3.92	3.73	3.90
2:B.S.	1998	13.55	12.60	13.34	12.13
	1997	7.083	5.72	5.83	6.15
	1996	11.42	9.61	8.30	9.87
3:Geo	1998	6.511	5.74	6.22	6.02
	1997	10.50	10.50	10.16	9.70
4:Gln.	1997	3.70	3.70	3.45	2.88
	1996	17.08	13.76	15.65	11.96
5:Lan.	1998	20.55	17.09	18.52	21.48
	1997	13.61	10.26	9.94	10.02
	1996	7.63	7.17	6.54	7.37
6:Mag.	1998	24.66	23.34	24.50	23.05
	1997	6.43	6.26	6.47	6.61
	1996	11.16	9.37	8.40	8.30
7:Oak.	1996	8.65	7.99	5.62	8.57
8:P.A.	1997	10.28	8.66	8.86	9.54
	1996	7.09	5.56	5.14	5.04
9:R.L.	1996	7.32	6.44	5.46	7.01
10:R.H.	1998	13.17	10.94	11.48	11.46
	1997	21.18	18.48	18.75	17.81
	1996	14.47	12.13	10.16	13.19
11:St.J.	1998	17.45	12.16	13.49	13.26
Mean		*11.41*	*9.79*	*9.74*	*9.96*

Notes: *Square root of the mean square prediction error (difference between predictor and target values for each genotype obtained by an iterative validation procedure). †1:All—A.V. Allain & Sons, 2:B.S.—Bon Secour, 3:Geo.—Georgia, 4:Gln.—Glenwood, 5:Lan.—Lanaux, 6:Mag.—Magnolia, 7:Oak.—Oaklawnhy, 8:P.A.—Palo Alto, 9:R.L.—Raceland, 10:R.H— Ronal Hebert, and 11:StJ.—Levert-St. John.

model approach produced errors of 11.406 Mg·ha⁻¹ compared with the mixed models mean values between 9.738 and 9.959 Mg·ha⁻¹. The model with the lowest root mean square error, however, varied by year and location. While superiority of the various mixed models over the fixed model was clearly evident in most experiments, the following exceptions were noted: Fixed Model and Mixed Model for Georgia in 1997 (error = 10.50) and for Glenwood in 1997 (error = 3.70) produced equal errors, Mixed AMMI produced larger errors than the Fixed Model at Lanaux in 1998 and at Magnolia in 1997. Mixed Shukla resulted in a larger error than that of the Fixed Model in 1997 at Magnolia.

The mixed AMMI model [3] produced the lowest errors in nine out of 23 location-year combinations. However, the other two mixed models, mixed ANOVA [1] and mixed Shukla [2], each produced the lowest errors in seven out of 23 tests. Mixed-model narrow inferences incorporate expected GEI effects. The fluctuation of these empirical results may be related to the small GEI observed in these trials, which is likely a direct result of the preoutfield, multilocation testing (Milligan, 1994). Successful varieties must display high yields across all tested environments to be advanced to the outfield trials. Hence, a screening for low GEI prior to testing in the outfield trials occurred.

Simultaneous comparisons of these models with regard to narrow inferences in METs involving a larger genotype and environment list were done using peanut yield data from EERA Manfredi-INTA, Argentina. The MET database involved 16 genotypes and 20 environments (year-location combination). The GEI was significant ($P < 0.001$). Figure 12.1 represents MSPEs for seven environments at one location (Manfredi). Comparisons among the fixed model (FM), the mixed ANOVA model, ANOVA with a two-way structure (MM_2W), the mixed model with heterogeneous GEI variance component by genotype (MM_SV), and a mixed model with a multiplicative interaction relative to an AMMI model with two multiplicative terms (MM_MI) were made. The results revealed important improvement in the narrow inference prediction when using a mixed AMMI model with respect to the fixed-model approach. The additional advantage of using a mixed AMMI model is that it is valid even when one has unbalanced data. Biplots to visualize the GEI patterns can be derived from the covariance parameters of the mixed model (Balzarini, 2000).

Thus, the use of mixed models to analyze advanced variety trials offers the potential to improve predictive precision at virtually no additional cost. It also enables the researcher to objectively integrate GEI stability measures with mean performance. More complex mixed models that may model within-environment covariance or consider environmental factors linked to variability of genotype response across environments are possible (Biarnes-Dumoulin et al., 1996; Kang and Magari, 1996; Cullis, et al., 1996).

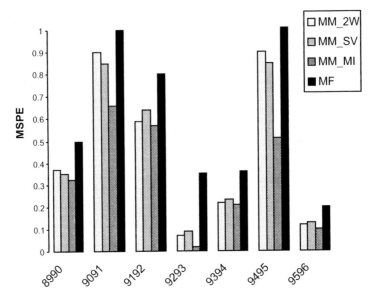

FIGURE 12.1. Narrow inference prediction accuracy of four models to analyze peanut METs. Environments represent seven test years at one location (Manfredi, Argentina). MSPE denotes mean square prediction error. The X axis represents environments.

REFERENCES

Balzarini, M. 2000. Biometrical models for predicting future performance in plant breeding. Doctoral Diss., Louisiana State University, Baton Rouge, LA.

Becker, H.C. and Leon, J. 1988. Stability analysis in plant breeding. *Plant Breeding* 101:1-23.

Bernardo, R. 1994. Prediction of maize single-cross performance using RFLPs and information from related hybrids. *Crop Sci.* 34:25-30.

Bernardo, R. 1995. Genetic models for predicting maize single-cross performance in unbalanced yield trial data. *Crop Sci.* 35:141-147.

Bernardo, R. 1996a. Best linear unbiased prediction of maize single-cross performance. *Crop Sci.* 36:50-56.

Bernardo, R. 1996b. Best linear unbiased prediction of the performance of crosses between untested maize inbreds. *Crop Sci.* 36:872-876.

Bernardo, R. 1999. Best linear unbiased predictor analysis. In J.G. Coors and S. Pandey (Eds.), *Genetics and exploitation of heterosis in crop* (pp. 269-276). ASA, CSSA, SSSA, Madison, WI.

Biarnes-Dumoulin, V., Denis J.B., Lejeune-Henaut, I., and Eteve G. 1996. Interpreting yield stability in pea using genotype and environmental covariates. *Crop Sci.* 36:115-120.

Crossa, J. 1990. Statistical analyses of multilocation trials. *Adv. Agronomy* 44: 55-85.

Cullis, B.R., Thompson, F.M., Fisher, J.A., Gilmour, A.R., and Thompson, R. 1996. The analysis of the NSW wheat variety database. II. Variance component estimation. *Theor. Appl. Genet.* 92:28-39.

Eberhart, S.A. and Russell, W.A. 1966. Stability parameters for comparing varieties. *Crop Sci* 6:36-40.

Finlay, K.W. and Wilkinson, G.N. 1963. The analysis of adaptation in a plant breeding programme. *Aust. J. Agric. Res.* 14: 742-754.

Gauch, H.G. Jr. 1988. Model selection and validation for yield trials with interaction. *Biometrics* 44:705-715.

Guillot, D.P., Milligan, S.B., Bischoff, K.P., Quebedeaux, K.L, Gravois, K.A., Garrison, D.D., Jackson W.R., and Waguespack, H.L. 1998. *1998 Outfield variety trials. Sugarcane Res. Annual Progress Report.* Louisiana State Univ. Agric. Center. Louisiana Agric. Exp. Stn., 88-101.

Henderson, C.R. 1975. Best linear unbiased estimation and prediction under a selection model. *Biometrics* 31:423-447.

Hill Jr., R.R. and Rosenberg, J.L. 1985. Models for combining data from germplasm evaluation trials. *Crop Sci* 25:467-470.

Kang, M.S. (Ed.). 1990. *Genotype-by-environment interaction and plant breeding.* Louisiana State University Agriculture Center, Baton Rouge, LA.

Kang, M.S. and Gauch, H.G. Jr. (Eds.), 1996. *Genotype-by-environment interaction.* CRC Press, Boca Raton, FL.

Kang, M.S. and Magari, R. 1996. New developments in selecting for phenotypic stability in crop breeding. In M.S. Kang and H.G. Gauch Jr. (Eds.), *Genotype-by-environment interaction* (pp. 1-14). CRC Press, Boca Raton, FL.

Lin, C.S. and Binns, M.R. 1994. Concepts and methods for analyzing regional trial data for cultivar and location selection. *Plant Breed. Rev.* 12:271-297.

Lin, C.S., Binns, M.R., and Lefkovitch, L.P. 1986. Stability analysis: Where do we stand? *Crop Sci.* 26:894-900.

Littell, R.C., Milliken, G.A., Stroup ,W.W., and Wolfinger, R.D. 1996. *SAS® system for mixed models.* SAS Institute, Cary, NC.

Magari, R. and Kang, M.S. 1997. SAS-STABLE: Stability analyses for balanced and unbalanced data. *Agron. J.* 89:929-932.

Magari, R., Kang, M.S., and Zhang, Y. 1997. Genotype by environment interaction for ear moisture loss rate in corn. *Crop Sci.* 37:774-779.

Milligan, S.B. 1994. Test site allocation within and among stages of a breeding program. *Crop Sci.* 34:1184-1190.

Oman, S.D. 1991. Multiplicative effects in mixed models analysis of variance. *Biometrika* 78 729:739.

Panter, D.M. and Allen, F.L. 1995. Using best linear unbiased predictions to enhance breeding for yield in soybean: I. Choosing parents. *Crop Sci.* 35:397-405.

Piepho, H.P. 1994. Best linear unbiased prediction (BLUP) for regional yield trials: A comparison to additive main effects multiplicative interaction (AMMI) analysis. *Theor. Appl. Genet.* 89:647-654.

Piepho, H.P. 1997. Analyzing genotype-environment data by mixed models with multiplicative effects. *Biometrics* 53:761-766.

Piepho, H.P. 1998a. Stability analysis using the SAS system. *Agron. J.* 91:154-160.

Piepho, H.P. 1998b. Empirical best linear unbiased prediction in cultivar trials using factor-analytic variance-covariance structures. *Theor. Appl. Genet.* 97:195-201.

Quebedeaux, K.L., Milligan, S.B., Martin, F.A., Garrison, D.D., Jackson W.R., and Waguespack, H. Jr. 1996. *1996 Outfield variety trials. Sugarcane Res. Annual Progress Report.* Louisiana State University Agriculture Center. Louisiana Agriculture Experimental Station, 77-98.

Quebedeaux, K.L., Milligan, S.B., Martin, F.A., Garrison, D.D., Jackson W.R., and Waguespack, H. Jr. 1997. *1997 Outfield variety trials. Sugarcane Res. Annual Progress Report.* Louisiana State University Agriculture Center. Louisiana Agriculture Experimental Station, 70-90.

SAS Institute. 1997. *SAS/STAT software: Changes and enhancements through release 6.12.* SAS Institute, Cary, NC.

Searle, S.R., Casella, G., and McCulloch, C.H. 1992. *Variance components.* John Wiley, New York.

Shukla, G.K. 1972. Some statistical aspects of partitioning genotype-environmental components of variability. *Heredity* 29:237-245.

Stroup, W.W. and Mulitze, D.K. 1991. Nearest neighbor adjust best linear unbiased prediction. *Am. Stat.* 45:194-200.

Yates, F. and Cochran, W.G. 1938. The analysis of groups of experiments. *J. Agric. Sci.* (Cambridge) 28:556-580.

Zobel, W.R., Wright, M.J., and Gauch, H.G. 1988. Statistical analysis of a yield trial. *Agron. J.* 80:388-393.

Index

Page numbers followed by the letter "f" indicate figures; those followed by the letter "t" indicate tables.